Studies in Fuzziness and Soft Computing

Volume 394

Series Editor

Janusz Kacprzyk, Systems Research Institute, Polish Academy of Sciences, Warsaw, Poland

The series "Studies in Fuzziness and Soft Computing" contains publications on various topics in the area of soft computing, which include fuzzy sets, rough sets, neural networks, evolutionary computation, probabilistic and evidential reasoning, multi-valued logic, and related fields. The publications within "Studies in Fuzziness and Soft Computing" are primarily monographs and edited volumes. They cover significant recent developments in the field, both of a foundational and applicable character. An important feature of the series is its short publication time and world-wide distribution. This permits a rapid and broad dissemination of research results.

Indexed by ISI, DBLP and Ulrichs, SCOPUS, Zentralblatt Math, GeoRef, Current Mathematical Publications, IngentaConnect, MetaPress and Springerlink. The books of the series are submitted for indexing to Web of Science.

More information about this series at http://www.springer.com/series/2941

Marie-Jeanne Lesot · Christophe Marsala
Editors

Fuzzy Approaches for Soft Computing and Approximate Reasoning: Theories and Applications

Dedicated to Bernadette Bouchon-Meunier

 Springer

Editors
Marie-Jeanne Lesot
Sorbonne Université, CNRS, LIP6
Paris, France

Christophe Marsala
Sorbonne Université, CNRS, LIP6
Paris, France

ISSN 1434-9922   ISSN 1860-0808   (electronic)
Studies in Fuzziness and Soft Computing
ISBN 978-3-030-54343-3   ISBN 978-3-030-54341-9   (eBook)
https://doi.org/10.1007/978-3-030-54341-9

This Springer imprint is published by the registered company Springer Nature Switzerland AG
The registered company address is: Gewerbestrasse 11, 6330 Cham, Switzerland

This book is dedicated to Bernadette Bouchon-Meunier, with warm regards and acknowledgements for all her contributions to the Soft Computing and Computational Intelligence community.

Foreword

This volume is a small token of appreciation for the entire broadly perceived fuzzy logic, but maybe more generally computer science and applied mathematics community of the whole world for Professor Bernadette Bouchon-Meunier for her many years of great scientific achievements and service to science and to the entire research community. I feel deeply honored and privileged to be able to write these couple of words, and one reason for this is that I have known her practically since the beginning of her illustrious research and scholarly career, and have always experienced her extraordinary consideration and a deep and loyal friendship.

Bernadette, Professor Bernadette Bouchon-Meunier, is a graduate of the Ecole Normale Superieure at Cachan, France, a very prestigious school, part of the famous French "grandes écoles", which adopt a very strict admission system, and maintain the highest academic standards. Bernadette then continued her education at the University Pierre and Marie Curie, known also as Paris 6, which is now Sorbonne Université, receiving her Ph.D. in Applied Mathematics and D.Sc. ("habilitation") in Computer Science. This excellent education, combined with a brilliant and original thinking, have certainly been important factors for her illustrious career in the future.

The academic career of Professor Bouchon-Meunier has been associated with both the CNRS (the French National Center for Scientific Research) and the University Pierre and Marie Curie, called some time also University Paris 6, where she started as a very young and promising researcher, and—through all steps of the academic promotions—had reached the highest ranks, both formally and in the sense of a huge appreciation for her class as a researcher and scholar by both her colleagues from the University and the entire world community.

My personal acquaintance and friendship with Bernadette started at the very beginning of her, and also my academic career. Throughout all those years, we have been meeting at the same conferences, participated at the same events, been involved in the same learned societies, etc. This all has implied great relations, professional and personal.

My constant deep appreciation for Bernadette has been manyfold but I will just mention some main aspects. First, I have greatly appreciated Bernadette's illustrious scientific achievements and original contributions. She has been the originator of many relevant scientific directions, for instance, a new approach to fuzzy questionnaires with wide applications in many "soft" and "hard" sciences, and new methods—notably using fuzzy logic—of approximate reasoning, machine learning, computational and artificial intelligence, followed by relevant applications in decision-making, data mining, risk analyses. forecasting, information retrieval, user modeling, emotional computing, to just name a few.

These brilliant scientific achievements have been ingeniously combined with Bernadette's organizational successes. More specifically, since the very beginning of her career, she has practically spared no effort to form strong research groups which would make it possible to do intensive research by reaching a critical mass. She has been extremely successful in this respect and is often considered by the research community as an example of an extremely effective and efficient scientist, not only at the French level but internationally. This is a great achievement, indeed.

Her brilliant scientific stature, and those extraordinary organizational abilities, has brought about great results. First, her research group has quickly become one of the world's top innovation centers in the broadly perceived fuzzy logic, and computational and artificial intelligence, with many visitors from all over the world. Second, she has attracted many younger collaborators, and her record of Ph.D. students has been growing year by year. What is important is that virtually all her Ph.D. students have gained much success in science, and also that she has been able to keep some of her best Ph.D. students in her group, notably the editors of this volume. She has been happy to see how these people gain a high stature and respect in science and continue the great job of running of such a world-class group she has founded.

Luckily enough, this deep worldwide appreciation for the scientific stature and brilliant achievements of Bernadette has been combined with an admiration of her personal qualities, consideration, and warmth. She has always had a kind word for everybody, has never refused to help, and—more generally—has always been a good spirit for our community. These great personal qualities are difficult to overestimate. They have certainly contributed to her many successes in international societies, recently as President of the IEEE Computational Intelligence Society.

The great qualities, professional and personal, of Professor Bernadette Bouchon-Meunier have been admired for years by the entire academic community. This has resulted in a natural way in so many contributions submitted to this volume by people from all over the world. This volume is clearly just a small token of appreciation for all what she has done for the community, for all of us, both particular persons, and science in general, and for her great personal qualities we all have always enjoyed. As they say, such great people do make the world better.

Academician Janusz Kacprzyk
Professor, Ph.D, D.Sc
Fellow of IEEE, IFSA
Systems Research Institute
Polish Academy of Sciences
Warsaw, Poland
kacprzyk@ibspan.waw.pl

Preface

This volume aims to celebrate all the fortunate, and still ongoing, years of scientific research Bernadette Bouchon-Meunier has carried out all along her career, among others through numerous collaborations: these collaborations have been as profuse as successful, involving a high multiplicity of researchers all across the world, as well as a very large diversity of research topics within the general domain of fuzzy logic and fuzzy systems.

Bernadette Bouchon-Meunier unquestionably belongs to the pioneers of fuzzy logic, she published her first paper in this domain as early as 1975, at a time when fuzzy logic was barely starting to flourish. She invited Lotfi Zadeh to France in 1977 to participate in the "International CNRS Symposium on recent developments of information theory and their applications" held in Cachan, prelude to a long-standing and deep friendship. Bernadette published more than 450 papers including 27 co-edited books and 5 co-authored books and she supervised more than 50 Ph.D. students in the domain of fuzzy logic and fuzzy systems, covering an extremely wide range of diverse topics, from multiple variants of approximate reasoning and non-classical logics to machine learning, soft computing, computational intelligence and very diverse applications such as affective computing, web document analysis or medical image processing, to name but a few. Bernadette's pioneering status has been "officially" acknowledged by the IEEE Computational Intelligence Society that bestowed her the Fuzzy System Pioneer Award in 2018, presented at the IEEE World Conference on Computational Intelligence (WCCI 2018, Rio de Janeiro).

Bernadette participated in the creation and structuring of the fuzzy community, in France and at an international level, in particular through the creation of the International Journal of Uncertainty, Fuzziness and Knowledge-Based Systems in 1993 and of the conference on Information Processing and Management of Uncertainty (IPMU) together with Ronald R. Yager in 1986. IPMU has taken place every two years since then, in multiple places across France, Spain, Italy, Germany or the Netherlands and has become a place-to-be for the community since its creation. At a national level, Bernadette also co-created the French national conference on fuzzy logic, named Rencontres Francophones sur la Logique Floue et ses Applications in 1995, whose first edition counted more than 200 registrations.

Bernadette also has an intense associative activity, at the European level in the EUSFLAT society and, at a worldwide level, within the IEEE Computational Intelligence Society. For the latter, since 2004 Bernadette assumed roles in multiple committees, such as the Conference, Nomination, Award, Distinguished Lecturer Program, Administrative or Women in Computational Intelligence committees, just to name a few. In 2020, Bernadette becomes the President of the Computational Intelligence Society that groups more than 7000 members.

From the research point of view, starting from the domain of information processing, Bernadette rapidly moved to integrating the newest propositions of the fuzzy logic emerging domain, in particular to manage uncertainties. To name a few of the multitude of her contributions, one can mention approximate reasoning, covering the whole range of deductive, inductive, abductive, analogical or interpolative forms, linguistic knowledge representation and extraction, in particular regarding modifiers and summaries, or soft computing, both in supervised and unsupervised settings, with a particular focus on similarity and entropy measures, for various types of data including tabular data, but also databases, images or time series. Bernadette's approach always favours interdisciplinary works, especially with cognitive sciences where collaboration with the domain of fuzzy logic has always proved to be very fruitful.

At the industrial level, Bernadette has also actively participated in technological transfers towards numerous companies for multiple applications, including electricity consumption monitoring, air traffic flow management, intelligent tutoring systems or ethno-political conflict detection and crisis detection for instance.

We are thus very glad that her collaborators and friends answered favourably our invitation for this book, offering an overview of the diversity of soft computing topics, from the theoretical and applicative points of view. The range of contributors highlights the variety of Bernadette's collaborations, thematically and geographically. We thank them warmly for having accepted to participate in this book.

Paris, France   Marie-Jeanne Lesot
February 2020   Christophe Marsala

Contents

Contributors

Inmaculada Alemán Doctoral program in biomedicine, Department of Legal Medicine, Toxicology and Physical Anthropology, University of Granada, Granada, Spain

Benjamín Bedregal Departamento de Informáticae Matemática Aplicada, Universidade Federal do Rio Grande do Norte, Natal, Brazil

Miguel Angel Bernal Sonora Institute of Technology, Ciudad Obregón, Mexico

James C. Bezdek School of Computing and Information Systems, The University of Melbourne, Melbourne, Australia

Humberto Bustince Department of Statistics, Computer Science and Mathematics, Public University of Navarra, Pamplona, Spain

Walmir Caminhas Department of Electronic Engineering, Federal University of Minas Gerais, Belo Horizonte, Brazil

B. Rosario Campomanes-Álvarez Data for Value Unit, CTIC Technological Centre, Asturias, Spain

Carmen Campomanes-Álvarez Vision Technologies Unit, CTIC Technological Centre, Asturias, Spain

Giulianella Coletti Department of Mathematics and Computer Sciences, Perugia, Italy

Oscar Cordón Andalusian Research Institute DaSCI, University of Granada, Granada, Spain

Sergio Damas Andalusian Research Institute DaSCI, University of Granada, Granada, Spain

Marcin Detyniecki AXA REV - Research Engineering and Vision, Paris, France; Sorbonne Universite, CNRS, Paris, France; Polish Academy of Science, IBS PAN, Warsaw, Poland

Graçaliz Dimuro Centro de Ciências Computacionais, Universidade Federal do Rio Grande, Rio Grande, Brazil

Alexander Dockhorn Institute for Intelligent Cooperating Systems, Otto-von-Guericke University, Magdeburg, Germany

Didier Dubois IRIT, Université Paul Sabatier, Toulouse, France

Francesc Esteva IIIA—CSIC, Bellaterra, Spain

Javier Fernandez Department of Statistics, Computer Science and Mathematics, Public University of Navarra, Pamplona, Spain

Lluis Godo IIIA—CSIC, Bellaterra, Spain

Fernando Gomide Department of Computer Engineering and Automation, University of Campinas, Campinas, Brazil

Thierry-Marie Guerra Polytechnic University Hauts-de-France, CNRS, UMR 8201, LAMIH, Valenciennes, France

Oscar Ibáñez Andalusian Research Institute DaSCI, University of Granada, Granada, Spain

László T. Kóczy Széchenyi István University, Gyor, Hungary; Budapest University of Technology and Economics, Budapest, Hungary

Olga Kosheleva University of Texas at El Paso, El Paso, TX, USA

Melinda Kovács Széchenyi István University, Gyor, Hungary

Vladik Kreinovich University of Texas at El Paso, El Paso, TX, USA

Rudolf Kruse Institute for Intelligent Cooperating Systems, Otto-von-Guericke University, Magdeburg, Germany

Florentin Kucharczak LIRMM – Université de Montpellier, Montpellier, France

Maria T. Lamata Universidad de Granada, E.T.S. de Ingenierías Informática y de Telecomunicación, Granada, Spain

Anne Laurent LIRMM, Université de Montpellier, CNRS, Montpellier, France

Andre Lemos Department of Electronic Engineering, Federal University of Minas Gerais, Belo Horizonte, Brazil

Ferenc Lilik Széchenyi István University, Gyor, Hungary

Kevin Loquin DMS Imaging, Montpellier, France

Luis Magdalena Escuela Técnica Superior de Ingenieros Informáticos, Universidad Politécnica de Madrid, Campus de Montegancedo, Boadilla del Monte, Madrid, Spain

Rubén Martos Doctoral program in biomedicine, Department of Legal Medicine, Toxicology and Physical Anthropology, University of Granada, Granada, Spain

Radko Mesiar Department of Mathematics and Descriptive Geometry, Slovak University of Technology, Bratislava, Slovakia

Szilvia Nagy Széchenyi István University, Gyor, Hungary

Marimuthu Palaniswami Department of Electrical and Electronic Engineering, The University of Melbourne, Melbourne, Australia

David A. Pelta Universidad de Granada, E.T.S. de Ingenierías Informática y de Telecomunicación, Granada, Spain

Irina Perfilieva University of Ostrava, Institute for Research and Applications of Fuzzy Modeling, Ostrava 1, Czech Republic

Olivier Pivert University of Rennes 1/IRISA, Lannion, France

Henri Prade IRIT, Université Paul Sabatier, Toulouse, France

Anca L. Ralescu EECS Department, University of Cincinnati, Cincinnati, OH, USA

Dan A. Ralescu Department of Mathematical Sciences, University of Cincinnati, Cincinnati, OH, USA

Punit Rathore Department of Electrical and Electronic Engineering, The University of Melbourne, Melbourne, Australia

Marek Z. Reformat Electrical and Computer Engineering, University of Alberta, Edmonton, Canada

Sandra Sandri LAC—INPE, SP, Brazil

Mika Sato-Ilic Faculty of Engineering, Information and Systems, University of Tsukuba, Tsukuba, Ibaraki, Japan

Chris Saxton Institute for Intelligent Cooperating Systems, Otto-von-Guericke University, Magdeburg, Germany

Grégory Smits University of Rennes 1/IRISA, Lannion, France

Olivier Strauss LIRMM – Université de Montpellier, Montpellier, France

Michio Sugeno Tokyo Institute of Technology, Tokyo, Japan

Brigita Sziová Széchenyi István University, Gyor, Hungary

Charles Tijus CHArt, University Paris 8, Paris, France

José Luis Verdegay Universidad de Granada, E.T.S. de Ingenierías Informática y de Telecomunicación, Granada, Spain

Ronald R. Yager Iona College, New Rochelle, NY, USA

Michael Zakharevich SeeCure Systems, Inc., Belmont, CA, USA

The Fuzzy Theoretic Turn

Michio Sugeno

Abstract This short essay characterizes Zadeh's contribution as the Fuzzy Theoretic Turn, recalling the past two turns in the human history: Kopernikanishe Wende (turn) by Kopernik in astronomy and the Linguistic Turn by Wittgenstein in philosophy.

Fuzziness, as a typical uncertainty seen in ordinary wordings in everyday life, is considered to be one of the three essential uncertainties that humans recognize. These are "Probability of Phenomena", "Fuzziness of Words" and "Vagueness of Consciousness". Among the three, a mathematical theory on "Probability of Phenomena" was initiated with the study by Pascal and Fermat in the latter half of the 17th century. The word "Probability" is an antonym of "certainty"; probability as the degree of uncertainty studied by Pascal is opposed to objective science where certainty is naturally required. However, through the axiomatization as probability measure by Kolmogorov in the beginning of the 20th century, the concept "probability" obtained its unwavering status in mathematics. After 300 hundred years since Pascal, the concept of "subjective probability" was presented with axiomatization by Savage; we can suppose that Pascal's original thought on probability stemmed from an idea like "subjective degree of uncertainty", if we look at his philosophy.

Modern science began with the methodology of science by Descartes in the 17th century. Since then, all rational sciences such as physics, chemistry, medical science, social science and even psychology have been based on this methodology. It is why Descartes is called "the father of modern rationalism". On the other hand, it is well known that Pascal criticized Descartes just when the Cartesian philosophy characterized by rationality, objectivism and universality was born. He said "we can recognize the truth not only by reason but also by sentiment". While the Cartesian philosophy is often called "l'esprit de géométrie", the philosophy of Pascal is called "l'esprit de finesse". As such, Pascal is even called "the father of modern non-rationalism". For Pascal, the spirit of probability (l'esprit de probabilité) must have

M. Sugeno (✉)
Tokyo Institute of Technology, Tokyo, Japan
e-mail: michio.sugeno@gmail.com

M.-J. Lesot and C. Marsala (eds.), *Fuzzy Approaches for Soft Computing and Approximate Reasoning: Theories and Applications*, Studies in Fuzziness and Soft Computing 394, https://doi.org/10.1007/978-3-030-54341-9_1

been related to his "l'esprit de finesse". We could say that our sprit considering a fuzzy set as a subjective set has precisely its root in "l'esprit de finesse" of Pascal.

What kinds of uncertainties are we aware of? With such a fundamental question, the author investigated the categories of uncertainty 30 years ago. To this aim, adjectives anyhow implying uncertainty were collected. There were found about 150 adjectives in Japanese, 160 in Chinese and 170 in English[1]). We can classify these adjectives into seven categories. Attaching modalities to categories, there are found (1) "Probability of Phenomena", (2) "Vagueness of Consciousness", (3) "Non-determinacy of Actions", (4) "Unclearness of Cognition", (5) "Fuzziness of Words", (6) "Uncertainty of Knowledge/Information" and (7) "Non-rationality of Logic", where "Phenomena" and "Consciousness" are categories, while "Probability" and "Vagueness" are modalities. If we more finely classify them, two more categories are found: (1') "Non-evidentness of Existence" and (7') "Incompleteness of Form", altogether nine categories. Among them, (1) "Probability of Phenomena", (2) "Vagueness of Consciousness" and (5) "Fuzziness of Words" are considered as three essential uncertainties which humans recognize as such. Why?

This question is resolved when we look at adjectives corresponding to these three categories. For examples in "Incompleteness of Form", we find adjectives such as "imperfect, improper, incomplete and incongruent"; most adjectives are associated with negative prefixes. On the other hand, in "Vagueness of Consciousness", we find "anxious, blurred, confusing and dubious"; most adjectives are not associated with negative prefixes. What does it mean? "Form" is usually complete in its nature, while "Consciousness" is often vague. Therefore, to express the uncertainty of "Form", we need to use adjectives with negative prefixes, but not the case about that of "Consciousness". These phenomena are commonly seen in English, French, Japanese and other languages. Based on these considerations, we could say that there exist three essential uncertainties."

Let us consider about "Fuzziness of Words". It is said that words have three main uncertainties: these are "Fuzziness", "Ambiguity" and "Generality". "Generality" is seen, for instance, when we say "Bring that book, please" or "Let's take a rest around there" where a book pointed by "that" and a place pointed by "around there" contain a kind of uncertainty. Among these uncertainties, "Fuzziness" is the most typical uncertainty, the next "Ambiguity" and then "Generality". We should notice that "Fuzziness" is much more general than "Ambiguity". For instance, when one says "Let's meet at 11 AM", the concept of "11 AM" is not crisp, implying "around 11 AM". As such, many wordings looking crisp are fuzzy in everyday life; it is a pity that only "Ambiguity" is discussed in natural language processing.

Next let us consider "Fuzziness of Words" from a view-point of our fuzzy theory. For example, consider the concept "small number". Then, it is found that this concept is never compatible with the Cartesian methodology of science. The methodology of science that Descartes presented in "Discours de la Méthode" consists of four principles such as (1) "evidentness", (2) "analysis", (3) "synthesis" and (4) "enumeration", where (1) "evidentness" implies that we should only deal with evident facts (or data)

---

[1]Later, the author recounted the number of adjectives and found about 430 in English, about 180 in French, about 200 in Spanish and also about 150 in Basque.

in science , (2) "analysis" implies that we should decompose a fact into elements, (3) "synthesis" implies that we should reconstruct a fact from its elements and (4) "enumeration" implies that we should consider many other similar facts in order to verify the results of the above analysis and synthesis. Here, we note that "evident" of Descartes means both "clear" and "distinct": key concepts in Cartesian philosophy. If we look at the definitions of a set, we can well understand the difference between "clear" and "distinct". Consider a set of natural numbers from 1 to 10. If a subset E is defined as $E = \{2, 4, 6, 8, 10\}$, then this definition is called extensional definition; E is defined by showing its elements. On the other hand, if E is defined as $E = \{n \mid n$ is even number$\}$, then this definition is called intentional definition; E is defined by giving the common property of its elements, where we note that the concept "even" is crisp in the sense of binary logic. "Extension" implies an element of a set, while "intention" implies the common property of the elements of a set. That is, a concept is said "clear" when its extensions are shown and said "distinct" when its intention is given. An intentional definition is more general than an extensional definition.

Now let us consider a fuzzy set S "Small (numbers)" as a subset of the above set: a set of natural numbers from 1 to 10. First we try to make an extensional definition. We may say that "1" is *certainly* small, and "2" is *almost certainly* small, "3" is *possibly* small, but we must wonder about "4". Therefore we cannot show the extensions of S; it is impossible to give an extensional definition of "Small". So we try to define it intentionally; $S = \{n \mid n$ is *small*$\}$. We soon recognize this as a tautology; "Small" is defined by itself. Of course, the predicate "small" is not crisp in the sense of binary logic. In fact there does not exist a common property of the concept "small". Therefore, S can be neither defined extensionally nor intentionally. The first principle in modern science is "evidentness". As we see, a fuzzy set S is not "evident" and, hence, contradicts the principles. A fuzzy set is, in this sense, never an object of science. In its first 20 years, fuzzy theory had been often criticized as "not science". Well, we can say that the criticism is right in the sense of Descartes. Then, how a fuzzy set was defined by Zadeh? It was like "$S = 1 / 1 + 0.8 / 2 + 0.5 / 3 + 0.2 / 4$". The numbers at the right hand sides of slashes are the extensions of an underlining set and the membership grades at the left hand sides are called the quantitative meaning of "Small". That is, the definition is based on extensions (those of an underling set) and intention (not qualitative but quantitative): in this sense, the definition is the combination of extensional definition and intentional definition. What else possible?

The fact that fuzzy theory concerning uncertainty of words was born in the 20th century would be not merely incidental. It is because the 20th century is called "The Century of Language". Wittgenstein is no doubt one of the most distinguished philosophers in the 20th century; he stated that all past problems of philosophy had been those of language. "The Century of Language" owes its name wholly to Wittgenstein. In the beginning of the 20th century, Russel and Whitehead dreamed to create an ideal language since ordinary language seemed so vague. Their dream was, however, over when Wittgenstein said "Ordinary language is perfect as it is". Let us look at two discourses in his "Philosophical Investigation". He says "How is the concept of a game bounded?

The concept "game" is a concept with blurred edge". As examples of "game", Wittgenstein showed "hopscotch (stone game) by children", "board-game" like chess or "Olympic games". This discourse is interpreted to state that the extensions of the concept "game" are not certain and, therefore, the concept "game" is not "clear". Further, let us look at the other discourse about "family resemblance" that "Family resemblance of a word: Consider for examples the proceedings that we call games.

You will not see something that is common at all". This second discourse is interpreted to state that the intention of the concept "game" is not distinct. That is, Wittgenstein stated that a word is not an "evident" existence. Following the concept of Wittgenstein, we may call the membership values of a fuzzy set as "family resemblance" degrees.

Kant (the most distinguished philosopher in German Idealism in the 19th century) stated that "recognition" is to define "phenomenon", but not the converse. He expressed such epistemology (his own epistemology) as "Kopernikanishe Wende (turn)"; Kopernik changed the Ptolemaic theory to the Copernican theory. In the 20th century, Rohty (a philosopher of neo-pragmatism in US) described the achievements of Wittgenstein as "The Linguistic Turn".

What Zadeh achieved is the third turn in science following the first turn in astronomy and the second turn in philosophy. It is, indeed, "fuzzy theoretic turn" by Zadeh. Mathematical theory for "Probability of Phenomena" was initiated in the latter half of the 17th century and that of "Fuzziness of Words" was born in the latter half of the 20th century. On the other hand, when we think that mathematical theory on "Vagueness of Consciousness" will never appear, we can truly understand the greatness of Zadeh. The subjectivity of Pascal seemed to submerge in the outside of the main stream of science in modern history. However, it never ceased. The return to the philosophy of Pascal has been realized in science by the birth of fuzzy theory. This is a revolutionary event to be called "the restitution of subjectivity in science". In this sense, it is in every way right to call it "The Fuzzy Theoretic Turn" following "Kopernikanishe Wende" and "The Linguistic Turn".

Indeed, it is very fortunate for us to have lived in the common ages with Zadeh and also to have shared the common researches with him.

Membership Functions

Didier Dubois and Henri Prade

Abstract The idea of a fuzzy set is formally modeled by a membership function that plays the same role as the characteristic function for an ordinary set, except that the membership function takes intermediary values between full membership and no membership. In this short note we first provide some references about the historical emergence of this notion, then discuss the nature of the scale for membership grades, and finally review their elicitation in relation with their intended meaning as a matter of similarity, uncertainty or preference.

1 Introduction

The notion of a fuzzy set stems from the observation made by Lotfi Zadeh more than fifty years ago in his seminal paper [80] that

> more often than not, the classes of objects encountered in the real physical world do not have precisely defined criteria of membership.

By "precisely defined", Zadeh means all-or-nothing, thus emphasizing the continuous nature of many categories used in natural language. This observation points out the gap existing between mental representations of reality and usual mathematical representations thereof, which are traditionally based on binary logic, precise numbers, differential equations and the like. Classes of objects referred to in Zadeh's quotation exist only through such mental representations, e.g., through natural language terms such as *high* temperature, *young* man, *big* size, etc., and also with nouns such as *bird*, *chair*, etc. Classical logic is too rigid to account for such categories where it appears that membership is a gradual notion rather than an all-or-nothing matter.

D. Dubois (✉) · H. Prade
IRIT, Université Paul Sabatier, 118 Route de Narbonne, 31062 Toulouse Cedex 9, France
e-mail: dubois@irit.fr

H. Prade
e-mail: prade@irit.fr

M.-J. Lesot and C. Marsala (eds.), *Fuzzy Approaches for Soft Computing and Approximate Reasoning: Theories and Applications*, Studies in Fuzziness and Soft Computing 394,
https://doi.org/10.1007/978-3-030-54341-9_2

5

In spite of the considerable interest for multiple-valued logics raised in the early 1920s by Łukasiewicz [53] and his school that developed logics with intermediary truth value(s), it was the American philosopher Black [8] who first proposed so-called "consistency profiles" (the ancestors of fuzzy membership functions), in 1937, in order to "characterize vague symbols."

As early as in 1946, the philosopher Abraham Kaplan argued in favor of the usefulness of the classical calculus of sets for practical applications. The essential novelty he introduces with respect to the Boolean calculus consists in, citing Kaplan [48] "entities which have a degree of vagueness characteristic of actual (empirical) classes". The generalization of the traditional characteristic function has been first considered by Weyl [75] in the same year; he explicitly replaces it by a continuous characteristic function. They both suggested calculi for generalized characteristic functions of vague predicates, and the basic fuzzy set connectives already appeared in these works.

Such a calculus has been presented by Kaplan and Schott [49] in more detail, and they called it the calculus of empirical classes (CEC). Instead of the notion of "property", Kaplan and Schott prefer to use the term "profile" defined as a type of quality. This means that a profile could refer to a simple property like *red, green*, etc. or to a complex property like *red and 20 cm long*, *green and 2 years old*, etc. They have replaced the classical characteristic function by an indicator which takes on values in the unit interval. These values are called weights from a given profile to a specified class. In the work of Kaplan and Schott, the notion of "empirical class" corresponds to the actual notion of "fuzzy set", and a value in the range of the generalized characteristic function (indicator, in their terminology) is already called by Kaplan and Schott a "degree of membership" (the same as the Zadehian grade of membership). Indicators of profiles are now called membership functions of fuzzy sets. Strangely enough it is the mathematician of probabilistic metric spaces, Karl Menger, who, in 1951, was the first to use the term "ensemble flou" (the French counterpart of "fuzzy set") in the title of a paper [61] of his, dealing with [0, 1]-valued fuzzy relations.

2 Mathematical Aspects of Membership Functions

A fuzzy set can be understood as a class equipped with an ordering of elements expressing that some objects are more inside the class than others. However, in order to extend the Boolean connectives, we need more than a mere relation in order to extend intersection, union and complement of sets, let alone inclusion. The set of possible membership grades has to be a complete lattice (the so-called L-fuzzy sets [40]) so as to capture union and intersection, and either the concept of residuation or an order-reversing function in order to express some kind of negation and implication.

2.1 Classical Versus Gradual Properties

A property A referring to an attribute a with domain D_a is said to be classical if A and its negation $\neg A$ make an ordinary partition of D_a. Let YA (resp. NA) denote the set of items having property A (resp. having property not A). Then the following properties hold:

$$\text{excluded-middle law: } YA \cup NA = D_a$$
$$\text{non-contradiction law: } YA \cap NA = \emptyset$$

However, these laws may fail in more general settings. For instance, if only the second (resp. the first) holds, A is said to be an intuitionistic (resp. paraconsistent) property. This situation is encountered with properties that are inherently gradual. Namely there may be items that fully satisfy neither A nor $\neg A$; however under a loose understanding of the property, they can be considered as satisfying both properties to some extent. Examples of gradual properties are numerous in natural languages. Clearly, properties such as *young, small, heavy, . . .* do not lead to a clear-cut binary partition of the domain. A predicate like *bald*, used in sorites paradoxes is another example of a gradual property. A clear test for checking whether a property A is gradual or not is to try to prefix it with the hedge "*very*". If "*very A*" makes sense, then there are natural situations where the property A is gradual. The above examples are gradual properties according to this test; while for instance, the property "*single*" is not gradual. In case of a noun rather than an adjective, fuzziness can model the idea that some instances of one noun are more typical than other ones, for instance a given animal can be a more or less typical *bird*.

2.2 Graduality and Partial Pre-orderings

Gradual properties naturally presuppose a preordering $\geq_A$ of the elements of D_a (which in turn induces a partial ordering in the set of items as well), in the sense that we can set, for any $u, v \in D_a$, $u \geq_A v$ iff the value u is at least as A as v. A first example is the case of general categories like *bird, chair*, and the like. The preordering then reflects an idea of typicality so that there are preferred instances in the class. For instance, *robin $>_{bird}$ penguin*, because penguins do not fly. In this sense, "*bird*" is not a classical category because when the agent claims some animal is a bird, this animal is more likely to be a robin than a penguin. As for $\neg A$, one can reasonably admit $u \geq_{\neg A} v$ if and only if $v \geq_A u$. Note that $\geq_A$ may be only partial (e.g., a hen is not clearly comparable with a duck as to their birdiness: the former cannot swim but is a poor flyer, the latter is a good flyer but most of the time it swims). Note that it is difficult to define numerical degrees of typicality. Although the idea of typicality is not necessarily thought as a matter of degree, the use of degrees of typicality has been advocated in the fuzzy set literature, with different ways of assessing them, see

[31, 32, 54, 77, 85]. However, according to a number of authors (e.g., Smith and Osherson [72], Freund [37]), extensions of concepts or categories cannot even have gradual boundaries.

Another example is the case of predicates referring to a numerical scale. For instance, for a = height, A = tall, and $D_a = [1.20, 2.20]$ meters, then $\geq_A$ is nothing but the natural order in the real interval $[1.20, 2.20]$, since the greater the height of a person, the taller he/she is; similarly with a = age, A = old and $D_a = [0, 150]$ years. But, this is not always the case. Remaining with a = age, take A = middle-aged, then the ordering on D_a is not in agreement with the natural order of ages. Namely we have $u \geq_A v$ if $u \geq v$ and both u and v are smaller than for instance 40 and greater than 30, and we also have $v \geq_A u$ if $u \geq v$ and both u and v are greater than for instance 50 and smaller than 60. While, for $40 \leq u, v \leq 50$, we may consider that both $u \geq_A v$ and $v \geq_A u$ hold. Relation $\geq_A$ is then partial (e.g., 35 is not comparable with 55, or 65 with respect to middle-aged). The view of a gradual property A just as an ordered structure $(D_a, \geq_A)$, advocated in [4, 35, 51, 52, 74], is very elegant, however very difficult to exploit for operational purposes when it comes to building a logic, due to the lack of *commensurateness* between two such preorderings pertaining to distinct properties.

2.3 Membership Scale: [0, 1] or Not

In contrast to the ordinal approach in the previous subsection, the use of a totally ordered membership scale for membership functions is based on an underlying assumption of commensurateness of grades of membership across properties. It provides a natural way of encoding a complete preorder. Although the unit interval $[0, 1]$ is the most commonly used scale in fuzzy set theory, it is not the only one that may make sense in practice. Observe that if a fuzzy set is defined on a finite universe, the number of grades needed for defining it is also finite. In fact, one may distinguish between the different options enumerated below [7].

1. The *qualitative finite setting*, with grades in a *finite* totally ordered scale: $L = \{\alpha_0 = 1 > \alpha_1 > \cdots > \alpha_{m-1} > 0\}$. This setting has a classificatory flavor, as we assign each item to a class in a finite totally ordered set thereof.
2. The *dense ordinal scale setting* using $L = [0, 1]$, seen as an ordinal scale. This has the advantage to have at one's disposal intermediary grades between any pair of grades.
3. The *denumerable setting*, using a scale of *powers* $L = \{\alpha^0 = 1 > \alpha^1 > \cdots > \alpha^i > \ldots, 0\}$, for some $\alpha \in (0, 1)$. This is isomorphic to the use of integers.
4. The *dense absolute setting*, for instance $L = [0, 1]$, seen as a genuine numerical scale equipped with product. The modeling of linguistic variables [82] on continuous measurable quantities requires such a scale. Toll sets [22] provides another example, where membership is a matter of cost for being a member in the set, hence valued on the positive real line.

2.4   Bipolar Scale or Not?

Another important issue is the presence or not of a "middle" element in the scale. This is related to the notion of bipolarity. A *bipolar* scale $(L, >)$ is a totally ordered set with a prescribed interior element **0** called *neutral*, separating the positive evaluations $\lambda > \mathbf{0}$ from the negative ones $\lambda < \mathbf{0}$. Mathematically, if the scale is equipped with a binary operation $\star$ (an aggregation operator), **0** is an idempotent element for $\star$, possibly acting as an identity. Classical examples are as follows:

- The most obvious quantitative bipolar scale is the (completed) real line equipped with the standard addition, where 0 is the neutral level. Isomorphic to it is the unit interval equipped with an associative uninorm [79] like $\frac{xy}{xy+(1-x)(1-y)}$ [14, 21]. Then the neutral point is 0.5, 0 plays the same role as $-\infty$ and 1 as $+\infty$ in the real line. Also the interval $[-1, 1]$ is often used as a bipolar scale.
- The simplest qualitative bipolar scale contains three elements: $\{-, \mathbf{0}, +\}$.

In such a bipolar scale, the negative side of the scale is the inverse mirror of the positive one. An item is evaluated on such a bipolar scale as being either positive or negative or neutral. It cannot be positive and negative at the same time. This is called a *univariate bipolar* framework. Fuzzy set theory often uses a bipolar membership scale, where the value 0.5 plays the role of a cross-over point between opposite fuzzy categories.

Another type of bipolar framework uses two distinct totally ordered scales L^+ and L^- for separately evaluating positive and negative information. This is the *bivariate unipolar* framework [28]. Here each scale is unipolar in the sense that the neutral level is at one end of the scale. Thus the meaning given to the bounds of the scale is an important issue. In a *positive* scale the bottom element is neutral. In a *negative* scale the top element is neutral. In the case of toll sets being valued on the positive real line, the scale is unipolar, as it is difficult to make sense of a cross-over point (even if it is mathematically possible to define one).

A bipolar scale can be viewed as the union of a positive and a negative scale $L^+ \cup L^-$ extending the ordering relations on each scale so $\forall \lambda^+ \in L^+, \lambda^- \in L^-, \lambda^+ > \lambda^-$. The symmetrization of finite unipolar scales is incompatible with associative operations [42]: only infinite bipolar scales seem to support such operations. See [27] for a general discussion of fuzzy set aggregation connectives in relation with the nature of the membership scale.

2.5   Beyond Scalar Membership Grades

Ten years after the invention of fuzzy sets, Zadeh [82] considered the possibility of iterating the process of changing membership grades into (fuzzy) sets, giving birth to interval-valued fuzzy sets, fuzzy set-valued (type 2) fuzzy sets, and more generally type n fuzzy sets. In fact, three other authors introduced interval-valued fuzzy sets independently in the same year [43, 46, 66]. While this idea is philosophically tempting and reflects the problematic issue of measuring membership grades of linguistic concepts, it has given birth to a large number of publications pertaining to variants of higher-order fuzzy sets. To name a few[1]: intuitionistic fuzzy sets, vague sets, hesitant fuzzy sets, soft sets, arbitrary combinations of the above notions (for instance, interval-valued intuitionistic fuzzy sets, fuzzy soft sets, etc.).

There are several concerns with these complexifications (rather than generalizations) of fuzzy sets to be pointed out. They shed some doubt on the theoretical or applied merits of such developments of fuzzy set theory:

- Several of these constructions reinvent existing notions under different sometimes questionable names. For instance vague sets are the same as intuitionistic fuzzy sets, and formally they are just a different encoding of interval-valued fuzzy sets (like using a pair of nested sets, versus an orthopair of disjoint sets [9]). Moreover the link between intuitionistic fuzzy sets and intuitionism hardly exists [16] or seems to be based on a confusion between *neither true nor false* and *unknown*, motivating the rejection of the excluded-middle law (see Atanassov [3] quotation of Brouwer). As for hesitant fuzzy sets, they were proposed already in 1975 under a different name [43]. Finally, soft sets are set-valued mappings, a notion that has been well-known for a long time in the theory of random sets, or can be viewed as plain relations.
- Algebraic structures for complex membership grades stemming from higher-order fuzzy sets are often special kinds of lattices. Actually, these higher-order fuzzy sets are special cases of L-fuzzy sets, hence often redundant from a mathematical point of view [44, 78].
- Some generalizations of fuzzy sets often underlie a misunderstanding [30]: are they special kinds of L-fuzzy sets or an approach to handling uncertainty about membership grades? The latter motivation is often put forward in the introduction of such papers, while the main text adopts an algebraic structure derived from L-fuzzy sets. Yet, handling uncertainty about complex logical formulas cannot be achieved in a truth-functional way by applying logical connectives.

The point is not to claim that such variants of fuzzy sets are necessarily misleading or useless. They often try to capture convincing intuitions. However, they are too often developed for their own sake, sometimes at odds with these intuitions. See [16] for a full-fledged discussion on intuitionistic fuzzy sets and interval-valued fuzzy sets, and [30] for the clash of intuitions between notions of bipolarity and uncertainty pervading intuitionistic fuzzy sets. Regarding soft sets, only outlined in the founding paper [62], they were originally meant as an extension of the alpha-cut mapping to non-nested sets, a concept more recently considered by several authors [29, 58, 67] in a more applied perspective. However, followers of the soft set trend often adopt

---

[1]We omit references here, for the sake of conciseness; readers can find a lot of them by searching for the corresponding key-words.

the set-valued mapping point of view without reference to cuts of fuzzy sets. For instance, the highly cited paper of Maji et al. [55] seems to consider soft sets in the algebraic framework of formal concept analysis [38] only.

3 Finding Membership Grades

The elicitation of membership grades is an important issue. The way to find them depends on the intended semantics and use of the fuzzy sets thye define. They may reflect similarity, uncertainty or preference, as pointed out more than 20 years ago [23]. In each case, the elicitation may rely on quantitative or qualitative views; see, e.g., [2] for an early survey.

3.1 Based on Similarity

In the measurement of the membership function associated to a linguistic term, say *tall* for instance, the membership $\mu_{tall}(h)$ is often constructed as a function of the distance between the value h and the closest height $\hat{h}$ that can be considered prototypical for the term, here *tall*, i.e., $\mu_{tall}(\hat{h}) = 1$, for instance,

$$\mu_{tall}(h) = f(d(h, \hat{h})) \tag{1}$$

where f is a non-negative, decreasing function such that $f(0) = 1$, for instance $f(u) = \frac{1}{1+u}$, and $d(h, \hat{h}) = \min\{d(h, x) : \mu_{tall}(x) = 1\}$, where d is a distance. Sudkamp [73] points out that conversely, given a possibility distribution π, the two-place function $\delta(x, y) = |\pi(x) - \pi(y)|$ is a pseudo-distance indeed.

A popular part of the fuzzy set literature deals with fuzzy clustering where gradual transitions between classes and their use in interpolation are the basic contribution of fuzzy sets. The idea that fuzzy sets would be instrumental to avoid too rough classifications was provided very early by Zadeh, along with Bellman and Kalaba [5]. They outline how to construct membership functions of classes from examples thereof. Intuitively speaking, a cluster gathers elements that are rather close to each other (or close to some core element(s)), while they are well-separated from the elements in the other cluster(s). Thus, the notions of graded proximity, similarity (dissimilarity) are at work in fuzzy clustering. With gradual clusters, the key issue is to define fuzzy partitions. The most widely used definition of a fuzzy partition is originally due to Ruspini [63], where the sum of membership grades of one element to the various classes is 1. This was enough to trigger the fuzzy clustering literature, that culminated with the numerous works by Bezdek and colleagues, with applications to image processing for instance [7]. Defining fuzzy Delaunay triangulations [36] is another example, in a planar point field, of taking into account the fuzziness of neighborhoods. See also the case of fuzzy sensors [60], where membership functions based on similarity are often to be determined in the context of fuzzy partitions.

Results of fuzzy clustering methods can indeed be interpreted as distance-based membership functions. Alternatively one may define a fuzzy set F from a crisp set A of prototypes of μ_{tall} and a similarity relation $S(x, y)$ on the height scale, such that $S(x, x) = 1$ (then $1 - S(x, y)$ is akin to a distance). Ruspini [64] proposes to define the membership function as a kind of upper approximation of A:

$$\mu_F(h) = \max_{u \in A} S(u, h).$$

Then A stands as the core of the fuzzy set F. We refer the reader to the survey by Türksen and Bilgic [76] for membership degree elicitation using measurement methods outside the possibility theory view, and more recently to papers by Marchant [56, 57].

Besides, the interpretation of a membership function $\mu_A(x)$ as a likelihood function $Prob(A|x)$ was pointed out early by Hisdal [45]. The idea is that $\mu_A(x)$ is all the greater as item x is more likely to be qualified by A, which relates probability to distance. This approach has been more recently elaborated in the setting of De Finetti's notion of coherence [11]. The connection between membership and likelihood functions actually pervades the probabilistic literature: the use of normal distributions as likelihood functions can be viewed as a way to define degrees of likelihood via the Euclidean distance between a given number and the most likely value (which in that case coincides with the mean value of the distribution). In the neuro-fuzzy literature, one often uses Gaussian membership functions of the form (1) with $f = e^{-x^2}$.

3.2 Fuzzy Sets as Possibility Distributions

In his paper introducing possibility theory, Zadeh [83] starts with the representation of pieces of information of the form 'X is A', where X is a parameter or attribute of interest and A is a fuzzy set on the domain of X, often representing a linguistic category (e.g., *John is Tall*, where $X = height\ (John)$, and A is the fuzzy set of *tall* heights for humans). The question is then, knowing that 'X is A', to determine what is the possibility distribution π_X restricting the possible values of X (also assuming we know the meaning of A, given by a $[0, 1]$-valued membership function μ_A). Then Zadeh represents the piece of information 'X is A' by the elastic restriction

$$\forall u \in U, \pi_X(u) = \mu_A(u)$$

where U is the universe of discourse on which X ranges. In the above example, U is the set of human heights. Thus, the degree of possibility that $X = u$ is measured by, hence equated to, the degree of compatibility $\mu_A(u)$ of the value u with the fuzzy set A. Originally, A is a *conjunctive* fuzzy set [84], the fuzzy set of all values more or less compatible with the meaning of A. Note however that π_X acts as a *disjunctive* restriction (X takes a single value in U). Here, μ_A is turned into a kind of likelihood function for X.

Note that the possibility distribution π takes its values in a unipolar scale where

- $\pi(u) = 0$ means that state u is rejected as impossible;
- $\pi(u) = 1$ means that state u is totally possible.

Possibility theory is qualitative if we use an ordinal setting, or a finite totally ordered scale, or a dense ordinal scale setting; it is quantitative if we use the unit interval $[0, 1]$ to measure possibility [25]. Moreover, just as membership functions of fuzzy linguistic terms are used to define possibility distributions, possibility distributions extracted from statistical information can be viewed as membership functions of fuzzy sets.

3.2.1 Qualitative Possibility Distributions

In this view, degrees of possibility $\pi(u)$ lie in a finite totally ordered set L [24]. Their elicitation is made easier by their qualitative nature as it comes down to assigning a possibility class to element u. In fact, it basically amounts to determining a well-ordered partition. Under a qualitative view, if we use a dense ordinal scale $L = [0, 1]$, the precise values of the degrees do not matter, only their relative position are important as expressing strict inequalities between possibility levels.

In a purely ordinal setting, a possibility ordering is a complete pre-order of states denoted by $\geq_\pi$, which determines a well-ordered partition $\{E_1, \ldots, E_k\}$ of U. It is the comparative counterpart of a possibility distribution π, i.e., $u \geq_\pi u'$ if and only if $\pi(u) \geq \pi(u')$. By convention E_1 contains the most plausible (or normal), or the most satisfactory (or acceptable) states, E_k the least plausible (or most surprising), or the least satisfactory ones, depending if we are modeling knowledge, or preferences. However as we use no scale, we again lose commensurateness and we cannot express full-fledged impossibility $\pi(u) = 0$ or full possibility $\pi(u) = 1$.

3.2.2 Quantitative Possibility Distributions: Imprecise Probabilities

Besides linguistic information, there is another situation where numerical membership functions are naturally obtained, and are interpreted as possibility measures: incomplete statistical information. A straightforward way of deriving a quantitative possibility distribution from incomplete statistical data in the form of a probability distribution m over a power set 2^U is to consider what Shafer [69] called the *contour function* of the random set m (actually, its one-point coverage function):

$$\pi_*(a) = \sum_{a \in E} m(E).$$

It is exactly a possibility distribution if the focal sets E with $m(E) > 0$ are nested. This view of possibility distributions and fuzzy sets as random sets was very early pointed out by Kampé de Feriet [47] and Goodman [41]. Without the nestedness condition, function π_* only yields a partial view of the data, as it is then not possible to reconstruct m from π_*. A variant of this approach to eliciting membership grades has been proposed in [50] under the name "label semantics": Instead of identifying a random set on universe U, the method defines a random set over a set of labels, from which one fuzzy set on U associated to each label is derived as a contour function. One merit of the approach is to obtain the membership functions of a whole term set.

Conversely, given a possibility distribution in π on U, there is a convex set of probability functions of the form $\mathscr{P}(\pi) = \{P : P(A) \leq \Pi(A), \text{ for all } A\}$ and the upper probability $P^*(A) = \sup_{P \in \mathscr{P}(\pi)} P(A) = \Pi(A) = \max_{u \in A} \pi(u)$ is a possibility measure. When the possibility distribution π is the membership function of a fuzzy interval, the set of probability measures $\mathscr{P}(\pi)$ dominated by its possibility measure Π takes the form $\{P : P(\pi_\alpha) \geq 1 - \alpha, \forall \alpha \in (0, 1]\}$, where $\pi_\alpha = \{x : \pi(x) \geq \alpha\}$, the α-cut of π, is a closed interval $[a_\alpha, b_\alpha]$ [12, 17].

Possibility distributions, when related to probability measures, are closely related to cumulative distributions. Namely, given a family $I_t = [a_t, b_t], t \in [0, 1]$ of nested intervals, such that $t < s$ implies $I_s \subset I_t$, $I_1 = \{\hat{x}\}$, and a probability measure P whose support lies in $[a_0, b_0]$, letting

$$\pi(a_t) = \pi(b_t) = 1 - P(I_t), t \in [0, 1]$$

yields a possibility distribution (it is the membership function of a fuzzy interval) that is compatible with P. Now, $1 - P(I_t) = P((-\infty, a_t)) + P((b_t, +\infty))$ making it clear that the possibility distribution coincides with a two-sided cumulative distribution function. Choosing $I_t = \{x : p(x) \geq t\}$ for $t \in [0, \sup p]$, where p is the density of P, of height $\sup p$, one gets the most specific possibility distribution compatible with P [33]. It has the same shape as p and $\hat{x}$ is the mode of p. It is the continuous counterpart of Eq. (2). It provides a faithful description of the dispersion of P.

In the finite case, there is a most specific possibility distribution π^* dominating a probability distribution encoding a linear order of elements, called *optimal transformation*, namely, considering the well-ordered partition $(E_1, \ldots E_n)$ induced by the probability distribution on U:

$$\forall a \in E_i, \pi^*(a) = \sum_{j \geq i} p_j \tag{2}$$

where $p_j = P(E_j)$ and $p_1 > p_2 > \ldots p_n$. Indeed one can check that $P(A) \in [N^*(A), \Pi^*(A)]$ and $P(\bigcup_{i=1}^j E_j) = \Pi^*(\bigcup_{i=1}^j E_j)$. The distribution π^* is known as the *Lorentz curve* of the vector $(p_1, p_2, \ldots, p_n)$. In fact, the main reason why this transformation is interesting is that, in the discrete as well as in the continuous case, it provides a systematic method for comparing probability distributions in terms of their relative peakedness (or dispersion). Namely, it has been shown that if π_p^* and π_q^* are optimal transformations of distributions p and q (sharing the same order of elements), and $\pi_p^* < \pi_q^*$ (the former is more informative than the latter), then $-\sum_{i=1}^n p_i \ln p_i < -\sum_{i=1}^n q_i \ln q_i$, and this property holds for all entropies [18]. For the continuous case, see Mauris [59].

3.2.3 Quantitative Possibility Distributions: Likelihood Functions

Another interpretation of numerical possibility distributions is the likelihood function in non-Bayesian statistics (Smets [70], Dubois et al. [19]). In the framework of an estimation problem, the problem is to determine the value of some parameter $\theta \in \Theta$ that characterizes a probability distribution $P(\cdot \mid \theta)$ over U. Suppose that our observations are summarized by the data set $\hat{d}$. The function $P(\hat{d} \mid \theta), \theta \in \Theta$ is not a probability distribution, but a likelihood function $\mathscr{L}(\theta)$: A value a of θ is considered as being all the more plausible as $P(\hat{d} \mid a)$ is higher, and the hypothesis $\theta = a$ will be rejected if $P(\hat{d} \mid a) = 0$ (or is below some relevance threshold). If we extend the likelihood of elementary hypotheses $\lambda(\theta) = c P(\hat{d} \mid \theta)$ (it is defined up to a positive multiplicative constant c [34]), viewed as a representation of uncertainty about θ, to disjunctions of hypotheses, the corresponding set-function Λ should obey the laws of possibility measures, i.e., $\Lambda(T) = \max_{\theta \in T} \lambda(\theta)$, in the absence of a probabilistic prior [13, 15].

We recover Shafer's proposal of a consonant belief function derived from likelihood information [69], more recently studied by Aickin [1]. Clearly, acquiring likelihood functions is one way of constructing possibility distributions. However, in this case possibility degrees are defined up to a multiplicative constant and, insofar as we stick to the likelihood interpretation, possibility distributions coming from different sources cannot be directly compared since it makes no sense to compare $P(\hat{d} \mid a)$ and $P(\check{d} \mid a)$ for two datasets $\hat{d}$ and $\check{d}$. This approach to possibility theory and membership functions on a ratio scale has not been very much explored yet.

3.3 Fuzzy Sets in Preference Evaluation

Fuzzy sets can be useful in decision sciences [20]. This is not surprising since decision analysis is a field where human-originated information is pervasive. While the suggestion of modeling fuzzy optimization as the (product-based) aggregation of an objective function with a fuzzy constraint first appeared in the last section of [81], the full-fledged seminal paper in this area was written by Bellman and Zadeh [6] in 1970, highlighting the role of fuzzy set connectives in criteria aggregation.

Then membership functions can be viewed as a variant of utility functions or rescaled objective functions, and optimized as such. The work of Giles [39] can be viewed as pertaining to the same trend, whereby a membership grade is interpreted in terms of pay-offs attached to the assertion of fuzzy statements. Savage acts are a key-notion in decision under uncertainty [68]. They represent decisions whose consequences depend on the state of the world, and can be viewed as fuzzy subsets of the state space since consequences are evaluated on an interval scale. Then, the expected utility of an act can be viewed as the probability of the fuzzy event *this act is the best*. Chen and Otto [10] also use the decision-theoretic framework for defining membership grades using the notion of certainty equivalent.

Using a method he developed later for his AHP (Analytic Hierarchy Process) approach to preference modeling and decision, Saaty [65] outlined an original method for determining membership degrees on the basis of data representing the *relative* membership of an element with respect to another element, using a scale of the form $\{1/n, \ldots, 1/2, 1, 2, \ldots, n\}$ where n is an integer (in practice $n = 9$). However the method requires the acquisition of relative membership grades (measuring how much an item belongs more to a fuzzy set than another one) and the enforcement of their consistency.

4 Concluding Remarks

This survey has outlined methods to obtain membership grades. A crucial step for properly defining a membership function in practice is to clarify its intended meaning in a given application and to choose an appropriate scale. Membership functions of linguistic terms are often quantified in terms of distance. However there are fuzzy sets that can be derived from frequentist information in the forms of possibility distributions modeling incomplete information or likelihood functions. Finally, utility functions may qualify as special kinds of membership functions whose semantics refer to the idea of cost or pay-off. Some of these views certainly need further developments. Another important related issue, not addressed here, is the identification of proper fuzzy set connectives for combining membership functions.

Dedication The authors are very glad to dedicate this chapter to Bernadette for her continuous and endless efforts to develop and promote the fuzzy set methodology in an open-minded way through her publications, and also for the launching of important international conferences she organized such as IPMU, which have been essential forums for exchanges between various approaches to uncertainty and vagueness, including fuzzy set theory, over more than four decades. Thank you so much, Bernadette.

References

1. Aickin, M.: Connecting Dempster-Shafer belief functions with likelihood-based inference. Synthese **123**(3), 347–364 (2000)
2. Aladenise, N., Bouchon-Meunier, B.: Acquisition de connaissances imparfaites: mise en évidence d'une fonction d'appartenance. Rev. Int. Syst. **11**(1), 109–127 (1997)

3. Atanassov, K.: Answer to D. Dubois, S. Gottwald, P., Hajek, J., Kacprzyk, H., Prade's paper "Terminological difficulties in fuzzy set theory—the case of intuitionistic fuzzy sets". Fuzzy Sets Syst. **156**, 496–499 (2005)
4. Basu, K., Deb, R., Pattanaik, P.K.: Soft sets: an ordinal formulation of vagueness with some applications to the theory of choice. Fuzzy Sets Syst. **45**, 45–58 (1992)
5. Bellman, R.E., Zadeh, L.A.: Decision making in a fuzzy environment. Manag. Sci. **17**, B141–B164 (1970)
6. Bellman, R.E., Kalaba, R., Zadeh, L.A.: Abstraction and pattern classification. J. Math. Anal. Appl. **13**, 1–7 (1966)
7. Benferhat, S., Dubois, D., Prade, H., Williams, M.-A.: A framework for iterated Belief revision using possibilistic counterparts to Jeffrey's rule. Fundam. Inform. **99**(2): 147-168 (2010)
8. Bezdek, J., Keller, J., Krishnapuram, R., Pal, N.: Fuzzy Models for Pattern Recognition and Image Processing. Kluwer (1999)
9. Black, M.: Vagueness. Phil. Sci. **4**, 427–455 (1937). Reprinted in Language and Philosophy: Studies in Method, Cornell University Press, Ithaca and London, pp. 23–58 (1949). Also in Int. J. Gener. Syst. **17**, 107–128 (1990)
10. Cattaneo, G., Ciucci, D.: Basic intuitionistic principles in fuzzy set theories and its extensions (A terminological debate on Atanassov IFS). Fuzzy Sets Syst. **157**(24), 3198–3219 (2006)
11. Chen, J.E., Otto, K.N.: Constructing membership functions using interpolation and measurement theory. Fuzzy Sets Syst. **73**, 313–327 (1995)
12. Coletti, G., Scozzafava, R.: Coherent conditional probability as a measure of uncertainty of the relevant conditioning events. In: Proceedings of ECSQARU'03, Aalborg, LNAI, vol. 2711, pp. 407–418. Springer-Verlag (2003)
13. Coletti, G., Scozzafava, R.: Conditional probability, fuzzy sets, and possibility: a unifying view. Fuzzy Sets Syst. **144**(1), 227–249 (2004)
14. Couso, I., Montes, S., Gil., P.: The necessity of the strong α-cuts of a fuzzy set. Int. J. Uncertain. Fuzziness Knowl. Based Syst. **9**(2), 249–262 (2001)
15. Dombi, J.: Basic concepts for a theory of evaluation: the aggregative operator. Eur. J. Oper. Res. **10**, 282–293 (1982)
16. Dubois, D., Perny, P.: A review of fuzzy sets in decision sciences: achievements, limitations and perspectives. In: Greco S., et al. (eds.) Multiple Criteria Decision Analysis. State of the Art Surveys, pp. 637–691. Springer (2016)
17. Dubois, D., Prade, H., Rannou, E.: An improved method for finding typical values. In: Proceedings 7th International Conference on Information Processing and Management of Uncertainty in Knowledge-Based Systems (IPMU'98), Paris, July 6–10, Editions EDK, Paris, pp. 1830–1837 (1998)
18. Dubois, D., Prade, H., Sandri, S.: On possibility/probability transformations. In: Lowen, R., Roubens, M. (eds.) Fuzzy Logic. State of the Art, pp. 103–112. Kluwer Academic Publishers, Dordrecht (1993)
19. Dubois, D., Prade, H.: New results about properties and semantics of fuzzy-set-theoretic operators. In: Wang, P.P., Chang, S.K. (eds.) Fuzzy Sets: Theory and Applications to Policy Analysis and Information Systems, pp. 59–75. Plenum Publishers (1980)
20. Dubois, D., Prade, H.: Possibility theory: qualitative et quantitative aspects. In: Gabbay, D., Smets, P. (eds.) Quantified Representation of Uncertainty et Imprecision. Handbook of Defeasible Reasoning et Uncertainty Management Systems, vol. 1, pp. 169–226. Kluwer Academic Publishers (1998)
21. Dubois, D., Prade, H.: Qualitative possibility theory and its applications to reasoning and decision under uncertainty. JORBEL (Belg. J. Oper. Res.) **37**(1–2), 5–28 (1997)
22. Dubois, D., Prade, H.: Toll sets and toll logic. In: Lowen, R., Roubens, M. (eds.) Fuzzy Logic: State of the Art, pp. 169–177. Kluwer Academic Publishers (1993)
23. Dubois, D.: Possibility theory and statistical reasoning. Comput. Stat. Data Anal. **51**, 47–69 (2006)
24. Dubois, D., Hüllermeier, E.: Comparing probability measures using possibility theory: a notion of relative peakedness. Int. J. Approx. Reason. **45**, 364–385 (2007)

25. Dubois, D., Prade, H.: The three semantics of fuzzy sets. Fuzzy Sets Syst. **90**, 141–150 (1997)
26. Dubois, D., Prade, H.: Possibility theory, probability theory and multiple valued logics: a clarification. Ann. Math. Artif. Intell. **32**, 35–66 (2001)
27. Dubois, D., Prade, H.: On the use of aggregation operations in information fusion processes. Fuzzy Sets Syst. **142**, 143–161 (2004)
28. Dubois, D., Prade, H.: An introduction to bipolar representations of information and preference. Int. J. Intell. Syst. **23**(8), 866–877 (2008)
29. Dubois, D., Prade, H.: Gradual elements in a fuzzy set. Soft. Comput. **12**, 165–175 (2008)
30. Dubois, D., Prade, H.: Gradualness, uncertainty and bipolarity: making sense of fuzzy sets. Fuzzy Sets Syst. **192**, 3–24 (2012)
31. Dubois, D., Prade, H., Rossazza, J.P.: Vagueness, typicality and uncertainty in class hierarchies. Int. J. Intell. Syst. **6**(2), 167–183 (1991)
32. Dubois, D., Moral, S., Prade, H.: A semantics for possibility theory based on likelihoods. J. Math. Anal. Appl. **205**, 359–380 (1997)
33. Dubois, D., Foulloy, L., Mauris, G., Prade, H.: Possibility/probability transformations, triangular fuzzy sets, and probabilistic inequalities. Reliab. Comput. **10**, 273–297 (2004)
34. Dubois, D., Gottwald, S., Hájek, P., Kacprzyk, J., Prade, H.: Terminological difficulties in fuzzy set theory—the case of "Intuitionistic Fuzzy Set". Fuzzy Sets Syst. **156**(3), 485–491 (2005)
35. Edwards, W.F.: Likelihood. Cambridge University Press, Cambridge, U.K. (1972)
36. Finch, P.D.: Characteristics of interest and fuzzy subsets. Inf. Sci. **24**, 121–134 (1981)
37. Förstner, W.: Uncertain neighborhood relations of point sets and fuzzy Delaunay triangulation. In: Förstner, W., Buhmann, J.M., Faber, A., Faber, P. (eds.) Proceedings of the Mustererkennung 1999, 21. DAGM-Symposium, Bonn, 15–17 Sept., Informatik Aktuell, pp. 213–222. Springer (1999)
38. Freund, M.: On the notion of concept I & II. Artif. Intell. **172**, 570–590, 2008 & **173**, 167–179 (2009)
39. Ganter, B., Wille, R.: Formal Concept Analysis. Springer-Verlag (1999)
40. Giles, R.: The concept of grade of membership. Fuzzy Sets Syst. **25**, 297–323 (1988)
41. Goguen, J.A.: L-fuzzy sets. J. Math. Anal. Appl. **8**, 145–174 (1967)
42. Goodman, I.R.: Fuzzy sets as equivalence classes of random sets. In: Yager, R. (ed.) Fuzzy Sets and Possibility Theory: Recent Developments, pp. 327–342. Pergamon Press, Oxford (1981)
43. Grabisch, M.: The Moebius transform on symmetric ordered structures and its application to capacities on finite sets. Discret. Math. **287**, 17–34 (2004)
44. Grattan-Guiness, I.: Fuzzy membership mapped onto interval and many-valued quantities. Z. Math. Logik. Grundladen Math. **22**, 149–160 (1975)
45. Gutiérrez García, J., Rodabaugh, S.E.: Order-theoretic, topological, categorical redundancies of interval-valued sets, grey sets, vague sets, interval-valued "intuitionistic" sets, "intuitionistic" fuzzy sets and topologies. Fuzzy Sets Syst. **156**(3), 445–484 (2005)
46. Hisdal, E.: Are grades of membership probabilities? Fuzzy Sets Syst. **25**, 325–348 (1988)
47. Jahn, K.U.: Intervall-wertige Mengen. Math. Nach. **68**, 115–132 (1975)
48. Kampé de Fériet, J.: Interpretation of membership functions of fuzzy sets in terms of plausibility and belief. In: Gupta, M., Sanchez, E. (eds.) Fuzzy Information and Decision Processes, pp. 93–98. North-Holland, Amsterdam (1982)
49. Kaplan, A.: Definition and specification of meanings. J. Phil. **43**, 281–288 (1946)
50. Kaplan, A., Schott, H.F.: A calculus for empirical classes. Methods **III**, 165–188 (1951)
51. Lawry, J.: Modelling and Reasoning with Vague Concepts. Springer (2006)
52. Lee, J.W.T.: Ordinal decomposability and fuzzy connectives. Fuzzy Sets Syst. **136**, 237–249 (2003)
53. Lee, J.W.T., Yeung, D.S., Tsang, E.C.C.: Ordinal fuzzy sets. IEEE Trans. Fuzzy Syst. **10**(6), 767–778 (2002)
54. Lesot, M.-J., Mouillet, L., Bouchon-Meunier, B.: Fuzzy prototypes based on typicality degrees. In: Reusch, B. (ed.) Proceedings of the International Conference on 8th Fuzzy Days, Dortmund, Germany, Sept. 29–Oct. 1, 2004. Advances in Soft Computing, vol. 33, pp. 125–138. Springer (2005)

55. Łukasiewicz, J.: Philosophical, remarks on many-valued systems of propositional logic (1930). In: Borkowski (ed.) Studies in Logic and the Foundations of Mathematics, pp. 153–179. North-Holland (1970) (Reprinted in Selected Works)
56. Maji, P.K., Biswas, R., Roy, A.R.: Soft set theory. Comput. Math. Appl. **45**(4–5), 555–562 (2003)
57. Marchant, T.: The measurement of membership by comparisons. Fuzzy Sets Syst. **148**, 157–177 (2004)
58. Marchant, T.: The measurement of membership by subjective ratio estimation. Fuzzy Sets Syst. **148**, 179–199 (2004)
59. Martin, T.P., Azvine, B.: The X-mu approach: fuzzy quantities, fuzzy arithmetic and fuzzy association rules. In: Proceedings of the IEEE Symposium on Foundations of Computational Intelligence (FOCI), pp. 24–29 (2013)
60. Mauris, G.: Possibility distributions: a unified representation of usual direct-probability-based parameter estimation methods. Int. J. Approx. Reason. **52**(9), 1232–1242 (2011)
61. Mauris, G., Foulloy, L.: A fuzzy symbolic approach to formalize sensory measurements: an application to a comfort sensor. IEEE Trans. Instrum. Measure. **51**(4), 712–715 (2002)
62. Menger, K.: Ensembles flous et fonctions aléatoires. C. R. l'Acad. Sci. Paris **232**, 2001–2003 (1951)
63. Molodtsov, D.: Soft set theory—first results. Comput. Math. Appl. **37**(4/5), 19–31 (1999)
64. Ruspini, E.H.: A new approach to clustering. Inform. Control **15**, 22–32 (1969)
65. Ruspini, E.H.: On the semantics of fuzzy logic. Int. J. Approx. Reason. **5**(1), 45–88 (1991)
66. Saaty, T.L.: Measuring the fuzziness of sets. J. Cybern. **4**(4), 53–61 (1974)
67. Sambuc, R.: Fonctions ϕ-floues. Application à l'aide au diagnostic en pathologie thyroidienne. Ph.D. Thesis, Univ. Marseille, France (1975)
68. Sanchez, D., Delgado, M., Villa, M.A., Chamorro-Martinez, J.: On a non-nested level-based representation of fuzziness. Fuzzy Sets Syst. **192**, 159–175 (2012)
69. Savage, L.J.: The Foundations of Statistics. Wiley (1954)
70. Shafer, G.: A Mathematical Theory of Evidence. Princeton University Press (1976)
71. Smets, P.: Possibilistic inference from statistical data. In: Ballester, A. (ed.) Proceedings of the 2nd World Conference on Mathematics at the Service of Man, Las Palmas (Spain) pp. 611–613 (1982)
72. Smets, Ph: Constructing the pignistic probability function in a context of uncertainty. In: Henrion, M., et al. (eds.) Uncertainty in Artificial Intelligence, vol. 5, pp. 29–39. North-Holland, Amsterdam (1990)
73. Smith, E.E., Osherson, D.N.: Conceptual combination with prototype concepts. Cognit. Sci. **8**, 357–361 (1984)
74. Sudkamp, T.: Similarity and the measurement of possibility. In: Actes Rencontres Francophones sur la Logique Floue et ses Applications (Montpellier, France), Toulouse: Cepadues Editions, pp. 13–26 (2002)
75. Trillas, E., Alsina, C.: A reflection on what is a membership function. Mathw. Soft Comput. **VI**(2–3), 201–215 (1999)
76. Türksen, I.B., Bilgic, T.: Measurement of membership functions: theoretical and empirical work. In: Dubois, D., Prade, H. (eds.) Fundamentals of Fuzzy Sets. The Handbooks of Fuzzy Sets, pp. 195–230. Kluwer Publ. Comp. (2000)
77. Vasudev Murthy, S., Kandel, A.: Fuzzy sets and typicality theory. Inform. Sci. **51**(1), 61–93 (1990)
78. Wang, G.-J., He, Y.-H.: Intuitionistic fuzzy sets and L-fuzzy sets. Fuzzy Sets Syst. **110**, 271–274 (2000)
79. Weyl, H.: Mathematics and logic- a brief survey serving as preface to a review of "the Philosophy of Bertrand Russell". Amer. Math. Month. **53**, 2–13 (1946)
80. Yager, R.R., Rybalov, A.: Uninorm aggregation operators. Fuzzy Sets Syst. **80**(1), 111–120 (1996)
81. Zadeh, L.A.: Fuzzy sets and systems. In: Fox, J. (ed.) Systems Theory; Proceedings of the Simposium on System Theory, New York, April 20–22, pp. 29–37 (1965). Polytechnic Press, Brooklyn, N.Y. (1966) (reprinted in Int. J. Gener. Syst. **17**, 129–138 (1990))

82. Zadeh, L.A.: The concept of a linguistic variable and its application to approximate reasoning. Inf. Sci. Part I **8**(3), 199–249; Part II **8**(4), 301–357; Part III **9**(1), 43–80 (1975)
83. Zadeh, L.A.: Fuzzy sets. Inf. Control **8**, 338–353 (1965)
84. Zadeh, L.A.: Fuzzy sets as a basis for a theory of possibility. Fuzzy Sets Syst. **1**, 3–28 (1978)
85. Zadeh, L.A.: PRUF—A meaning representation language for natural languages. Int. J. Man-Mach. Stud. **10**, 395–460 (1978)
86. Zadeh, L.A.: A note on prototype theory and fuzzy sets. Cognition **12**(3), 291–297 (1982)

The Evolution of the Notion of Overlap Functions

Humberto Bustince, Radko Mesiar, Graçaliz Dimuro, Javier Fernandez, and Benjamín Bedregal

Abstract In this chapter we make a review of the notion of overlap function. Although originally developed in order to determine up to what extent a given element belongs to two sets, overlap functions have widely developed in the last years for very different problems. We recall here the motivation that led to the introduction of this new notion and we discuss further theoretical developments that have appeared to deal with other types of problems.

1 The Origin of the Notion of Overlap Function

In many different real situations it is necessary to determine to which one of a set of classes a considered instance belongs to. In the case of disjoint crisp classes, it may be assumed that the instance belongs to one or another class, but not to more than one class at the same time. However, it may happen that, even for crisp classes,

H. Bustince (✉) · J. Fernandez
Department of Statistics, Computer Science and Mathematics, Public University of Navarra, Pamplona, Spain
e-mail: bustince@unavarra.es

J. Fernandez
e-mail: fcojavier.fernandez@unavarra.es

R. Mesiar
Department of Mathematics and Descriptive Geometry, Slovak University of Technology, Bratislava, Slovakia
e-mail: mesiar@math.sk

G. Dimuro
Centro de Ciências Computacionais, Universidade Federal do Rio Grande, Rio Grande, Brazil
e-mail: gracalizdimuro@furg.br

B. Bedregal
Departamento de Informáticae Matemática Aplicada, Universidade Federal do Rio Grande do Norte, Natal, Brazil
e-mail: bedregal@dimap.ufrn.br

M.-J. Lesot and C. Marsala (eds.), *Fuzzy Approaches for Soft Computing and Approximate Reasoning: Theories and Applications*, Studies in Fuzziness and Soft Computing 394, https://doi.org/10.1007/978-3-030-54341-9_3

the boundary separating one class from another is not clearly defined for the expert, due to the lack of information or the imprecision on the available data. Finally, it also happens commonly that classes are not clearly defined by their own nature, and hence several or even all the instances may belong up to some extent to two or more of the classes [1, 4, 15].

This situation arises, for instance, in some image processing problems (see [22]). Consider an image where we want to separate an object from the background. A possible approach to deal with this problem is to represent the object by means of an appropriate fuzzy set and the background by means of another fuzzy set [5]. These fuzzy sets can be built, for instance, on the referential set of the considered possible intensities and need not be disjoint in the sense of Ruspini. Obviously, the result of the separation procedure will be worse or better according to the accuracy of the fuzzy sets used to represent the background and the object. To build the membership functions that define these fuzzy sets it is necessary to know the exact property that provides a characterization for those pixels which belong to the object (or the background). This property determines the expression of the membership function associated to the fuzzy set representing the object (background) (see [8, 9]). But, in most of the real cases, the expert is not able to provide a specific expression for this function. Furthermore, there exist some pixels that the expert assign to the object or to the background without any doubt, but there are other pixels which the expert can not assign with certitude either to the object or to the background. For these last pixels the value of the membership function is not accurately known, since the pixels belong, up to some extent, to both classes (object and background).

Overlap functions were first introduced to deal with this kind of problems [6]. Overlap functions provide a mathematical model in such a way that the outcome of the function can be interpreted as the representation of up to what extent the considered instance belongs to both classes simultaneously. In this way, overlap functions are also related to the notion of overlap index and in fact they can be used to build the latter, see [24]. Moreover, the concept of overlap has also been extended to a more general situation in [21], with three or more classes involved. In this chapter, we start reviewing results for the bivariate case and later we also consider the general n-dimensional case.

In this chapter we make an overview of some aspects of the history of the notion of overlap functions and their applications. The structure of the chapter is as follows. We start recalling the original definition of overlap function. Then, we comment the link between overlap functions and triangular norms. We also discuss the extension to any dimension of the definition of overlap functions and we finish with some applications.

2 The Mathematical Definition of Overlap Functions

As we have said, the concept of overlap as a bivariate aggregation operator was first introduced in [6] to deal with the problem of measuring up to what extent a given object (instance) belongs simultaneously to two classes. All the definitions and results in this section can be found in that reference. The original definition reads as follows.

Definition 3.1 A mapping $G_O : [0, 1]^2 \to [0, 1]$ is an overlap function if it satisfies the following conditions:

(G_O1). G_O is symmetric.
(G_O2). $G_O(x, y) = 0$ if and only if $xy = 0$.
(G_O3). $G_O(x, y) = 1$ if and only if $xy = 1$.
(G_O4). G_O is increasing in both variables.
(G_O5). G_O is continuous.

Regarding this definition, it is worth to mention that continuity arises precisely from the applicability of the notion of overlap functions in image processing. Note that in most of the image processing algorithms it is desired that slight variations in the intensity of the pixels do not lead to large variations in the final outcome, and this is precisely what continuity reflects. Furthermore, note that associativity is not a requirement in the definition. This, on the one hand, provides more flexibility than other functions such as t-norms, for instance, and, on the other hand, together with the other properties in the definition, allows overlap functions to be greater that the minimum.

There are many possible examples of overlap functions. For instance, $G_O(x, y) = (\min(x, y))^p$ for $p > 0$ or $G_O(x, y) = xy$ are overlap functions. Moreover, the class of bivariate overlap functions, $\mathscr{G}$, is convex and overlap functions can be characterized by the following result.

Theorem 3.1 *The mapping* $G_O : [0, 1]^2 \to [0, 1]$ *is an overlap function if and only if*

$$G_O(x, y) = \frac{f(x, y)}{f(x, y) + h(x, y)}$$

for some $f, h : [0, 1]^2 \to [0, 1]$ *such that*

(1) *f and h are symmetric;*
(2) *f is non decreasing and h is non increasing;*
(3) *$f(x, y) = 0$ if and only if $xy = 0$;*
(4) *$h(x, y) = 0$ if and only if $xy = 1$;*
(5) *f and h are continuous functions.*

Theorem 3.1 allows the definition of interesting families of overlap functions.

Corollary 3.1 *Let f and h be two functions in the setting of the previous theorem. Then, for $k_1, k_2 \in]0, \infty[$, the mappings*

$$G_S^{k_1,k_2}(x, y) = \frac{f^{k_1}(x, y)}{f^{k_1}(x, y) + h^{k_2}(x, y)}$$

define a parametric family of overlap functions.

Corollary 3.2 *In the same setting of Theorem 3.1, let us assume that G_O can be expressed in two different ways:*

$$G_O(x, y) = \frac{f_1(x, y)}{f_1(x, y) + h_1(x, y)} = \frac{f_2(x, y)}{f_2(x, y) + h_2(x, y)}$$

for any $x, y \in [0, 1]$ and let M be a bivariate continuous aggregation function that is homogeneous of order one. Then, if we define $f(x, y) = M(f_1(x, y), f_2(x, y))$ and $h(x, y) = M(h_1(x, y), h_2(x, y))$ it also holds that

$$G_O(x, y) = \frac{f(x, y)}{f(x, y) + h(x, y)} .$$

3 Overlap Functions Versus Triangular Norms

Note that from Definition 3.1 it follows straightforwardly that overlap functions are particular instances of aggregation functions without divisors of zero (i.e., such that $G(x, y) \neq 0$ if $x \neq 0$ and $y \neq 0$) or one (i.e., such that $G(x, y) \neq 1$ if $x \neq 1$ and $y \neq 1$). In fact, there exists a close relation between overlap functions and the well-known class of triangular norms [2, 6]. Recall that the latter are commonly used in order to represent the intersection between fuzzy sets and one of their crucial properties is that of associativity.

Theorem 3.2 ([2]) *Let G_O be an associative overlap function. Then G_O is a t-norm.*

Theorem 3.3 ([6]) *Let T be a continuous triangular norm without divisors of zero. Then T is an overlap function.*

However, when we are dealing with only two classes, associativity is not a natural requirement, as only two variables are involved. In this sense, overlap functions can be considered as an alternative tool to triangular norms. In particular, this approach has been successfully developed to build new classes of residual implication functions replacing the t-norm by an overlap function. Furthermore, in the same way as t-conorms can be defined by considering the dual of t-norms, it is possible to define the concept of grouping function as the dual (with respect to some strong negation N) of an overlap function.

Definition 3.2 [7] A mapping $G_G : [0, 1]^2 \to [0, 1]$ is a grouping function if it satisfies the following conditions:

(G_G1). G_G is symmetric.
(G_G2). $G_G(x, y) = 0$ if and only if $x = y = 0$.
(G_G3). $G_G(x, y) = 1$ if and only if $x = 1$ or $y = 1$.
(G_G4). G_G is increasing in both variables.
(G_G5). G_G is continuous.

Combining the notions of overlap and grouping functions, it is possible for instance to define preference structures in the same way as Fodor and Roubens [18], but replacing t-norms and t-conorms by overlap and grouping functions, respectively [7]. More specifically, given a fuzzy preference relation R, we can define the degree of indifference between alternatives i and j as:

$$I_{ij} = G_O(R_{ij}, R_{ji})$$

and the degree of incomparability between the same alternatives as

$$J_{ij} = 1 - G_G(R_{ij}, R_{ji})$$

where G_O is an overlap function and G_G is a grouping function. Note that in this case, the considered strong negation is Zadeh's negation. This approach was further analyzed in [7], where its application to decision making problems was widely discussed.

4 n-Dimensional Overlap Functions

Since overlap functions are not assumed to be associative, their extensions to the n-dimensional case with $n > 2$ requires a specific definition. On the other hand, in many problems it is natural to deal with more than two classes. These considerations led the authors in [21] to propose the following definition of n-dimensional overlap function. The results in this section can be found in that reference.

Definition 3.3 An n-dimensional aggregation function $G_O : [0, 1]^n \longrightarrow [0, 1]$ is an n-dimensional overlap function if and only if:

1. G_O is symmetric.
2. $G_O(x_1, \ldots, x_n) = 0$ if and only if $\prod_{i=1}^{n} x_i = 0$.
3. $G_O(x_1, \ldots, x_n) = 1$ if and only if $\prod_{i=1}^{n} x_i = 1$.
4. G_O is increasing.
5. G_O is continuous.

Note that, taking into account this definition, an object c that belongs to three classes C_1, C_2 and C_3 with degrees $x_1 = 1$, $x_2 = 1$ and $x_3 = 0.5$ will not have the maximum degree of overlap since condition (3) of the previous definition is not satisfied. Even more, if the degrees are $x_1 = 1$, $x_2 = 1$ and $x_3 = 0$, from the second condition we will conclude that the n-dimensional degree of overlapping of this object into the classification system given by the classes C_1, C_2 and C_3 will be zero. This is the reason why this first extension of the original idea of overlap proposed in [6] has been called n-*dimensional* overlap. Note that this definition is closely related to the idea of intersection of n classes, and in this sense, it follows also one of the inspirations behind the original idea of overlap function.

Example 3.1 It is easy to see that the following aggregation functions are n-dimensional overlap functions:

- The minimum powered by p. $G_O(x_1, \ldots, x_n) = \min_{1 \le i \le n}\{x_i^p\} = \left[\min_{1 \le i \le n}\{x_i\}\right]^p$ with $p > 0$. Note that for $p = 1$ we recover an averaging aggregation function (i.e., an aggregation functions which is bigger than or equal to the minimum and smaller than or equal to the maximum at every point).
- The product. $G_O(x_1, \ldots, x_n) = \prod_{i=1}^{n} x_i$.
- The Einstein product aggregation operator. $EP(x_1, \ldots, x_n) = \frac{\prod_{i=1}^{n} x_i}{1 + \prod_{i=1}^{n}(1-x_i)}$.
- The sinus induced overlap. $G_O(x_1, \ldots, x_n) = \sin \frac{\pi}{2} (\prod_{i=1}^{n} x_i)^p$ with $p > 0$.

The characterization results already introduced for bivariate overlap functions may be extended in a straight way for n-dimensional overlap functions. In particular, we remark the following result.

Proposition 3.1 *Let $A_n : [0, 1]^n \longrightarrow [0, 1]$ be an aggregation function. If A_n is averaging, then A_n is an n-dimensional overlap function if and only if it is symmetric, continuous, has zero as absorbing element and satisfies $A_n(x, 1, \ldots, 1) \neq 1$ for any $x \neq 1$.*

In fact, we have the following theorem.

Theorem 3.4 *Let $G_1, \ldots G_m$ be n-dimensional overlap functions and let $M : [0, 1]^m \longrightarrow [0, 1]$ be a continuous aggregation function such that if $M(x) = 0$ then $x_i = 0$ for some i and $M(x) = 1$ only if $x_i = 1$ for some i. Then the aggregation function $G : [0, 1]^n \longrightarrow [0, 1]$ defined as $G(x) = M(G_1(x), \ldots, G_m(x))$ is an n-dimensional overlap function.*

Notice that, since any averaging aggregation function M being continuous satisfies the conditions of the previous Theorem, it is possible to conclude that any continuous averaging aggregation of n-dimensional overlap functions is also an n-dimensional overlap function. Also note that it is also possible to consider $m = 1$, so, for instance, the power of a 2-dimensional overlap function is again an overlap function.

Example 3.2 As an illustrative case, consider that of OWA operators (see [3]). Let $W = (w_1, \ldots, w_n) \in [0, 1]^n$ be a weighting vector (i.e., such that $w_1 + \cdots + w_n = 1$). Then, the only OWA operator which is also an n-dimensional overlap function is that for which $w_n = 1$. That is, the minimum. On the other hand, for any continuous t-norm T with no zero divisors, its n-ary form is always an n-dimensional overlap function.

For the case of Kolmogorov-Nagumo means we can state the following result.

Proposition 3.2 *Let $f : [0, 1] \rightarrow [-\infty, 0]$ be a continuous increasing bijection, that is, $f :]0, 1] \rightarrow] - \infty, 0]$ is an increasing bijection such that $\lim_{x \rightarrow 0} f(x) = -\infty$, so by abuse of notation we define $f(0) = -\infty$. Then the function: $G_O(x_1, \ldots, x_n) = f^{-1}(\frac{1}{n} f(x_1) + \cdots + \frac{1}{n} f(x_n))$ is an n-dimensional overlap function.*

5 Applications of Overlap Functions

As we have already said, the first application of overlap functions was in image processing, and more specifically, in image thresholding problems. However, along the last years, the notion of overlap function has displayed a strong versatility and has been applied to many different situations and problems. Among them we can mention:

- Decision making problems [17].
- Forest fire detection [20].
- Classification problems [16, 23, 24].
- Approximate reasoning [19].

The number of different applications is growing very fast, as these functions have shown themselves very useful in order to build new aggregation functions which are well adapted to specific data. In this sense, it is worth to mention the construction of fuzzy measures (used as the basis to obtain Choquet integrals and hence averaging aggregation functions) using overlap functions and overlap indexes obtained from them. See in particular, [24] for more details.

Besides, recently there have been discussed relevant results on the development of implication functions defined in terms of overlap functions [9, 10, 12, 13, 25]. Also other classes of aggregation functions have been defined in terms of overlap functions [26]. Finally, also extensions of the notions of overlap function from a theoretical point of view have been considered [8]. Some important properties of overlap and grouping functions, such as migrativity, homogeinity or Archimedeanity, were studied in [2, 11, 27]. Additive generators of overlap functions were introduced in [14].

6 Conclusions

In this chapter we have made a review of the evolution of the notion of overlap function. This notion is being nowadays an active topic of interest for researchers, as overlap functions can be used instead of t-norms, for instance, in applications where associativity is not a must. For this reason, recent works are interested both in theoretical aspects of overlap functions (including the possible definition of implication functions, etc.) as well as in specific applications in image processing or optimization.

Acknowledgements This work has been partially supported by research project PID2019-108392GB-I00 of the Spanish Government, by Brazilian Funding Agency CNPq (Proc. 307781/2016-0), and by project APVV-14-0013.

References

1. Amo, A., Montero, J., Biging, G., Cutello, V.: Fuzzy classification systems. Eur. J. Oper. Res. **156**, 459–507 (2004)
2. Bedregal, B.C., Dimuro, G.P., Bustince, H., Barrenechea, E.: New results on overlap and grouping functions. Inf. Sci. **249**, 148–170 (2013)
3. Beliakov, G., Bustince, H., Calvo, T.: A Practical Guide to Averaging Functions. Springer (2016)
4. Bustince, H., Pagola, M., Barrenechea, E.: Construction of fuzzy indices from fuzzy DI-subsethood measures: application to the global comparison of images. Inf. Sci. **177**(3), 906–929 (2007)
5. Bustince, H., Barrenechea, E., Pagola, M.: Image thresholding using restricted equivalence functions and maximizing the measures of similarity. Fuzzy Sets Syst. **158**(5), 496–516 (2007)
6. Bustince, H., Fernandez, J., Mesiar, R., Montero, J., Orduna, R.: Overlap functions. Nonlinear Anal. Theory Methods Appl. **72**(3–4), 1488–1499 (2010)
7. Bustince, H., Pagola, M., Mesiar, R., Hullermeier, E., Herrera, F.: Grouping, overlap, and generalized bientropic functions for fuzzy modeling of pairwise comparisons. IEEE Trans. Fuzzy Syst. **20**(3), 405–415 (2012)
8. De Miguel, L., Gómez, D., Rodríguez, J.T., Montero, J., Bustince, H., Dimuro, G.P., Sanz, J.A.: General overlap functions. Fuzzy Sets Syst. (in press)
9. Caoa, M., Hua, B.Q., Qiao, J.: On interval (G, N)-implications and (O, G, N)-implications derived from interval overlap and grouping functions. Int. J. Approx. Reason. **100**, 135–160 (2018)
10. Dimuro, G.P., Bedregal, B., Santiago, R.H.N.: On (G, N)-implications derived from grouping functions. Inf. Sci. **279**, 1–17 (2014)
11. Dimuro, G.P., Bedregal, B.: Archimedean overlap functions: the ordinal sum and the cancellation, idempotency and limiting properties. Fuzzy Sets Syst. **252**, 39–54 (2014)
12. Dimuro, G.P., Bedregal, B.: On residual implications derived from overlap functions. Inf. Sci. **312**, 78–88 (2015)
13. Dimuro, G.P., Bedregal, B., Bustince, H., Jurio, A., Baczyński, M., Miś, K.: QL-operations and QL-implication functions constructed from tuples (O, G, N) and the generation of fuzzy subsethood and entropy measures. Int. J. Approx. Reason. **82**, 170–192 (2017)
14. Dimuro, G.P., Bedregal, B., Bustince, H., Asiain, M.J., Mesiar, R.: On additive generators of overlap functions. Fuzzy Sets Syst. **287**, 76–96 (2016)
15. Dubois, D., Ostasiewicz, W., Prade, H.: Fuzzy sets: history and basic notions. In: Fundamentals of Fuzzy Sets. Kluwer, Boston, MA (2000)
16. Elkano, M., Galar, M., Sanz, J., Bustince, H.: Fuzzy rule-based classification systems for multi-class problems using binary decomposition strategies: on the influence of n-dimensional overlap functions in the fuzzy reasoning method. Inf. Sci. **332**, 94–114 (2016)
17. Elkano, M., Galar, M., Sanz, J., Schiavo, P.F., Pereira, S., Dimuro, G.P., Borges, E.N., Bustince, H.: Consensus via penalty functions for decision making in ensembles in fuzzy rule-based classification systems. Appl. Soft Comput. **67**, 728–740 (2018)
18. Fodor, J., Roubens, M.: Fuzzy preference modelling and multicriteria decision support. In: Theory and Decision Library. Kluwer Academic Publishers, Boston, MA (1994)
19. Garcia-Jimenez, S., Bustince, H., Huellermeier, E., Mesiar, R., Pal, N.R., Pradera, A.: Overlap indices: construction of and application to interpolative fuzzy systems. IEEE Trans. Fuzzy Syst. **23**(4), 1259–1273 (2015)

20. Garcia-Jimenez, S., Jurio, A., Pagola, M., De Miguel, L., Barrenechea, E., Bustince, H.: Forest fire detection: a fuzzy system approach based on overlap indices. Appl. Soft Comput. **52**, 834–842 (2017)
21. Gómez, D., Rodriguez, J.T., Montero, J., Bustince, H., Barrenechea, E.: n-Dimensional overlap functions. Fuzzy Sets Syst. **287**, 57–75 (2016)
22. Jurio, A., Bustince, H., Pagola, M., Pradera, A., Yager, R.R.: Some properties of overlap and grouping functions and their application to image thresholding. Fuzzy Sets Syst. **229**, 69–90 (2013)
23. Lucca, G., Sanz, J., Dimuro, G.P., Bedregal, B., Bustince, H.: A proposal for tuning the α parameter in $C_\alpha C$-integrals for application in fuzzy rule-based classification systems. Natl. Comput. (in press). https://doi.org/10.1007/s11047-018-9678-x
24. Paternain, D., Bustince, H., Pagola, M., Sussner, P., Kolesárová, A., Mesiar, R.: Capacities and overlap indexes with an application in fuzzy rule-based classification systems. Fuzzy Sets Syst. **305**, 70–94 (2016)
25. Qiao, J., Hu, B.Q.: The distributive laws of fuzzy implications over overlap and grouping functions. Inf. Sci. **438**, 107–126 (2018)
26. Qiao, J., Hu, B.Q.: On the migrativity of uninorms and nullnorms over overlap and grouping functions. Fuzzy Sets Syst. **346**, 1–54 (2018)
27. Qiao, J., Hu, B.Q.: On homogeneous, quasi-homogeneous and pseudo-homogeneous overlap and grouping functions. Fuzzy Sets Syst. (in press)

Interpolative Reasoning: Valid, Specificity-Gradual and Similarity-Based

Marcin Detyniecki

Abstract In this paper, a recall of the similarity-based view of interpolative reasoning considered as an analogical scheme of reasoning introduced by Bernadette Bouchon-Meunier and her colleagues. Based on this, an extension of this replacement is presented that finely corrects the validity by choosing a similar observation that non only guarantees convexity, but also a gradual behaviour with respect to a specifity measure.

1 Introduction

Work on fuzzy rule interpolation date from the early 90s, raised by a Lotfi Zadeh's plenary lecture at the conference NAFIPS'91 where he mentioned the problem of sparse fuzzy rule base. Since then and still, this topic is the subject of much research. Even in the classical logic, the technique of *knowledge interpolation* [8] has been proposed as a solution to knowledge-intensive computations of classical algorithms. The general idea of *knowledge interpolation* is described as follows: "carrying out computations of similar nature to find solution that balances two extremes, in order to build up a solution that falls between them". For the authors, two important issues must be addressed: *how to measure the distances between two symbols* and *how to carry out the actual interpolation*. These issues are also important issues for a fuzzy interpolation method.

A lot of approaches in fuzzy rule interpolation in the literature refer to an analogical framework, or a similarity-based framework - promoted among others by Bernadette Bouchon-Meunier. Among these approaches, two main principles can be distinguished: either the interpolation process is considered as a one-step process

M. Detyniecki (✉)
AXA REV - Research Engineering and Vision, Paris, France
e-mail: marcin.detyniecki@axa.com

Sorbonne Universite, CNRS, Laboratoire d'Informatique de Paris 6, LIP6, Paris, France

Polish Academy of Science, IBS PAN, Warsaw, Poland

© The Editor(s) (if applicable) and The Author(s), under exclusive license to Springer Nature Switzerland AG 2021
M.-J. Lesot and C. Marsala (eds.), *Fuzzy Approaches for Soft Computing and Approximate Reasoning: Theories and Applications*, Studies in Fuzziness and Soft Computing 394, https://doi.org/10.1007/978-3-030-54341-9_4

like in the classical interpolation (see for instance [11, 12, 15]), or it is considered as a two-steps process, one step dedicated to the location of the interpolated conclusion, the other one to the precision and specificity of the interpolated conclusion (for instance [2]).

Several works aimed to find a general framework to compare the numerous existing approaches since the pioneer work of [11]. For instance, a way to compare methods consists in checking a set of desirable properties that were brought out by [4] or by [10]. Some of these properties are for instance the validity of the conclusion (the conclusion should be a fuzzy set) or the monotonicity of the deduced solution (if the observation is more specific than the antecedent, then the interpolated set should also be more specific than the conclusion). Another framework aiming at comparing existing approaches was proposed by [13]. The problem of fuzzy interpolation is considered from an analytical point of view. Then, the aim is to represent algorithmic approaches by a fuzzy function.

Interpolative reasoning is concerned with sparse fuzzy rule-based systems and its aim is to provide the most convenient conclusion corresponding to an observation by considering only the two rules, which do not directly apply to that observation. Since we lack information to properly infer the conclusion, we use a set of assumptions, which as a result constraint the problem strongly enough so that a valid conclusion can be calculated.

Among the suited properties, there is the monotonicity with respect to observation-conclusion specificity. Based on our current knowledge no existing approach is able to be responsive, gradual with respect to specificity and produce valid solutions. Our approach is a combination of framing a previous work [5] in the analogical framework, which allowed us to improve the proposed validity correction.

This paper is organised as follows. First in Sect. 2, a recall of the analogical schema is presented. This analogical framework allows us to identify, in Sect. 3, the properties and the constraints imposed to a fuzzy interpolation approach in order to be a valid reasoning, and not just a generalised arithmetic computation. These general axioms along other technical ones as described in Sect. 4, are the theoretical pillars making the similarity-based interpolative reasoning a unique viable solution. Finally, a conclusion is drawn and some still open works are presented.

2   Interpolative Reasoning

Let X and Y be two ordered universes of values. We denote $[0, 1]^X$ (resp. $[0, 1]^Y$) the set of all the fuzzy subsets on X (resp. Y). For all fuzzy subset $A \in [0, 1]^X$, we denote for all $x \in X$, $A(x)$ the membership degree of x to A. Even if X and Y can be any ordered sets, in this paper, for the sake of simplicity, we consider that both X and Y are subsets of $\mathbb{R}$, and we consider $\preceq$ an order relation on the fuzzy sets of $\mathbb{R}$. Interpolative reasoning can be formalised as follow.

Let $A_1, \ldots, A_n$ be $n > 1$ fuzzy sets on X such that for any $i < n$, $A_i \preceq A_{i+1}$, and for all $1 \leq i < n$, $A_i \cap A_{i+1} = \emptyset$ for a given intersection of fuzzy sets. Let $B_1, \ldots, B_n$ be $n > 1$ fuzzy sets on Y such that for all $i < n$, $B_i \preceq B_{i+1}$.

It is well-known that a fuzzy decision rule R defines a link between a given $A \in [0, 1]^X$ and a given $B \in [0, 1]^Y$. That link is provided by an expert, or inferred by machine learning. If we consider $R_1, \ldots, R_n$ be n rules defined[1] as:

$$
\begin{aligned}
R_1 &: A_1 \rightarrow B_1 \\
R_2 &: A_2 \rightarrow B_2 \\
&\ldots \\
R_n &: A_n \rightarrow B_n
\end{aligned}
\tag{1}
$$

Each rule R_i is equivalent to a decision rule: "if X is A_i, then Y is B_i". Reasoning with such a rule-based system consists in building up the solution B_* corresponding to a given observation $A_* \in [0, 1]^X$ according to the rules.

In the case of A_* is not equal to one A_i, using the generalised modus ponens is not suitable as the involved premises do not cover the whole description space X. That lead to construct a total uncertain solution B_* bringing out no useful information to draw a valid decision. Interpolative reasoning has been proposed to solve such kind of problems.

Interpolative reasoning is concerned with sparse fuzzy rule-based systems and its aim is to provide the most convenient conclusion B_* corresponding to an observation $A_* \in [0, 1]^X$ by considering only the two rules R_i and R_{i+1} where $A_i \preceq A_* \preceq A_{i+1}$. Thus interpolative reasoning is to deduce a description B_* on $[0, 1]^Y$ for any description A_* on $[0, 1]^X$ according to the two rules R_i and R_{i+1}.

For the sake of simplicity, in the following, we consider that $i = 1$ and focus on the rules R_1 and R_2 where $A_1 \preceq A_* \preceq A_2$.

2.1 Similarity-Based Interpolative Reasoning

Approximate reasoning aims at handling imprecise rules and graduality [17]. Depending on the deduction that should be build there exist kinds of fuzzy rules, implicative or conjunctive rules, as gradual rules [14]. In our setting, interpolative reasoning aims at offering a good way to model human process of reasoning by handling graduality. In that spirit, general framework has been introduced that encompasses similarity-based interpolation methods as an analogical reasoning process [2]. Here, the construction of B_* is based on a concept of resemblance [6] and lies on the knowledge of two similarity measures: S_X a similarity measure on $[0, 1]^X \times [0, 1]^X$, and S_Y a similarity measure on $[0, 1]^Y \times [0, 1]^Y$.

---

[1] A very restrictive version of fuzzy decision base rules, however, it enables us to lighten the notations and focus only on the important properties we want to highlight.

The analogical schema has been applied to several reasoning schema: rule-base reasoning, inductive reasoning, gradual reasoning, and so on [2]. A particular form of that analogical schema is the interpolative reasoning as explained hereafter.

An approach to use sparse fuzzy rules under hypotheses of graduality and analogy has been introduced previously [1, 3, 7]. That approach is based on the following hypothesis (or axioms):

- There exists a gradual behaviour of the variables which are involved in both premise and conclusion of the rules. For instance, with a rule of the form (R) "if u is A then v is B", the value of v increases if the value of u increases (or decreases according to the kind of graduality).
- The knowledge about v is analogous to the knowledge about u with respect to accuracy and uncertainty. Thus, a precisiation of information regarding u entails a precisiation of information concerning v.
- Concerning the deduction of B_* related to A_*, the imprecision of B_* should depend not only on the imprecision of A_* but also on the one of A_1, A_2, B_1, and B_2.

Interpolative reasoning is an application of the analogical schema when the description B_* is deduced in the following way.

An intermediate solution B' is deduced from R_1 such that B' is related to B_1 as A_* is related to A_1 (i.e. $S_X(A_1, A_*) \leq S_Y(B_1, B')$). Another intermediate solution B'' is deduced from R_2 such that B'' is related to B_2 as A_* is related to A_2 (i.e. $S_X(A_2, A_*) \leq S_Y(B_2, B'')$). Finally, the description B_* associated to A_* is deduced by aggregating B' and B'', for instance by means of a weighted aggregation (the closeness of A_* to A_1 or to A_2 could be used).

A direct application of viewing interpolative reasoning as an analogical scheme leads to the Bouchon-Meunier's et al. approach [1]. In this method, B' is constructed by means of two seminal elements: its *location* on the universe of definition, and its *shape* around is location.

As a basic concept, the location of a fuzzy set of $\mathbb{R}$ is a real number that represents its place [1, 3, 7]. For fuzzy sets given as a triangular or a trapezoidal fuzzy interval, the location can be defined as the middle of its kernel.[2] The location of B_* is linearly interpolated from the locations of A_*, A_1, A_2, B_1, and B_2. Once the location of B_* has been determined, the approach focuses on the comparison of the shape of A_* against the shapes of A_1 and A_2. The aim here is to use the differences in shape of A_* against A_1 and A_2 to build B' from B_1 and B'' from B_2.

Concerning the shape, the *distinguishability* measure plays the role of the similarity that enables the implementation of the analogical schema of reasoning. In consequence the analogical construction of the shape consists in two steps:

- Let A' be the translated of A_* such that A' and A_1 have the same location. A' is compared to A_1 thanks to the shape distinguishability of their shapes. That shape distinguishability is thus used to modify B_1 in order to build the fuzzy set B' that

---

[2]Any approach used to defuzzify a fuzzy set (for instance, those used in fuzzy control) can be used to define the location of a fuzzy set. In particular, some approaches could be used in case of non-triangular or non-trapezoidal fuzzy sets.

has the same shape distinguishability against B_1 than A_* has against A_1. That fuzzy set B' is translated in order that its location coincide with the previously computed location of B_*.

- Let A'' be the translated of A_* such that A'' and A_2 have the same location. In the same way, A'' is compared to A_2 thanks to their shape distinguishability in order to build B'' from B_2. That fuzzy set B'' is translated in order that its location coincide with the computed location of B_*.

In this way, analogical schema for the deduction of the solution is respected: B' and B_1 (resp. B'' and B_2) have the same shape distinguishability than A' and A_1 (resp. A'' and A_2).

In order to obtain the final interpolated set we need to determine the shape of B_* from the shapes of B' and B''. The authors use a weighted aggregation that enables to take into account the position of A_* between A_1 and A_2: the closer A_* and A_1 (resp. A_2), the more important is the contribution of B' (resp. B'') in the aggregation to build B_*.

3 Similarity Based Axioms for Interpolative Reasoning

The use of the similarity-based reasoning framework implies a set of constraints, which translates into a number of suitable properties for a correct reasoning.

As mentioned in the introduction, several works [4, 10, 13] have aimed to establish a lists of desirable properties and/or a general framework to compare the numerous existing approaches.

In the following, we identify the recurrent important properties. We will focus on the interpretation of these properties, for a more complete presentation please refer to the papers mentioned above.

3.1 Behaviour When Observation Equals a Premise of a Known Rule

When an observation matches exactly a premise of a rule, we expect that the interpolated (i.e. inferred) conclusion corresponds exactly to the conclusion of that rule. Formally if $A_* = A_i$ then $B_* = B_i$.

This constraint corresponds to the limit condition when the position of the observation and one premise coincide and the similarity of the two equals 1 (i.e. the two fuzzy sets are identical). We expect, as for the generalised modus ponens [16], the conclusion to be identical to the rule's conclusion, and this for all degrees of truth, and not only in the Boolean case. This property is also known as compatibility with the rule-base [10].

This behaviour can also be required, with a different interpretation, from a pure interpolation perspective. In this case, we consider it as limit conditions of an inter- polation approach.

3.2   Homogeneity of Observation-Premise Specificity with Respect to Conclusions

The homogeneity of observation-premise specificity with respect to conclusions is a pivotal property differentiating similarity-based approaches from others [1, 3, 4, 7]. It is based on the knowledge of a specificity measure as explained hereafter.

If an observation is more specific than the premises of the rules then the conclu- sion should be more specific than the conclusions of the rules [4]. It is a suitable property, because if conclusion specificity is independent from premise specificity, then, as seen for all methods ignoring it, only the position of the observation is mean- ingful. Producing unresponsive behaviours such as obtaining the same interpolated conclusions for a very precise observation and for a very vague one.

The use of a similarity measure leads to a natural and meaningful solution to integrate this property, since it is the way to compare the specificities of the different sets. Notice that it is clearly not the only one, since most the approaches verify it.

3.3   Monotonicity with Respect to Observation-Conclusion Specificities

In this part, we focus on a very important property that an interpolative reasoning approach should fulfil: the property of monotonicity with respect to observation- conclusion specificities. A very similar property is described by Jenei [10]. The idea is that more specific observations should lead to a more specific interpolated solutions. Unlike the previous one, this constraint does not take directly into account the specificity of the premises.

This constraint is closely related to the homogeneity of observations-premise specificity (Sect. 3.2). In fact, a constraint stating that "*if the premise's specificity and observation's specificity are equal then the specificity of the interpolated conclusion should be equal to the specificity of rule's conclusion*" would imply that the two properties are equivalent.

3.4 Valid Convex Conclusions

Although in the similarity-based framework it is assumed that for any observation we can infer a conclusion, this is not really required. In fact, in theory we could imagine that the sets are not similar enough to infer anything meaningful.

In practice, because of the nature of the interpolative problem, the methods always produce an inferred conclusion and it is expected that the membership function describing the fuzzy sets to be valid. From a purely interpolation perspective this is an obvious constraint, since there is no reason that an interpolation does not exist.

It is important to notice that this constraint turn out to be not obvious to be fulfilled in particular due to its interaction with the specificity monotonicity property required in Sect. 3.2.

3.5 Graduality with Respect to Rule's Proximity

Although this required property has not been identified by any author, it is regularly present in proposed approaches. The idea is that *"the "closer" an observation is to a premise of a rule, the more similar the inferred conclusion should be to that rule's conclusion"*. In fact, behind the interpolation there is the idea of a weighted combining of two rules framing the observation.

In most existing approaches, this "graduality" is formulated in terms of linear equations, which as it is presented in the following leads to non-valid solutions (i.e. a solution that could not be interpreted as a fuzzy set). Similarity-based approaches, not necessarily linear, overcome this difficulty without major difficulties.

4 Success and Limitations of Similarity-Based Interpolative Reasoning

The main reason of success of similarity-based interpolative reasoning, is that by being designed inside the similarity-based framework (Sect. 2.1), they automatically respect the constraints (Sect. 3) and thus are sound from an approximate reasoning perspective.

Another fundamental but hidden reason is that the framework separates the computation of the specificity and the position in such a way that a double linear interpolation is possible.

In order to better understand advantages and limitations of these approaches, let us assume that a specificity measure is given. This measure provides a real positive value describing any fuzzy set. Moreover, let an estimation of the position of any fuzzy set be given. This estimation is based again on a measure providing real positive

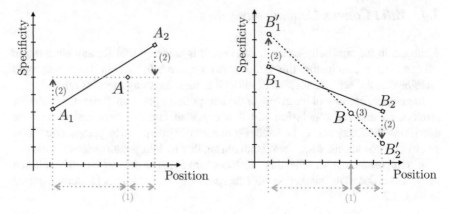

Fig. 1 Two specificity-position graphs on which any interpolative scenario can be represented. In particular, here an example of similarity-based similarity is illustrated: (1) the position is interpolated, (2) the similarities are exploited to obtain analogous specificity variations, (3) a linear interpolation between the new specificities is computed

values. In that case, the situation of any interpolative case can be represented in a space defined by the two axis *position* and *specificity* as shown in Fig. 1.

The specificity measure translates into a numerical value the shape of a membership function. An example of such a measure is the area under the curve. The position can be estimated by several means, for instance, the centre of gravity.

In the following, it is important to remember that any conclusion drawn from a particular configuration is independent of the specificity and positions measures, since for any of such a measure, as long as it is not trivial, such a situation can be built.

On the graph on the left of Fig. 1 points A_1 and A_2 represent the specificity and positions of premises of the framing rules. On the graph on the right we have two same descriptions of conclusions B_1 and B_2. The plain lines between these points represent a perfect linear interpolation where the *specificity* changes linearly with the *position*.

As illustrated on the figure, the specificity of A may not lie on the plain line. This implies that it is impossible to have a interpolation that respects *linearly* the graduality with respect to rule's proximity. In fact, at the interpolated position of the position (1), independently of how the inferred set is obtained, there is only one possible specificity that respects a linear constraint, and this would be on the plain line joining B_1 and B_2.

If responsive (i.e. non trivial) solutions are needed, the homogeneity of observations-premise specificity should be respected according to conclusions. In that case, a particular handling is needed, such as a "second-layer interpolation" [9] that will allow to manage the specificity differences. The particular handling is achieved in the similarity-based approaches by "shifting" the conclusions (on Fig. 1 step (2)) in such a way that the shifted conclusions are as *similar* with the original conclusions as the observation was with the corresponding premise. Then a linear interpolation between the shifted conclusions can be obtained (on Fig. 1 step (3)), so that it respects all the constraints mentioned in the previous section.

4.1 Non-convex Solutions

Based on what has been said above, it easy to see when *linear* interpolative approaches are about to fail. In fact, the only case they can work is when the specificity is obtained based on the position of the conclusion, independently on how it was obtained (ignoring suitable property Sect. 3.2). Since all the methods try to include the specificity in their calculations, in a way or another as for instance via alpha-cuts, it is not difficult to see that they should fail for some particular cases.

One of these cases producing a non-viable solution, specifically a non-convex ones, happens when the premises are rather imprecise, the conclusions and observation are very precise.

To simplify the computations we can even suppose that A_1 and A_2 (resp. B_1 and B_2) have the same specificity. In that case, when interpolating, since the observation A is more precise than the premises A_i, we expect to obtain a proportionally more precise B than the rule's conclusion B_i. And since the B_i are already very precise, the interpolation approaches try to reduce the conclusion's specificity going below zero, which is translated in practice by the construction of a solution that can be interpreted as a regular fuzzy set (that kind of solution is denoted "non-valid solution").

In terms of specificity, the non-convex solutions are obtained when for a fuzzy set an impossible negative specificity was interpolated.

4.1.1 Convexity correcting solutions

Since it is known that non-convex solutions may appear, most interpolative approaches of the Literature referenced in the introduction have produced post processing procedures that "repair the convexity" of the result.

Most often this is done by re-ordering the parameter defining the fuzzy set so that it becomes a valid solution. In that case, of course we obtain a convex solution but the interpolation does not make sense anymore.

In fact, let us assume that an interpolation for a particular observation and a set of rules gives as inferred conclusion a precise singleton. Let us also assume that in that same situation there is a new observation A'_* which is more precise than the previous one. In that case, as stated by the homogeneity constraint, an even more specific solution is expected but it is impossible to build it up.

Thus, in that case, interpolative approaches compute a non-valid solution, which is transformed into a convex solution by a post-processed "repair" procedure. However, for all known approaches, this solution is strictly less precise than what has been inferred in the previous situation, which not only goes against the required properties for the interpolation, but also is counter-intuitive.

4.2 Valid and Specificity-Gradual Solution

To avoid the non-valid solutions, in [5] authors, based on the analogical schema, proposed the compelling idea of replacing the observation by a similar but less specific substitute. As we observed in the previous section *all* the currently known proposed methods leads to some non-suitable behaviour. The reasons behind are, in the one hand, the replacement of the observation for both rules when just one of the rules similarities leads to a non convex solution, and in the other hand, the symmetric treatment for construction of the new observation.

Based on the descriptions associated to Fig. 1, we propose a new solution, which is responsive and linearly monotone with respect to the specificity. This time instead of a non valid B' there will be a similar to observation A'_* (analogously for a B'' non valid, a similar A''_*) each defined by:

$$A'_* = \begin{cases} (b-u, b, b+k-k') & \text{if } u > q - q' \text{ and } v \leq k - k' \\ (b-q+q', b, b+k-k') & \text{if } u \leq q - q' \text{ and } v \leq k - k' \\ (b-q+q', b, b+v) & \text{if } u \leq q - q' \text{ and } v > k - k' \end{cases} \qquad (2)$$

$$A''_* = \begin{cases} (b-u, b, b+l-l') & \text{if } u > h - h' \text{ and } v \leq l - l' \\ (b-h+h', b, b+l-l') & \text{if } u \leq h - h' \text{ and } v \leq l - l' \\ (b-h+h', b, b+v) & \text{if } u \leq h - h' \text{ and } v > l - l' \end{cases} \qquad (3)$$

With this new definition of the most similar observation allowing a valid solution, we obtain the suited behaviour in all cases.

4.2.1 Observing the Gradual Specificity: Illustration

In Fig. 2, we present two very similar rule interpolation configurations: one with symmetric fuzzy sets and one with asymmetric ones. In both cases we study the influence of the increase of specificity of an observation.

We immediately notice that in the symmetric case, increasing the specificity implies an augmentation of the specificity of the interpolated conclusion, while in the asymmetric case the interpolated conclusion specificity is blocked. This is due to the fact that in latter there is a replacement of the observation to avoid non-convex conclusions.

Now, in Fig. 3, we kept the same configuration and use the proposed solution. We notice that the behaviour in the symmetric fuzzy sets scenario is the same as

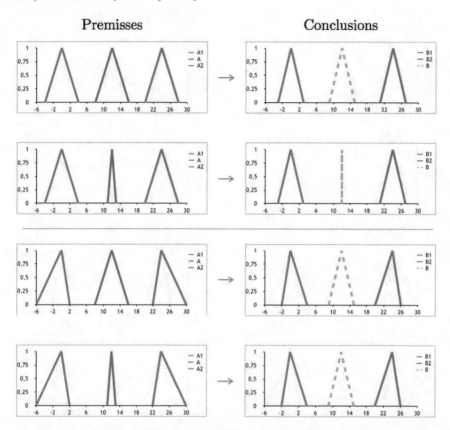

Fig. 2 Two very similar configurations with minor differences: above, we have symmetric fuzzy sets, while, below, we have asymmetric ones. In both cases we study the influence of the same increase of specificity of an observation. In the symmetric case the interpolation is responsive to specificity increase, while in the asymmetric case the interpolated conclusion specificity is blocked by the observation replacement-strategy proposed in [5]

for the one proposed in [5]. However this time, in the non asymmetric scenario, the behaviour of the specificity is coherent (i.e. the specificity reduction is not blocked by the convexity correction).

5 Conclusions

For more than twenty years the fuzzy community has been addressing the problem known as "interpolation" in the context of sparse rules. It took then years to realise that the similarity-based reasoning, promoted among others by Bernadette Bouchon-

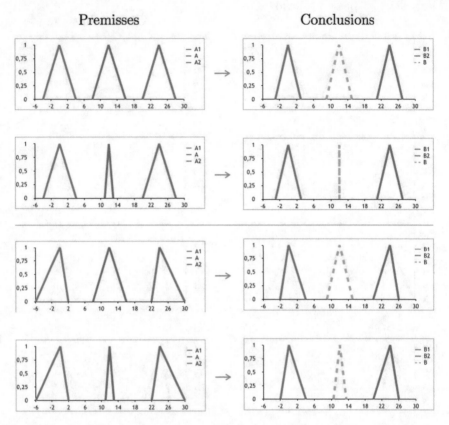

Fig. 3 The improved convexity correction leads to the suited behaviour, to be compared with the one shown in Fig. 2. We have again the two very similar configurations with minor differences: above symmetric fuzzy sets, while, below, asymmetric ones. This time in both cases the increase of specificity of an observation is reflected on the inferred conclusion

Meunier, can lead to a theoretical framework, in which interpolation is a valid and viable reasoning-based solution—as shown in this paper.

As most of other approaches, in practice, the similarity-based interpolation can lead to non convex fuzzy sets. Based on the similarity framework it has been proposed an intuitive an coherent way to produce valid fuzzy sets. The main idea is to replace the observation leading to invalidity by a similar observation.

In this paper, an extension of this replacement is presented that finely corrects the validity by choosing a similar observation that non only guarantees convexity, but also a gradual behaviour with respect to specifity: a property identified in the literature as very suited.

We believe this new improved convexity correction—and in particular its general behaviour as studied in this paper—opens up new perspectives for correct repair mechanisms for other approaches not only similarity-based ones.

References

1. Bouchon-Meunier, B., Delechamp, J., Marsala, C., Mellouli, N., Rifqi, M., Zerrouki, L.: Analogy and interpolation in the case of sparse rules. In: Eurofuse, SIC'99, pp. 132–136. Budapest, Hungary (1999)
2. Bouchon-Meunier, B., Delechamp, J., Marsala, C., Rifqi, M.: Several forms of fuzzy analogical reasoning. In: Proceedings of the IEEE International Conference on Fuzzy Systems, FUZZ-IEEE'1997, pp. 45–50. Barcelona, Spain (1997)
3. Bouchon-Meunier, B., Delechamp, J., Marsala, C., Rifqi, M.: Analogy as a basis of various forms of approximate reasoning. In: Bouchon-Meunier, B., Yager, R.R., Zadeh, L.A. (eds.) Uncertainty in Intelligent and Information Systems, vol. 20, pp. 70–79. World Scientific (2000)
4. Bouchon-Meunier, B., Dubois, D., Marsala, C., Prade, H., Ughetto, L.: A comparative view of interpolation methods between sparse fuzzy rules. In: IFSA'01 World Congress, pp. 2499–2504. Vancouver, Canada (2001)
5. Bouchon-Meunier, B., Esteva, F., Godó, L., Rifqi, M., Sandri, S.: A principled approach to fuzzy rule base interpolation using similarity relations. In: EUSFLAT–LFA 2005, pp. 757–763. Barcelona, Spain (2005)
6. Bouchon-Meunier, B., Rifqi, M., Bothorel, S.: Towards general measures of comparison of objects. Fuzzy Sets Syst. **84**(2), 143–153 (1996)
7. Bouchon-Meunier, B., Rifqi, M., Marsala, C.: Interpolative reasoning based on graduality. In: Proceedings of the IEEE International Conference on Fuzzy Systems, FUZZ-IEEE'2000, pp. 483–487. San Antonio, USA (2000)
8. Chatterjee, N., Campbell, J.A.: Knowledge interpolation: a simple approach to rapid symbolic reasoning. Comput. Artif. Intell. **17**(6), 517–551 (1998)
9. Detyniecki, M., Marsala, C., Rifqi, M.: Double-linear fuzzy interpolation method. In: Proceedings of the IEEE International Conference on Fuzzy Systems, FUZZ-IEEE'2011, pp. 455–462. IEEE, Taipei, ROC (2011)
10. Jenei, S.: Interpolation and extrapolation of fuzzy quantities revisited-an axiomatic approach. Soft. Comput. **5**(3), 179–193 (2001)
11. Kóczy, L.T., Hirota, K.: Rule interpolation in approximate reasoning based fuzzy control. In: IFSA'91, pp. 89–92. Brussels, Belgium (1991)
12. Kóczy, L.T., Hirota, K.: Interpolative reasoning with insufficient evidence in sparse fuzzy rule bases. Inf. Sci. **71**(1–2), 169–201 (1993)
13. Perfilieva, I., Dubois, D., Prade, H., Esteva, F., Godó, L., Hodakova, P.: Interpolation of fuzzy data: analytical approach and overview. Fuzzy Sets Syst. **192**, 134–158 (2012)
14. Ughetto, L., Dubois, D., Prade, H.: Implicative and conjunctive fuzzy rules: a tool for reasoning from knowledge and examples. In: Proceeding of the Sixteenth National Conference on AI (AAAI'99), pp. 214–219. Orlando, Fl, USA (1999)
15. Wu, Z.Q., Masaharu, M., Shi, Y.: An improvement to Kóczy and Hirota's interpolative reasoning in sparse fuzzy rule bases. Int. J. Approx. Reason. **15**(3), 185–201 (1996)
16. Zadeh, L.A.: Outline of a new approach to the analysis of complex systems and decision processes. IEEE Trans. Syst. Man Cybern. **3**, 28–44 (1973)
17. Zadeh, L.A.: A theory of approximate reasoning. Mach. Intell. **9**, 149–194 (1979)

A Similarity-Based Three-Valued Modal Logic Approach to Reason with Prototypes and Counterexamples

Francesc Esteva, Lluis Godo, and Sandra Sandri

Abstract In this paper we focus on the application of similarity relations to formalise different kinds of graded approximate reasoning with gradual concepts. In particular we extend a previous approach that studies properties of a kind of approximate consequence relations for gradual propositions based on the similarity between both prototypes and counterexamples of the antecedent and the consequent. Here we define a graded modal extension of Łukasiewicz's three-valued logic Ł$_3$ and we show how the above mentioned approximate consequences can be interpreted in this modal framework, while preserving both prototypes and counterexamples.

1 Introduction

Indistinguishability relations, also known as *fuzzy similarity relations*, are suitable graded generalisations of the classical notion of equivalence relation that go back to Zadeh, Trillas and colleagues [31–33, 38].

Definition 1 Let W be a universe and let $*$ be a t-norm. A binary fuzzy relation $S : W \times W \longrightarrow [0, 1]$ is called an *indistinguishability relation* with respect to $*$ (or an $*$-indistinguashability relation, or $*$-similarity relation) if, for any $x, y, z \in W$, the following properties hold:

- Reflexivity: $S(x, x) = 1$,
- Symmetry: $S(x, y) = S(y, x)$,
- $*$-Transitivity: $S(x, y) * S(y, z) \leq S(x, z)$.

F. Esteva · L. Godo (✉)
IIIA—CSIC, Campus de la UAB, 08193 Bellaterra, Spain
e-mail: godo@iiia.csic.es

F. Esteva
e-mail: esteva@iiia.csic.es

S. Sandri
LAC—INPE, SJ Campos, SP 12227-010, Brazil
e-mail: sandra.sandri@inpe.br

M.-J. Lesot and C. Marsala (eds.), *Fuzzy Approaches for Soft Computing and Approximate Reasoning: Theories and Applications*, Studies in Fuzziness and Soft Computing 394,
https://doi.org/10.1007/978-3-030-54341-9_5

Additionally, if the following property holds, then S is called a *strict* indistinguishability relation:

• Strictness: $S(x, y) = 1$ iff $x = y$,

Actually, this definition is a generalization of Zadeh's concept of fuzzy similarity relations introduced in [38], that corresponds to the particular case of min-indistinguishability relations, that is, when one takes $* = \min$ above. Actually, min-indistinguishability relations are a very particular class of fuzzy relations since all the cuts of a min-similarity relation turn out to be classical equivalence relations, and hence it defines a partition tree (a partition for each cut of the relation, and in which the partition in one level is a refinement of the partition in the level above it). This does not hold for a $*$-similarity with $* \neq \min$.

Fuzzy similarity relations have been extensively studied from a mathematical point of view, see for example [2, 11, 15, 16, 24–27, 34] and its applications cover many different topics like classification, analogical reasoning, preferences, interpolation, morphology, etc. In this paper we are interested in the application of similarity relations to formalise different kinds of graded approximate reasoning with gradual concepts. This approach dates back to Ruspini [30] where he proposes a semantics for fuzzy sets based on the idea of prototypes and similarity relations. Along this line, there have been a number of contributions towards the logical formalisation of (graded) approximate inference patterns of the kind "if φ then *approximately ψ*", either as graded consequence relations and with unary or binary modalities [18, 20, 21, 36, 37], but always dealing with classical propositions φ and ψ. In a recent paper [19], this approach has been extended to cope with vague concepts (or propositions) based on the similarity between both prototypes and counterexamples of the antecedent and the consequent. This approach is a natural generalization for Łukasiewicz's three-valued logic Ł$_3$ of the notion of logical consequence that preserves truth-degrees ($\models^{\leq}$). In this paper, we define a graded modal extension of Łukasiewicz's three-valued logic Ł$_3$, where modalities $\Diamond_a \varphi$ stand for *approximately φ to the degree at least a*, where a is a value in a finite scale $G \subset [0, 1]$, and we show how one can interpret in this modal framework the above mentioned approximate consequences preserving both prototypes and counterexamples.

The paper is structured as follows. After this short introduction, Sects. 2 and 3 are devoted to briefly recall material from [19]. Namely, in Sect. 2 we overview a logical approach to reason with vague concepts represented by examples and counterexamples based on the three-valued Łukasiewicz logic Ł$_3$, while in Sect. 3 we characterize three similarity-based graded notions of approximate logical consequence among vague propositions. Finally, in Sect. 4 we formally define a multi-modal extension of Ł$_3$ to capture reasoning about the approximate consequences, and prove its completeness. We end up with some conclusions.

2 Prototypes, Counter-Examples and Borderline Cases: A Simple 3-Valued Model for Gradual Properties

2.1 A 3-Valued Approach

A vague property, in the sense of gradual, is characterized by the existence of borderline cases; that is, objects or situations for which the property only partially applies.

In a recent paper [19], the authors investigate how a logic for vague concepts can be defined starting from the most basic description of a vague property or concept φ in terms of two subsets of the set Ω of situations: the set of *examples* –situations where φ definitely applies– and the set of *counterexamples* –situations where φ does not apply for sure–, denoted $[\varphi^+]$ and $[\varphi^-]$ respectively.

The consistency condition $[\varphi^+] \cap [\varphi^-] = \emptyset$ is assumed to hold. Further, the remaining set of situations $[\varphi^\sim] = \Omega \setminus ([\varphi^+] \cup [\varphi^-])$ are assumed to be those where φ only partially applies, that is, the set of borderline cases. In such a scenario, one is led to a three-valued framework, where for each situation $w \in \Omega$, either $w \in [\varphi^+]$, $w \in [\varphi^-]$ or $w \in [\varphi^\sim]$.[1]

To define a logic to deal with these 3-valued concepts, a first question is how to define the prototypes and counter-examples of compound concepts from their basic constituents. Let us consider a language with four connectives: conjunction ($\wedge$), disjunction ($\vee$), negation ($\neg$) and implication ($\rightarrow$). The rules for $\wedge$, $\vee$ and $\neg$ seem clear to be given as follows:

$$[(\varphi \wedge \psi)^+] = [\varphi^+] \cap [\psi^+], \quad [(\varphi \wedge \psi)^-] = [\varphi^-] \cup [\psi^-],$$
$$[(\varphi \vee \psi)^+] = [\varphi^+] \cup [\psi^+], \quad [(\varphi \vee \psi)^-] = [\varphi^-] \cap [\psi^-],$$
$$[(\neg\varphi)^+] = [\varphi^-], \qquad\qquad [(\neg\varphi)^-] = [\varphi^+].$$

The case for $\rightarrow$ is not as straightforward as for the previous connectives since several choices can be considered. For instance, if w is a borderline case for both φ and ψ then it can be considered as an example for $\varphi \rightarrow \psi$ rather than a borderline case of $\varphi \rightarrow \psi$. The former choice leads to the well-known three-valued Łukasiewicz logic, while the latter would lead to Kleene's three-valued logic as done in [35] (see e.g. [14] for a relevant discussion on three-valued logical representations of imperfect information). In this paper, we follow [19] and will use the three-valued Łukasiewicz logic $Ł_3$ as base logic to reason with vague concepts, we will thus stick to the following rule for $\rightarrow$:

$$[(\varphi \rightarrow \psi)^+] = [\varphi^-] \cup [\psi^+] \cup ([\varphi^\sim] \cap [\psi^\sim]), \quad [(\varphi \rightarrow \psi)^-] = [\varphi^+] \cap [\psi^-].$$

---

[1]It is worth noticing that in this 3-valued model, the set $[\varphi^\sim]$ is not meant to represent the situations the agent does not know whether φ applies or not; rather it is meant to represent the situations where the concept only partially applies, or equivalently, the situation that are borderline cases for the concept φ (see [17] for a discussion on this topic).

2.2 A Refresher on 3-Valued Łukasiewicz Logic Ł₃

Let us briefly recall the formal logical framework of the 3-valued Łukasiewicz logic Ł₃, see e.g. [13, 23]. Let Var denote a (finite) set of atomic concepts, or propositional variables, from which compound concepts (or formulas) are built using the connectives $\wedge, \vee, \rightarrow$ and $\neg$. We will denote the set of formulas by $\mathscr{L}_3(Var)$, or simply by $\mathscr{L}_3$ in case of no doubt. Further, we identify the set of all possible situations Ω with the set of all evaluations w of atomic concepts Var into the truth set $\{0, 1/2, 1\}$, that is, $\Omega = \{0, 1/2, 1\}^{Var}$, with the following intended meaning: for every $\alpha \in Var$, $w(\alpha) = 1$ means that w is an example of α (resp. w is a model of α in logical terms), $w(\alpha) = 0$ means that w is a counterexample of α (resp. w is a counter-model of α), and $w(\alpha) = 1/2$ means that w is borderline situation for α, i.e. it is neither an example nor a counterexample. According to the previous discussion, truth-evaluations w will be extended to compound concepts according to the semantics of Ł₃, defined by the following truth-functions[2]: for all $x, y \in \{0, 1/2, 1\}$,

$$x \wedge y = \min(x, y), \quad x \vee y = \max(x, y), \quad x \rightarrow y = \min(1, 1 - x + y), \quad \neg x = 1 - x.$$

In Ł₃, strong conjunction and disjunction connectives can be defined from $\rightarrow$ and $\neg$ as follows: for all $\varphi, \psi \in \mathscr{L}_3$, $\varphi \otimes \psi := \neg(\varphi \rightarrow \neg\psi)$ and $\varphi \oplus \psi := \neg\varphi \rightarrow \psi$.[3] Actually, for each concept $\varphi \in \mathscr{L}_3$, three related *Boolean* concepts can be defined using the connective $\otimes$:

$$\varphi^+ := \varphi \otimes \varphi, \quad \varphi^- := (\neg\varphi) \otimes (\neg\varphi) = (\neg\varphi)^+, \quad \varphi^\sim := \neg\varphi^+ \wedge \neg\varphi^-,$$

with the following semantics:

$w(\varphi^+) = 1$ if $w(\varphi) = 1$, $w(\varphi^+) = 0$ otherwise;
$w(\varphi^\sim) = 1$ if $w(\varphi) = 1/2$, $w(\varphi^\sim) = 0$ otherwise;
$w(\varphi^-) = 1$ if $w(\varphi) = 0$, $w(\varphi^-) = 0$ otherwise.

Therefore, if for $\star \in \{+, -, \sim\}$ we let $[\varphi^\star] = \{w \in \Omega \mid w(\varphi^\star) = 1\}$, then $[\varphi^+]$, $[\varphi^-], [\varphi^\sim]$ capture respectively the (classical) sets of examples, counterexamples and borderline cases of φ.

The usual notion of logical consequence in 3-valued Łukasiewicz logic is defined as follows: for any set of formulas $\Gamma \cup \{\varphi\}$,

$$\Gamma \models \varphi \quad \text{if,} \quad \text{for any evaluation } w, w(\psi) = 1 \text{ for all } \psi \in \Gamma, \text{ then } w(\varphi) = 1.$$

It is well known that this consequence relation can be axiomatized by the following axioms and rule (see e.g. [13]):

(Ł1) $\varphi \rightarrow (\psi \rightarrow \varphi)$
(Ł2) $(\varphi \rightarrow \psi) \rightarrow ((\psi \rightarrow \chi) \rightarrow (\varphi \rightarrow \chi))$

---

[2]We use the same symbols of connectives to denote their corresponding truth-functions.
[3]One could take $\rightarrow$ and $\neg$ as the only primitive connectives since $\wedge$ and $\vee$ can be defined from $\rightarrow$ and $\neg$ as well: $\varphi \wedge \psi = \varphi \otimes (\varphi \rightarrow \psi)$ and $\varphi \vee \psi = (\varphi \rightarrow \psi) \rightarrow \psi$.

(Ł3) $(\neg\varphi \rightarrow \neg\psi) \rightarrow (\psi \rightarrow \varphi)$
(Ł4) $(\varphi \vee \psi) \rightarrow (\psi \vee \varphi)$
(Ł5) $\varphi \oplus \varphi \leftrightarrow \varphi \oplus \varphi \oplus \varphi$

(MP) The rule of modus ponens: $\dfrac{\varphi, \quad \varphi \rightarrow \psi}{\psi}$

This axiomatic system, denoted Ł$_3$, is strongly complete with respect to the above semantics; that is, for a set of formulas $\Gamma \cup \{\varphi\}$, $\Gamma \models \varphi$ iff $\Gamma \vdash \varphi$, where $\vdash$, the notion of proof for Ł$_3$, is defined from the above axioms and rule in the usual way.

Remark: In the sequel we will restrict ourselves on considerations about logical consequences from a *finite* set of premises. In such a case, if $\Gamma = \{\varphi_1, \ldots, \varphi_n\}$ then $\Gamma \models \psi$ iff $\varphi_1 \wedge \ldots \wedge \varphi_n \models \psi$, and hence it will be enough to consider premises consisting of a single formula.

2.3 Dealing with both Prototypes and Counter-Examples: The Logic Ł$_3^{\leq}$

It is evident that, for any formulas φ, ψ, $\varphi \models \psi$ can be equivalently expressed as $[\varphi^+] \subseteq [\psi^+]$. This makes clear that $\models$ is indeed the consequence relation that preserves the examples of concepts. Similarly one can also consider the consequence relation that preserves counterexamples [19]. Namely, one can contrapositively define a falsity-preserving consequence as:

$$\varphi \models^C \psi \text{ if } \neg\psi \models \neg\varphi, \text{ that is,}$$
$$\text{if for any evaluation } w, w(\psi) = 0 \text{ implies } w(\varphi) = 0.$$

Unlike classical logic, in 3-valued Łukasiewicz logic it is not the case that $\varphi \models \psi$ iff $\neg\psi \models \neg\varphi$. As we have seen that the former amounts to require $[\varphi^+] \subseteq [\psi^+]$, while the latter, as shown next, amounts to require $[\psi^-] \subseteq [\varphi^-]$. Clearly these conditions, in general, are not equivalent, except when φ and ψ do not have borderline cases, that is, when $[\varphi^+] \cup [\varphi^-] = [\psi^+] \cup [\psi^-] = \Omega$.

Equivalently, $\varphi \models^C \psi$ holds iff $[\psi^-] \subseteq [\varphi^-]$, and iff for any evaluation $v \in \Omega$, $v(\varphi) \geq 1/2$ implies $v(\psi) \geq 1/2$, or in other words, $[\varphi^+] \cup [\varphi^\sim] \subseteq [\psi^+] \cup [\psi^\sim]$. Now we define the consequence relation that preserves both examples and counterexamples in the natural way.

Definition 2 $\varphi \models^{\leq} \psi$ if $\varphi \models \psi$ and $\varphi \models^C \psi$, that is, if $[\varphi^+] \subseteq [\psi^+]$ and $[\psi^-] \subseteq [\varphi^-]$.

Note that $\varphi \models^{\leq} \psi$ iff, for any $w \in \Omega$, $w(\varphi) \leq w(\psi)$, that justifies the use of the superscript $\leq$ in the symbol of consequence relation. Indeed, the consequence relation $\models^{\leq}$ is known in the literature as the *degree-preserving* companion of $\models$, as opposed to the *truth-preserving* consequence $\models$, that preserves the truth-value '1', see e.g. [3].

$\models^{\leq}$ can also be axiomatized by taking as axioms those of Ł₃ and the following two inference rules:

$$(Adj) : \frac{\varphi, \ \psi}{\varphi \wedge \psi} \qquad (rMP) : \frac{\varphi, \ \vdash \varphi \rightarrow \psi}{\psi}$$

The resulting logic is denoted by $Ł_3^{\leq}$, and its notion of proof is denoted by $\vdash^{\leq}$. Notice that (rMP) is a weakened version of modus ponens, called restricted modus ponens, since $\varphi \rightarrow \psi$ has to be a theorem of Ł₃ for the rule to be applicable.

Therefore, $Ł_3^{\leq}$ (and its semantical counterpart $\models^{\leq}$) appears as a more natural logical framework to reason about concepts described by examples and counterexamples than the usual three-valued Łukasiewicz logic Ł₃.

3 A Similarity-Based Refined Framework

In the previous section we have discussed a logic for reasoning about vague concepts described in fact as 3-valued fuzzy sets. A more fine grained representation, moving from 3-valued to [0, 1]-valued fuzzy sets, can be introduced by assuming the availability of a (fuzzy) similarity relation $S : \Omega \times \Omega \rightarrow [0, 1]$ among situations. Indeed, for instance, assume that all examples of φ are examples of ψ, but some counterexamples of ψ are not counterexamples of φ. Hence, we cannot derive that ψ follows from φ according to $\models^{\leq}$. However, if these counterexamples of ψ greatly resemble to counterexamples of φ, it seems reasonable to claim that ψ follows *approximately* from φ.

Actually, starting from Ruspini's seminal work [30], a similar approach has already been investigated in the literature in order to extend the notion of entailment in classical logic in different frameworks and using formalisms, see e.g. [21]. Following this line, here we recall from [19] a graded generalization of the $\models^{\leq}$ in the presence of similarity relation S on the set of 3-valued Łukasiewicz interpretations Ω, that allows to draw approximate conclusions.

Since, by definition $\varphi \models^{\leq} \psi$ if both $\varphi \models \psi$ and $\varphi \models^C \psi$, that is, if $[\varphi^+] \subseteq [\psi^+]$ and $[\psi^-] \subseteq [\varphi^-]$, it seems natural to define that ψ is an approximate consequence of φ to some degree $a \in [0, 1]$ when every example of φ is similar (at least to the degree a) to some example of ψ, as well as every counterexample of ψ is similar (to at least to the degree a) to some counterexample of φ. In other words, this means that to relax $\models^{\leq}$ we propose to relax both $\models$ and $\models^C$. This idea is formalized next, where we assume that a ∗-similarity relation $S : \Omega \times \Omega \rightarrow [0, 1]$ is given. Moreover, for any subset $A \subseteq \Omega$ and value $a \in [0, 1]$ we define its *a-neighborhood* as

$$A^a = \{w \in \Omega \mid \text{there exists } w' \in A \text{ such that } S(w, w') \geq a\}.$$

Definition 3 For any pair of formulas φ, ψ and for each degree $a \in [0, 1]$, we define the graded consequence relations $\models_a, \models_a^C$ and $\models_a^{\leq}$ as follows:

(i) $\varphi \models_a \psi$ if $[\varphi^+] \subseteq [\psi^+]^a$
(ii) $\varphi \models_a^C \psi$ if $[\psi^-] \subseteq [\varphi^-]^a$
(iii) $\varphi \models_a^{\leq} \psi$ if both $[\varphi^+] \subseteq [\psi^+]^a$ and $[\psi^-] \subseteq [\varphi^-]^a$.

Taking into account that for any formula χ we have $[(\neg\chi)^+] = [\chi^-]$, it is clear that both $\models_a^C$ and $\models_a^{\leq}$ can be expressed in terms of $\models_a$. Namely, $\varphi \models_a^C \psi$ iff $\neg\psi \models_a \neg\varphi$ and $\varphi \models_a^{\leq} \psi$ iff $\varphi \models_a \psi$ and $\neg\psi \models_a \neg\varphi$.

The consequence relations $\models_a$'s are very similar to the so-called approximate graded entailment relations defined in [18] and further studied in [21]. The main difference is that in [18] the authors consider classical propositions while here we consider three-valued Łukasiewicz propositions. Nevertheless, as shown in [19], one can prove very similar characterizing properties for the $\models^a$'s. In the following theorem we assume the language is built from a *finite* set of propositional variables Var, and for each evaluation $w \in \Omega$, $\overline{w}$ denotes the following proposition:

$$\overline{w} = (\bigwedge_{p \in Var: \, w(p)=1} p^+) \wedge (\bigwedge_{p \in Var: \, w(p)=1/2} p^{\sim}) \wedge (\bigwedge_{p \in Var: \, w(p)=0} p^-).$$

So, $\overline{w}$ is a (Boolean) formula which encapsulates the complete description provided by w. Moreover, for every $w' \in \Omega$, $w'(\overline{w}) = 1$ if $w' = w$ and $w'(\overline{w}) = 0$ otherwise.

Theorem 1 *([19]) The following properties hold for the family $\{\models_a\}_{a \in [0,1]}$ of graded entailment relations on $\mathscr{L}_3$ induced by a $*$-similarity relation S on Ω:*

(i) *Nestedness: if $\varphi \models_a \psi$ and $b \leq a$, then $\varphi \models_b \psi$*
(ii) *$\models_1$ coincides with $\models$, while $\models \subsetneq \models_a$ if $a < 1$. Moreover, if $\psi \not\models \perp$, then $\varphi \models_0 \psi$ for any φ.*
(iii) *Positive-preservation: $\varphi \models_a \psi$ iff $\varphi^+ \models_a \psi^+$*
(iv) *$*$-Transitivity: if $\varphi \models_a \psi$ and $\psi \models_b \chi$ then $\varphi \models_{a*b} \chi$*
(v) *Left-OR: $\varphi \vee \psi \models_a \chi$ iff $\varphi \models_a \chi$ and $\psi \models_a \chi$*
(vi) *Restricted Right-OR: for all $w \in \Omega$, $\overline{w} \models_a \varphi \vee \psi$ iff $\overline{w} \models_a \varphi$ or $\overline{w} \models_a \psi$*
(vii) *Restricted symmetry: for all $w, w' \in \Omega$, $\overline{w} \models_a \overline{w'}$ iff $\overline{w'} \models_a \overline{w}$*
(viii) *Consistency preservation: if $\varphi \not\models \perp$ then $\varphi \models_a \perp$ only if $a = 0$*
(ix) *Continuity from below: If $\varphi \models_a \psi$ for all $a < b$, then $\varphi \models_b \psi$.*

Conversely, for any family of graded entailment relations $\{\vdash_a\}_{a \in [0,1]}$ on $\mathscr{L}_3$ satisfying the above properties, there exists a $*$-similarity relation S such that $\vdash_a$ coincides with $\models_a$ for each $a \in [0, 1]$.

Actually, the above properties also indirectly characterize $\models_a^{\leq}$ since, in the finite setting, $\models_a$ (and thus $\models_a^C$ as well) can be derived from $\models_a^{\leq}$ in the following sense: $\varphi \models_a \psi$ holds iff for every $w \in \Omega$ such that $w(\varphi) = 1$ there exists $w' \in \Omega$ such that $w(\psi) = 1$ and $\overline{w} \models_a^{\leq} \overline{w'}$.

However, a nicer characterization of $\models_a^{\leq}$ can be obtained if we extend the language of Ł$_3$ with the truth-constant $\frac{1}{2}$.

Lemma 1 *([19]) For any formulas in the expanded language, the following conditions hold:*

- $\varphi \models_a \psi$ *iff* $\varphi \models_{\bar{a}}^{\leq} \psi \vee \frac{1}{2}$
- $\varphi \models_a^C \psi$ *iff* $\varphi \wedge \frac{1}{2} \models_{\bar{a}}^{\leq} \psi$.

As a consequence, we have that $\varphi \models_{\bar{a}}^{\leq} \psi$ *iff* $\varphi \models_{\bar{a}}^{\leq} \psi \vee \frac{1}{2}$ *and* $\varphi \wedge \frac{1}{2} \models_{\bar{a}}^{\leq} \psi$.

From this, one can get the following representation for the $\models_{\bar{a}}^{\leq}$ consequence relations.

Theorem 2 *([19]) Let* $\{\vdash_{\bar{a}}^{\leq}\}_{a \in G}$ *be a set of consequence relations on the expanded language satisfying the following conditions:*

- $\varphi \vdash_{\bar{a}}^{\leq} \psi$ *iff* $\neg\psi \vdash_{\bar{a}}^{\leq} \neg\varphi$
- $\varphi \vdash_{\bar{a}}^{\leq} \psi$ *iff* $\varphi \vdash_{\bar{a}}^{\leq} \psi \vee \frac{1}{2}$ *and* $\varphi \wedge \frac{1}{2} \vdash_{\bar{a}}^{\leq} \psi$
- *The set of relations* $\{\vdash_a\}_{a \in G}$, *where* $\varphi \vdash_a \psi$ *is defined as* $\varphi \vdash_{\bar{a}}^{\leq} \psi \vee \frac{1}{2}$, *satisfy the conditions (i)-(ix) of Theorem 1.*

Then, there exists a similarity relation $S : \Omega \times \Omega \to G$ *such that* $\vdash_{\bar{a}}^{\leq} = \models_{\bar{a}}^{\leq}$, *for any* $a \in G$.

4 A Multi-modal Approach to Reason About the Similarity-Based Graded Entailments

As we have seen in Sect. 3, the three kinds of similarity-based graded entailments $\models_a$, $\models_a^C$ and $\models_{\bar{a}}^{\leq}$ are based on an idea of *neighbourhood* of a set of interpretations (see Def. 3) . Indeed, for instance, given a $*$-similarity relation S on the set of interpretations Ω, the idea of a graded entailment $\varphi \models_a \psi$ is to replace the classical constraint that the set $[\varphi^+]$ of models (prototypes) of φ have to be included in the set $[\psi^+]$ of models (prototypes of ψ, to a more relaxed condition in the sense that $[\varphi^+]$ needs only to be included in the neighbourhood of $[\psi^+]$ of radius a (i.e. each model of φ has to be at least a-similar to some model of ψ). Similarly with the other graded entailments $\models_a^C$ and $\models_{\bar{a}}^{\leq}$.

In the framework of relational semantics for modal logic, such neighbourhoods can be nicely captured by a certain class of generalized Kripke frames of the form (W, S), where W is a set of possible worlds and S a $*$-similarity relation on W. Then S induces a nested set of binary accessibility relations on W: for each value a, we can define the accessibility relation S_a among interpretations (or worlds in the modal logic terminology) in the natural way: $(w, w') \in S_a$ if $S(w, w') \geq a$, that is, S_a is the a-cut of S. In fact, for each value a, (W, S_a) is a classical Kripke frame, and the semantics of a corresponding possibility modal formula $\Diamond_a \varphi$ exactly corresponds to the notion of a-neighbourhood of (the set of models of) φ (see for instance [20] for the case of dealing with classical propositions.)[4]

In this section we will define a multi-modal logic over 3-valued Łukasiewicz logic Ł$_3$ expanded with the truth-constant $\frac{1}{2}$, and we will see how the above notions of graded entailment can be faithfully captured in this logic. Actually, this 3-valued modal logic is a richer framework with a more expressive power. To avoid unnecessary complications, we will make the following assumptions: all $*$-similarity relations S will take values in a finite set $G \subset [0, 1]$, containing 0 and 1, and $*$ will be a *finite* t-norm operation on G, that is, $(G, *, \leq, 1, 0)$ will be a finite totally ordered semigroup. In this way, we keep our language finitary and will avoid the use of an infinitary inference rule to cope with Property (ix) of Theorem 1. Then, we will expand the propositional language of Ł$_3$ with the (finite) set of modal operators $\{\Box_a : a \in G\}$ ($\Diamond_a$ will be used as abbreviations of $\neg\Box_a\neg$) by means of the usual rules. We will denote the modal language by $\mathscr{L}_\Box$.

For a given finite t-norm $(G, *, 1, 0)$ as above, the semantics will be given by $*$-similarity Kripke frames (W, S), where W is a set of possible worlds and $S : W \times W \to G$ is a *strict* $*$-similarity relation. A 3-valued $*$-similarity Kripke model is a structure $M = (W, S, e)$, where (W, S) is as above and $e : W \times Var \to \{0, 1/2, 1\}$ is a 3-valued evaluation of propositional variables for each possible world. An evaluation $e(w, \cdot)$ is extended to arbitrary formulas using the truth-functions of Ł$_3$ with the following special stipulations:

- $e(w, \overline{1/2}) = 1/2$
- $e(w, \Box_a\varphi) = \min\{e(w', \varphi) \mid (w, w') \in S_a\}$
- $e(w, \Diamond_a\varphi) = \max\{e(w', \varphi) \mid (w, w') \in S_a\}$

where, as already mentioned, $S_a = \{(w, w') \mid w, w' \in W, S(w, w') \geq a\}$. The corresponding notions of satisfiability, validity and consequence are respectively as follows:

- $(M, w) \models \varphi$ if $e(w, \varphi) = 1$.
- $M \models \varphi$ if $(M, w) \models \varphi$ for every $w \in W$.

Given a similarity scale $(G, *, 1, 0)$, we will denote by $\mathsf{SK}(G)$ the class of similarity Kripke models $M = (W, S, e)$ where S is a G-valued $*$-strict similarity relation on W. Then, we can finally define the notion of consequence relative to $\mathsf{SK}(G)$: for any set of formulas $\Gamma \cup \{\varphi\}$,

- $\Gamma \models_{\mathsf{SK}(G)} \varphi$ if, for every model $M = (W, S, e) \in \mathsf{SK}(G)$ and world $w \in W$, $(M, w) \models \psi$ for every $\psi \in \Gamma$ implies $(M, w) \models \varphi$.

Next, we aim at providing a complete aximatization for this 3-valued modal logic. We start by observing that each modality $\Box_a$ is interpreted in a Kripke model

---

[4]Note that this is not to be confused with the so-called *neighbourhood semantics* (also known as Scott-Montague semantics) for modal logics (see e.g. [12]), a more general semantics that, instead of using relational frames (W, R) consisting of a set W of worlds and an accessibility relation R, it is based on neighborhood frames (W, N), where N is a neighborhood function $N : W \to 2^{2^W}$ assigning to each possible world of W a set of subsets of W.

$M = (W, S, e)$ by the a-cut S_a of the similarity relation, i.e. each graded modality has associated a crisp accessibility relation S_a. Thus, we have to look at what properties these crisp relations S_a have. It is clear that S_a is a reflexive and symmetric relation for every $a \in G$. Moreover, due to the $*$-transitivity property of S, the following transitivity-like inclusions hold for any $a, b \in G$: $S_a \circ S_b \subseteq S_{a*b}$, where $\circ$ denotes usual composition of relations. Hence, if $a * a = a$, the relation S_a is also transitive. Therefore, each operator $\Box_a$ is a sort of 3-valued KTB modality, as the following well-known axioms are valid in every Kripke model for all $a \in G$:

$(K_a)\ \Box_a(\varphi \to \psi) \to (\Box_a\varphi \to \Box_a\psi)$
$(T_a)\ \Box_a\varphi \to \varphi$
$(B_a)\ \varphi \to \Box_a\Diamond_a\varphi.$

while the following generalized form of Axiom 4 is also valid for every $a, b \in G$:

$(4_{a,b})\ \Box_{a*b}\varphi \to \Box_b(\Box_a\varphi)$, or equivalently $\Diamond_a(\Diamond_b\varphi) \to \Diamond_{a*b}\varphi.$

Thus, if $a * a = a$ (e.g. for $a = 0$ or $a = 1$ at least), then $\Box_a$ can be considered as 3-valued S5 modality as it satisfies many-valued versions of Axioms K, T, B and 4 axioms.

Moreover, since the S_a's relations form a nested set, in the sense that $S_a \subseteq S_b$ if $a \geq b$, then any Kripke model validates the following axiom:

$(N_{a,b})\ \Box_b\varphi \to \Box_a\varphi$, if $a \geq b.$

Finally, let us observe the boundary properties of the relations S_a (when $a = 0$ or $a = 1$), namely: $S_0 = W \times W$ and $S_1 = \{(w, w) \mid w \in W\}$.

These properties and existing results for many-valued modal logics in general [4] and for some modal extensions of finite-valued Łukasiewicz logics Ł_n [1] lead to the definition of the following axiomatic system.

Definition 4 Given a finite similarity scale $(G, *, 1, 0)$, the axiomatic system $\text{Ł}_3^\Box(G)$ consists of the following groups of axioms and rules, where the subindices a, b run over G:

(Ł$_3$) axioms of Ł$_3$,
(bk) $\neg(\overline{1/2}) \leftrightarrow \overline{1/2}$
$(K_a)\ \Box_a(\phi \to \psi) \to (\Box_a\phi \to \Box_a\psi),$
$(A_a^\wedge)\ (\Box_a\varphi \wedge \Box_a\psi) \to \Box_a(\varphi \wedge \psi),$
$(A_a^c)\ \Box_a(\overline{1/2} \to \varphi) \leftrightarrow (\overline{1/2} \to \Box_a\varphi),$
$(A_a^\oplus)\ (\Box_a\varphi \oplus \Box_a\varphi) \leftrightarrow \Box_a(\varphi \oplus \varphi),$

$(T_a)\ \Box_a\phi \to \phi$
$(B_a)\ \varphi \to \Box_a\Diamond_a\varphi,$
$(4_{a,b})\ \Box_{a*b}\varphi \to \Box_b(\Box_a\varphi)$

$(N_{a,b})\ \Box_b\varphi \to \Box_a\varphi$, if $b \leq a$
$(C_1)\ \varphi \to \Box_1\varphi$

(MP) from φ and $\varphi \to \psi$ derive ψ

(Nec) from φ derive $\square_a \varphi$

We will denote by $\vdash_{Ł_3^\square(G)}$ the notion of proof in $Ł_3^\square(G)$ defined as usual from the above axioms and rules.

The first group of axioms $(Ł_3)$-$(A_a^\oplus)$ corresponds to the axiomatization of each $\square_a$ for the minimal modal logic over $Ł_3$ with the truth constant $\overline{1/2}$ with respect to the semantics with crisp accessibility relations, see [4]. The second group aims at capturing the three characteristic properties of the relations S_a above mentioned, namely reflexivity, symmetry and $*$-transitivity, see e.g. [1]. Finally, $(N_{a,b})$ captures the nestedness of the graded operators, while (C_1) aims at capturing the particular behaviour of the modal operator for the extremal value $a = 1$: $\square_1 \varphi$ collapses with φ itself since for any strict similarity $S(w, w') = 1$ iff $w = w'$.

Now we can prove that the axiomatic system $Ł_3^\square(G)$ is indeed complete with respect to the intended semantics, that is, the class of similarity Kripke models $\mathsf{SK}(G)$.

Theorem 3 $Ł_3^\square(G)$ *is complete w.r.t. the class of models* $\mathsf{SK}(G)$, *that is, for any set of formulas* $\Gamma \cup \{\varphi\}$, $\Gamma \models_{\mathsf{SK}(G)} \varphi$ *iff* $\Gamma \vdash_{Ł_3^\square(G)} \varphi$.

Proof (Sketch) We have to prove that if $\Gamma \nvdash_{Ł_3^\square(G)} \varphi$ there is a model $M \in \mathsf{SK}(G)$ and a world $w \in W$ such that $(M, w) \models \psi$ for every $\psi \in \Gamma$, but $(M, w) \nvDash \varphi$.

Actually, we will make the proof in two steps. In the first step we will build a canonical model $M^c = (W^c, S^c, e^c)$, where S^c is a (non-necessarily strict) $*$-similarity relation on W^c, and a world $v \in W^c$ such that $(M^c, v) \models \psi$ for every $\psi \in \Gamma$, but $(M^c, v) \nvDash \varphi$. For this, we basically use the results and proofs of [1] where the authors define KD45 and S5-like modal logics over a k-valued Łukasiewicz logic $Ł_k$.

Consider for each modal formula φ its propositional counterpart φ^* by treating subformulas of the form $\square_a \psi$ for some $a \in G$ as new propositional variables. So formally, we can consider a propositional language $\mathscr{L}_3(Var^*)$ built from the extended set of variables $Var^* = Var \cup \{(\square_a \psi)^* \mid \psi \in \mathcal{L}_3(Var), a \in G\}$, and let Ω^* the set of $Ł_3$-evaluations over the fomulas from $\mathscr{L}_3(Var^*)$. Then, the canonical model $M^c = (W^c, S^c, e^c)$ is defined as follows:

- $W^c = \{w \in \Omega^* \mid \forall \phi^* \in \Lambda : w(\phi^*) = 1\}$ with $\Lambda = \{\phi^* \mid \vdash_{Ł_3} \phi\}$;
- $S_a^c = \{(w_1, w_2) \in \Omega^* \times \Omega^* \mid \forall \phi \in \mathscr{L}_3(Var) : \text{if } w_1((\square_a \phi)^*) = 1 \text{ then } w_2(\phi^*) = 1\}$;
- $S^c(w_1, w_2) = \max\{a \in G \mid (w_1, w_2) \in S_a^c\}$, for all $w_1, w_2 \in W^c$;
- $e^c(w, p) = w(p)$ for each variable $p \in Var$.

Then, using the same techniques in [1], we can show that the fundamental *Truth Lemma* for the canonical model M^c, namely, for any modal formula φ and any $w \in W^c$, it holds that

$$e^c(w, \varphi) = w(\varphi^*).$$

Moreover, following [1] one can show that, so defined, S^c is reflexive, symmetric and $*$-transitive, and that $S^c(w, w') = 1$ iff $w = w'$ thanks to Axiom $(C1)$.[5] However, even if S_0^c turns out to be an equivalence relation, we cannot guarantee that $S_0^c = W^c \times W^c$ and hence we do not know whether $M^c \in \mathsf{SK}(G)$.

To remedy this problem, we need a further step. For this we need a previous lemma that shows that deductions in $Ł_3^\square(G)$ can be reduced to propositional deductions in $Ł_3$. The proof is rather standard and it is omitted.

Lemma 2 *For any set of modal formulas* $\Gamma \cup \{\varphi\}$, $\Gamma \vdash_{Ł_3^\square(G)} \varphi$ *iff* $\Gamma^* \cup \Lambda \vdash_{Ł_3} \varphi^*$ *where* Γ^* *is defined as above.*

Continuing with the main proof, assume $\Gamma \nvdash_{Ł_3^\square(G)} \varphi$. Then by the above lemma and propositional completeness of $Ł_3$, there is a $Ł_3$-evaluation v such that $v(\Gamma^*) = v(\Lambda) = 1$ and $v(\varphi^*) < 1$. By definition, $v \in W^c$, and thus we can consider its equivalence class w.r.t. S_0^c, i.e. $W^0 = \{w \in W^c \mid (v, w) \in S_0^c\}$. Then, an easy computation shows that $M^0 = (W^0, S^0, e^0)$, where $S^0 = S^c$ and $e^0 = e^c$, belongs to the class $\mathsf{SK}(G)$, and by the Truth Lemma it preserves the evaluations. In particular, we have $e^0(v, \psi) = v(\psi^*) = 1$ for each $\psi \in \Gamma$, while $e^0(v, \varphi) = v(\varphi^*) < 1$. In other words, we have shown that $\Gamma \nvDash_{\mathsf{SK}(G)} \varphi$. This finishes the proof.

Finally, we can show that the graded entailments defined by similarity reasoning over $Ł_3$ can be faithfully captured in the multi-modal logic $Ł_3^\square(G)$. Intuitively, consider an instance of a graded entailment like "$\varphi \models_a \psi$", whose intended meaning is "for all $w \in \Omega$, if $w \models \varphi$ then there exists w' such that $S(w, w) \geq a$ and $w' \models \psi$". To encode this condition in the modal framework, we use the universal modality $\square_0$ to model the "for all $w \in \Omega$", while the rest of condition "if $w \models \varphi$ then there exists w' such that $S(w, w) \geq a$ and $w' \models \psi$" can be naturally encoded by the formula "$\varphi \to \Diamond_a \psi$". Therefore, we can encode "$\varphi \models_a \psi$" by the $Ł_3^\square(G)$-formula $\square_0(\varphi \to \Diamond_a \psi)$. In summary, the translations are as follows:

Graded consequences	$Ł_3^\square(G)$-formulas
$\varphi \models_a \psi$	$\square_0(\varphi \to \Diamond_a \psi)$
$\varphi \models_a^C \psi$	$\square_0(\neg\varphi \to \Diamond_a \neg\psi)$, or $\square_0(\square_a \psi \to \varphi)$
$\varphi \models_a^\leq \psi$	$\square_0((\varphi \to \Diamond_a \psi) \wedge (\square_a \psi \to \varphi))$

5 Conclusions and Dedication

In this paper we have first recalled an approach towards considering graded approximate entailments between vague concepts (or propositions) based on the similarity

---

[5]By definition, $(w, w') \in S_1^c$ iff for all φ, $w(\square_1\varphi) = 1$ implies $w'(\varphi) = 1$, equivalent to the condition: for all φ, $w(\square_1\varphi) \leq w'(\varphi)$. But by definition of W^c, $w(\square_1\varphi) = w(\varphi)$, thus $(w, w') \in S_1^c$ iff for all φ, $w(\varphi) \leq w'(\varphi)$. Since S_1^c is symmetric, then $(w, w') \in S_1^c$ iff for all φ, $w(\varphi) = w'(\varphi)$, therefore, iff $w = w'$.

between both prototypes and counterexamples of the antecedent and the consequent, presented in [19]. This approach is a natural generalization for Łukasiewicz's three-valued logic Ł$_3$ of the notion consequence that preserves truth-degrees. Then, we have provided a modal logic formalisation, by defining a similarity-based graded modal extension of Ł$_3$, and have shown how this modal framework is expressive enough to accommodate reasoning about instances of the above approximate entailments.

This small paper on similarity-based reasoning is our humble tribute to Bernadette Bouchon-Meunier for her dedication to the field of approximate reasoning. Similarity relations have been one of the many nuclear research topics for Bernadette, as it is witnessed by her numerous and relevant contributions on this subject; here we cite only some representative works of hers relating together the notion of fuzzy similarity with analogical reasoning [6, 7, 10], similarity measures [5, 28, 29], and interpolation of fuzzy rules [8, 9, 22]. Her own remarkable studies and her relentless efforts to bring the computational intelligence and approximate reasoning community together have substantially deepened our knowledge and will continue to impact generations of computer scientists to come.

Acknowledgements The authors are thankful to an anonymous reviewer for his/her careful reading of the manuscript. Esteva and Godo acknowledge partial support by the Spanish FEDER/MINECO project TIN2015-71799-C2-1-P.

References

1. Blondel, M., Flaminio, T., Schockaert, S., Godo, L., De Cock, M.: On the relationship between fuzzy autoepistemic logic and fuzzy modal logics of belief. Fuzzy Sets Syst. **276**, 74–99 (2015)
2. Boixader, D., Jacas, J., Recasens, J.: Fuzzy equivalence relations: advanced material. In: Dubois, D., Prade, H. (eds.) Fundamentals of Fuzzy Sets, pp. 261–290. Kluwer, Dordrecht (2000)
3. Bou, F., Esteva, F., Font, J.M., Gil, A., Godo, L., Torrens, A., Verdú, V.: Logics preserving degrees of truth from varieties of residuated lattices. J. Log. Comput. **19**(6), 1031–1069 (2009)
4. Bou, F., Esteva, F., Godo, L., Rodríguez, R.: On the minimum many-valued modal logic over a finite residuated lattice. J. Log. Comput. **21**(5), 739–790 (2011)
5. Bouchon-Meunier, B., Coletti, G., Lesot, M.J., Rifqi, M.: Towards a conscious choice of a similarity measure: a qualitative point of view. In: 10th European Conference on Symbolic and Quantitative Approaches to Reasoning with Uncertainty (ECSQARU 2009), Lecture Notes in Computer Science 5590, pp. 542–553. Springer, Verona, Italy (2009)
6. Bouchon-Meunier, B., Delechamp, J., Marsala, C., Mesiar, R., Rifqi, M.: Fuzzy deductive reasoning and analogical scheme. In: Proceedings of the EUSFLAT-ESTYLF Joint Conference, Palma de Mallorca, Spain, pp. 22–25 (1999)
7. Bouchon-Meunier, B., Delechamp, J., Marsala, C., Rifqi, M.: Several forms of fuzzy analogical reasoning. In: 6th IEEE International Conference on Fuzzy Systems, Fuzz'IEEE'97, pp. 45–50. IEEE, Barcelona, Spain(1997)
8. Bouchon-Meunier, B., Esteva, F., Godo, L., Rifqi, M., Sandri, S.: A principled approach to fuzzy rule base interpolation using similarity relations. EUSFLAT–LFA 2005, pp. 757–763. Barcelona, Spain (2005)
9. Bouchon-Meunier, B., Rifqi, M., Marsala, C.: Interpolative reasoning based on graduality. In: Fuzz-IEEE 2000—9th IEEE International Conference on Fuzzy Systems, pp. 483-487. IEEE, San Antonio, USA (2000)

10. Bouchon-Meunier, B., Valverde, L.: A fuzzy approach to analogical reasoning. Int. J. Soft Comput. **3**(3), 141–147 (1999)
11. Castro, J.L., Klawonn, F.: Similarity in fuzzy reasoning. Mathware Soft Comput (1996)
12. Chellas, B.F.: Modal Logic. Cambridge University Press (1980)
13. Cignoli, R., D'Ottaviano, I.M.L., Mundici, D.: Algebraic Foundations of Many-Valued Reasoning. Trends in Logic, vol. 7. Kluwer, Dordrecht (1999)
14. Ciucci, D., Dubois, D., Lawry, J.: Borderline vs. unknown: comparing three-valued representations of imperfect information. Int. J. Approx. Reason. **55**(9), 1866–1889 (2014)
15. Demirci, M.: On many-valued partitions and many-valued equivalence relations. Int. J. Uncert. Fuzz. Knowl. Based Syst. **11**(2), 235–254 (2003)
16. Demirci, M.: Representations of the extensions of many-valued equivalence relations. Int. J. Uncer. Fuzz. Knowl. Based Syst. **11**(3), 319–342 (2003)
17. Dubois, D., Esteva, F., Godo, L., Prade, H.: An information-based discussion of borderline cases in categorization: six scenarios leading to vagueness. In: Cohen, H., Lefebvre, C. (eds.) Handbook of Categorization in Cognitive Science (Second Edition), pp. 1029–1051. Elsevier (2017)
18. Dubois, D., Prade, H., Esteva, F., Garcia, P., Godo, L.: A logical approach to interpolation based on similarity relations. Int. J. Approx. Reason. **17**, 1–36 (1997)
19. Dutta, S., Esteva, F., Godo, L.: On a three-valued logic to reason with prototypes and counterexamples and a similarity-based generalization. In: O.Lucas et al. (eds.) Proceedings of CAEPIA 2016 Advances in Artificial Intelligence, LNAI 9868, pp. 498–508. Springer, Berlin (2016)
20. Esteva, F., Garcia, P., Godo, L., Rodríguez, R.O.: A modal account of similarity-based reasoning. Int. J. Approx. Reason. **16**, 235–260 (1997)
21. F. Esteva, L. Godo, R. O. Rodríguez, T. Vetterlein., Logics for approximate and strong entailments. Fuzzy Sets Syst. **197**, 59–70 (2012)
22. Esteva, F., Rifqi, M., Bouchon-Meunier, B., Detynicki, M.: Similarity-based fuzzy interpolation method. In: Proceedings of the International conference on Information Processing and Management of Uncertainty in knowledge-based systems, IPMU 2004, pp. 1443–1449. Perugia, Italy (2004)
23. Hájek, P.: Methamatematics of Fuzzy Logic. Trends in Logic, vol. 4. Kluwer, Dordrecht (1998)
24. Höhle, U.: Many-valued equalities, singletons and fuzzy partitions. Soft Comput. **2**(3), 134–140 (1998)
25. Jacas, J.: Similarity relations—the calculation of minimal generating families. Fuzzy Sets Syst. **35**, 151–162 (1990)
26. Ovchinnikov, S.: Similarity relations, fuzzy partitions and fuzzy orderings. Fuzzy Sets Syst. **40**, 107–126 (1991)
27. Recasens, J.: Indistinguishability Operators—Modelling Fuzzy Equalities and Fuzzy Equivalence Relations. Studies in Fuzziness and Soft Computing 260, Springer (2011)
28. Rifqi, M., Bouchon-Meunier, B.: Set-theoretic similarity measures. In: Proceedings of Knowledge-based Intelligent Information Engineering Systems and Allied Technologies (KES'2002), pp. 879–884. Cremona, Italy (2002)
29. Rifqi, M., Detyniecki, M., Bouchon-Meunier, B.: Discrimination power of measures of resemblance. In: Proceedings of the 10th International Fuzzy Systems Association World Congress (IFSA 2003), Istanbul, Turkey (2003)
30. Ruspini, E.H.: On the semantics of fuzzy logic. Int. J. Approx. Reason. **5**, 45–88 (1991)
31. Trillas, E.: Assaig sobre les relacions dÕindistingibilitat. In: Proceedings Primer Congrés Català de Lògica Matemàtica, pp. 51–59. Barcelona (1982)
32. Trillas, E., Valverde, L.: An inquiry into indistinguishability operators. Theory Decis. Lib. 231–256. Reidel, Dordrecht (1984)
33. Trillas, E., Valverde, L.: On implication and indistinguishability in the setting of fuzzy logic. In: Kacprzyk, J., Yager, R.R. (eds.), Management Decision Support Systems Using Fuzzy Sets and Possibility Theory, pp. 198–212. Verlag TUV Rheinland Koln (1984)
34. Valverde, L.: On the structure of F-indistinguishability operators. Fuzzy Sets Syst. **17**, 313–328 (1985)

35. Vetterlein, T.: Logic of prototypes and counterexamples: possibilities and limits. In: Alonso, J.M. et al. (eds.), Proceedings of IFSA-EUSFLAT-15, pp. 697–704. Atlantis Press (2015)
36. Vetterlein, T.: Logic of approximate entailment in quasimetric spaces. Int. J. Approx. Reason. **64**, 39–53 (2015)
37. Vetterlein, T., Esteva, F., Godo, L.: Logics for approximate entailment in ordered universes of discourse. Int. J. Approx. Reason. **71**, 50–63 (2016)
38. Zadeh, L.A.: Similarity relations and fuzzy orderings. Inform. Sci. **3**, 177–200 (1971)

Analogy

Charles Tijus

Abstract Among cognitive processes, although literally based on logically false propositions, analogical and metaphorical reasoning are the most used and useful thinking for communicating, understanding, discovering, problem-solving and learning. The topic of this chapter about analogy and metaphor, is to address the kind of computation linking a target category to a source category that belongs to another domain that might be able to support reasoning properties based on the fallacy of the falsity of propositions, on imperfection, imprecision and approximation, gradualness, vagueness, fuzziness, uncertainty and implicit plausibility of likeness. Because notable advances in the computation of analogies are from Bernadette Bouchon-Meunier's work with her team: the fuzzy logic computation of analogical reasoning and schemes, we examine how such modeling of the hu-man computation of analogies can be used in turn to model the machine computa-tion of analogies.

1 Introducing Analogy as One of Two Main Ways of Thinking

Cognitive processes of both human and machine are for understanding and for decision-making. These processes are cognitive investigations about external physical things, including people and oneself; about who they are, what they are (structure and functioning), about the events and actions in which they participate, as well as ascribing intention to other agents [1, 2]. Although a cognitive system requires information from the external physical world, investigations about external things are firstly based on what we already know about them: on their internal representation, including the kind of things they are and the categories to which they belong. This inference-making about what things are and about how they behave are made for interaction, for decision-making about our judgments about them, about actions

C. Tijus (✉)
CHArt, University Paris 8, 2, rue de la Liberté 93526 St Denis Cedex 02, Paris, France
e-mail: tijus@univ-paris8.fr

© The Editor(s) (if applicable) and The Author(s), under exclusive license to Springer Nature Switzerland AG 2021
M.-J. Lesot and C. Marsala (eds.), *Fuzzy Approaches for Soft Computing and Approximate Reasoning: Theories and Applications*, Studies in Fuzziness and Soft Computing 394, https://doi.org/10.1007/978-3-030-54341-9_6

we might perform on them or with them. Classically, there are four kinds of inferences of these cognitive processes—deduction, induction, analogy and metaphor [3, 4]—that we propose to be reduced to two main kinds: deduction and induction as one kind and analogy and metaphor as another kind.

Inference-making of the first kind is literal, which means that the target domain of investigation is domain specific constrained: restricted to the categories of this domain, restricted to the intension and extension of these categories and to the relations among these categories. In that case, things under investigation are supposedly known; as in computer programming for which a literal value is a value given explicitly in the source code of the algorithm for a given variable. Such literal inference-making classically stands for both deduction and induction: either from category to exemplar (i.e. deduction: what is known about a category can be attributed to subordinate categories and finally to exemplars of these categories): Cats meow, Maine Coon is a kind of cat, Felix is a Maine Coon, Felix should meow. It stands also from exemplars to category (i.e. induction: what is known about exemplars of a category could be attributed to this category): Felix is a cat and meows, Tom is also a cat and meows: cats might meow. In that case, the top-down (from category to exemplars) or bottom-up (from exemplars to category) processes are about things known to belong to the same given category or, for more complex thinking and reasoning, to the same specific domain.

Contrary to deduction and induction that provide "true" sentences, Inference-making of the second kind is unliteral (i.e. not literal; not literally comparable). Things under the cognitive investigation are cross-domain constrained: they are of different categories, usually of exclusive domains: inference-making is from singular to singular, from an exemplar of a given category to exemplar of another category.

Such unliteral inference-making is found with analogical propositions based on comparison (e.g. the nucleus of the atom is the sun of the solar system) but also with metaphorical propositions [5] based on attributive categorization (e.g. John is a cat): John, a troubadour, sings to beg for a little love. Felix, a cat, meows to beg for food. John's voice sounds like meowing: propositions such as "John is meowing" or "John is a cat" are analogies. Such associations, like "Juliet is my sun" (i.e. she brings me joy and light) in from Romeo's diary of Shakespeare's "Romeo and Juliet," are based on some fuzzy resemblance, on analogy or metaphor (upon, according to/so to speak). According to Plato, analogies are founded on a reasoning based on "an argument from the similarity of things in some ways inferring their similarity in others" and on a computation based on "partial agreement, likeness or proportion between things". Although such analogically-based reasoning provides sentences that do not have evidential support and are logically "false", this kind of reasoning appears to be the most prominent kind of human of thinking [6] and maybe the one that machines will be using in the very near future for understanding, thinking and communicating:

- For analogical reasoning that is based on comparison: [A: the nucleus] is to [B: the Atom] what [C: the sun] is to [C: the solar system]. If one knows the relation between the sun and the solar system, s/he can infer the relation between the nucleus and the atom.

- For metaphorical reasoning that is based on attributive categorization [B: singing] is to [A: John] what [D: meowing] is to [C: Felix]. If one knows that Felix is meowing to beg, the conclusion is that John is begging. Thus, Felix is the source of the analogy, while John is the target of the analogy.

Note that to produce an analogy or a metaphor, the computation is from-target-to-source. For example, to produce "Juliet is my sun" from Juliet Shakespeare had to find a likeness source. In order to understand the likeness of the source, the cognitive investigation of the listener is a from-source-to-target computation. Thus, to understand "Juliet is my sun" from "sun," one must find the likeness to attribute to Juliet.

Note also that because analogy is based on comparison (x is equivalent to), source and target can permute. For example, "the nucleus of the atom is the sun of the solar system" is equivalent to "the sun of the solar system is the nucleus of the atom." Conversely, because metaphor is based on implication (x is a kind of y), source and target cannot permute. "John is a cat," the reverse is not true: "Cats are not as John" [C].

As a matter of fact, there is a very challenging scientific and technological issue to discover and model the what and how of the thinking processes that are able to produce and understand analogies. The kind of computation linking a target category to a source category that belongs to another domain might be able to support reasoning properties such as imperfection, imprecision and approximation, gradualness, vagueness, fuzziness, uncertainty and implicit plausibility of likeness. Notable advances in the computation of analogies are from Bernadette Bouchon-Meunier's work with her team: the fuzzy logic computation of analogical reasoning and schemes [7–16].

2 The Necessity of Fuzzy Logic Computation of Analogical Reasoning and Schemes

The first main advance of Bernadette Bouchon-Meunier (BBM) about analogical reasoning is due to the hint of using the main principle of fuzzy logic: i.e. the gradualness of membership function. So, as crisp sets representing precise and certain descriptions of objects might be regarded as particular cases of fuzzy sets [10], tautology as identity (the sun is the sun), analogy as comparison (the nucleus of the atom is the sun of the solar system) and antilogy as metaphor (Juliet is the sun of Romeo) are fuzzy sets of special kinds. Identity is computed as a particular case of analogy. Analogy is computed as a particular case of metaphor, with transitivity, asymmetry and irreflexivity relations.

The first main advance of BBM on analogical reasoning is due to her insight that the main principle of fuzzy logic might be the central core of analogical thinking and reasoning. There is a graduality of membership function that can be used both for "John is a man" and "John is a cat"; certainty (John is a man) being a special case of uncertainty (John is a cat). So, as crisp sets that represent precise and certain descriptions of objects, they are to be regarded as particular cases of fuzzy sets [10]. Tautology as identity (the sun is the sun), analogy as comparison (the nucleus of the atom is as the sun of the solar system) and antilogy as metaphorical contradiction (Juliet is the sun although Juliet is not the sun) are all fuzzy sets of special kinds. Identity is computed as a particular case of analogy. Analogy is computed as a particular case of metaphor, with transitivity, asymmetry and irreflexivity relations.

The second advance of BBM's team about analogical and metaphorical reasoning is embedded by the first one. This advance relates to tautology that appears to be very useful in daily life activities,—as well as antilogy (In London, even when it's not raining, it's raining!) -, to provide useful information. However, according to classic logic, tautologies such as "the sun is the sun" and "Paris is Paris" are per se uninformative. Similarly, according to the Grice's maxims of pragmatics [17], although they respect quality (truthful, supported by evidence) and manner (avoiding obscurity and ambiguity), tautologies violate two other maxims: quantity (to be as informative as possible) and relation (be pertinent). Here again with the BBM approach, tautologies and antilogies can be seen as special cases of graduality of certainty-uncertainty, in contradt to a full or null membership. As will be seen in the next section, the same categorization-based cognitive process appears to be a good candidate for the computing all of these forms of metaphorical reasoning [18].

The third advance is that analogical reasoning and metaphorical reasoning are cross-domains: a target object T (Tenor), is investigated from the point of view of a source that is used as a V (vehicle) for transmission of meanings. Having similes among the same category (tautology), among different categories of the same domain (analogy), among different categories across domains, or among inter-domains categories (metaphor), T and V can be computed according to the evaluation of their "closeness" through fuzzy modifiers in order to measure their similarity [16]. The advance is that closeness of two objects (i.e. moon and sun) as a semantic distance (very close, close, far, very far) can be described through fuzzy sets that can manage approximation. Closeness approximation depends on to T and V role (the moon-sun closeness in "the moon is a sun" being of a different value in "the sun is a moon"), according to context (the moon-sun closeness in "tonight the moon is a sun" being of a different value in "today the moon is a sun"), and motive (the moon-sun closeness in "in your drawing, the moon is a sun" being of a different value in "in the sky, the moon is a sun").

Moving on, we go further to develop (i) what analogy is and what analogy is not, (ii) analogical reasoning as being metaphorical reasoning, (iii) the powerful use of approximation and imprecision by the brain using analogies and metaphors, (iv) the categorical human resolution of analogies and metaphors through fuzzy inference-making and finally (v) models of solving analogies and metaphors; a section on artificial cognition could mimic human cognition for producing and understanding analogical and metaphorical thinking.

3 What Analogy is and What Analogy is Not

Analogy might play an important role in epistemology, history of art, scientific discoveries and innovation, but also in the methods of doing art, science and techniques. There are many historical narratives about the emergence of new ideas, of discoveries and of problem solving. Analogy is generally described in the form "A is to B as C is to D" (A:B::C:D); where the source is some kind of substitute for the target for thinking and reasoning [19]. For instance, a well-known narrative is about how Archimedes found a solution to know whether the crown of King Hiero of Athens was really made out of pure gold, or if it was contaminated with cheap silver. After a long day of worrying, he decided to relax with a warm bath. When he entered the tub, he noticed the water level rising. This was something he knew, but now he suddenly realized that the water displacement was proportionate to the volume of the immersed part of his body. Then he put a weight of gold equal to the crown in a bowl filled with water. Next, the gold was removed and replaced by the crown. Silver would increase the bulk of the crown and cause the bowl to overflow. Thus, his body was a cognitive substitute for the crown to solve the problem.

In everyday life, analogies are used and solved either for symbolic representation of things with verbalization, or when acting with physical things, or both. For instance, a verbal analogical problem solving in the form of "The railway is to the train as the sea is to the boat" (railway: train:: sea: boat) while the corresponding physical analogical problem solving is having the engineer in the restless train.

Problem solving of verbal scholar analogies such as "The railway is to the train as the sea is to boat" (railway: train:: sea: boat). Problems have three terms, to solve for the fourth (e.g. A: train:: sea: boat) or have one set of two terms and their relation (railway: train:: C: D) to find the one that has the simile relation among other sets of objects (wheels: car; sea: boat; passengers: bus).

Note first that, in contrast to the real world problems, these scholar problems have simile relations that are academic (e.g. synonym, antonym, part-to-whole, category/type, object-to-function, performer to related action, cause and effect, degree of intensity, and symbol and representation) and that the relation is given. For instance, in a real world problem, an engineer is asked, to correct and assure passengers comfort because "this train is a boat" making the relation under investigation implicit. It could be in that case that the train behaves as a boat because the train's rails are to the passengers what the rough seas would be for the passengers of a boat. However, the relation could include other elements (e.g. due to the wind, due to the mountain shape as waves, and due to the wheels). Thus, the analogy is not in the form "A is to B as C is to D" but rather in the form "A is like C".

Secondly, as a matter of facts, not a single relationship but rather many are candidates for concluding the analogical solution. In addition, the conclusive solution can be made of a, set of relations with their interactions.

Thirdly, analogy is supposedly done based on literal similarity comparisons [13, 20, 21]. However, between two natural knowledge domain, A and B, it is hard for individuals to evaluate the similarity between A and B. It is hard to evaluate the size of the intersection (A ∩ B) of their feature sets as well as the alignment of these features: which feature in A matches a given feature in B. Such correspondence may also be at a given level of the decomposition tree of attributes but not in others (e.g. while red in A and blue in B do not match at the value level; they match as being a color). Such correspondence may also be at one of the dimensions of the domain description (e.g. dimension of surface, of structure, of function, of procedure, of dynamic behaviour).

Fourthly, based on the comparison of A and C, the analogy in the form "A is to B as C is to D" can be permuted as "C is to D as A is to B." We reasoned that such inter-domain comparisons require precise structural alignment and mapping [16] that must be hard to find. The equivalence between "A is to B" and "C is to B" should be rare, making analogical relations much more oriented from target to source than from source to target. This also means that analogy is from complex to simple. From this point of view, analogy is simplexity [22], given that meanings in everyday life language provide instructions to build understandable points of view [23]. Thus, many analogies appear to be assertions that have a metaphorical form, which is to say, not reversible: if New York is a big apple, a big apple is not New York.

This metaphor asymmetry is found with classical analogical problem solving, for instance between two isomorphic sub-domains of algebra and physics [24]. When students, who learned one of these subtopics and are familiar with that source, are presented with a target problem based on the unfamiliar but analogous domain (as in metaphor) the source-to-target transfer is asymmetric. Students who had learned arithmetic (source) were very likely to transfer to physics (target). In contrast, students who had learned physics (source) almost never exhibited transfer to the isomorphic algebra problem (target). If physics is recognized as algebra, algebra is not recognized as physics.

For studying the underlying cognitive processes of analogy resolution, an alternative of using already known natural domains is to build up unknown experimental isomorphic micro domains. The building up of experimental domains assert the analogical match of the two A and B descriptions, while minimizing the noise of complement sets of A's features that are not in B (A–B) and the reverse (B–A).

The prominent Cognitive science work of Herbert Simon and collaborators [25], as well as followers [26], is based on using domain-free puzzles such as the Tower of Hanoi (TOH) and the build up of term-to-term isomorphs. Thus, there is full similarity between two A and B TOH isomorphs that are done that way. The same space problem, the same transitions from state to state, each state in A is having its analog in B and the same minimum number of moves for reaching the goal state. Thus, the two TOH isomorphs have the same deep structure, but a different surface appearance. A persistent experimental result in the literature is that much exploration is often involved in solving some of the isomorphs, whose problem spaces are identical, but are packaged differently according to their surface properties. For instance, the classical 3-disks-TOH (e.g. A) is made of disks of permanent size that change place while a possible isomorphic problem (e.g. B) is made of disks of permanent place that change size. Although 7 moves are enough to solve both problems, some of these are easy to solve with 11 moves on average (A) while others require up to 120 moves (B). Some isomorphs (as B) of the same problem space take 16 times as long to solve as other isomorphs. The difficulty varies by a factor of 16, depending on the surface characteristics.

Studies of analogical transfer between isomorphs of these well defined puzzles [27, 28] show that solving a particular problem does not help solving an isomorphic one, except when conditions of transfer are based on generalization and internalization. On the contrary, from externalization according to surface properties and context, learning cannot be transferred when the problem content provides information that can be perceived and used without being explicitly interpreted and formulated.

For instance, there are problems for which the solution requires discovering that each winning state is made of an even number. What are the conditions for transferring this solution to another isomorphic target problem? How much of the "even number" concept can be generalized to the same problem but have different surface properties? Here again, it is found that the analogical gain of solving isomorphic problems is asymmetric.

Let's examine two problems. The first one is about numbers. The second one is about tokens. When the problem that is solved first is solely with numbers, the discovery that number 4 is a winning state can be learned and used as a source for the isomorphic target problem with a number of tokens because four tokens is recognized as number 4, it is a winning state. Conversely, when the source problem is the problem with tokens, it is not helpful to solve the analogous target problem solely with numbers. The state number 4 in the target problem is not recognized as equivalent to four tokens in the source problem.

Thus, because learning is based on generalization and deep structure, a problem that favors generalization and internalization, such as the problem with solely numbers, is a profitable source while a problem based on externalization through its surface properties, such as the problem with tokens, will impair transfer; except for other problems of same surface resemblance [28]. As physics is recognized as algebra, four tokens is recognized as number 4. As algebra is not recognized as physics, number 4 is not recognized as the equivalent of four tokens (one can externally manipulate four tokens without explicitly counting them).

In summary, as noticed by BBM [9], suppose that the price of a house is to be determined. This can be done according to several criteria such as its size, its state, or its location. But because each criterion cannot be independently evaluated, we need references to evaluate the co-occurrences or correlation of attribute values. The needs of references could be satisfied with a large amount of data about home prices and values. A more simple and appropriate reasoning to evaluate the particular target house is to look for a known particular house that can be used as an intra-domain source for this case-based reasoning: "this house is a large villa: let's compare it with a large villa we know". The question under investigation is therefore the resemblance of source and target that might be based on comparing their corresponding features. Thus, the target house H1 (T) can be compared to the source house H2 (S) on properties that can be aligned and matched [21]. Within this intra domain comparison of houses, most of the features, and therefore criteria, can be found in the (T ∩ S) intersection of their feature sets. It could be that some features (e.g. a swimming pool) belong to the (T-not S) set while others (e.g. a tennis court) belong to the (not T-S) set and that alignment might be difficult, but most of the reasoning is among the (T ∩ S) set of features that can be compared. In addition, target house H1 might serve as a source for house H2 and vice-versa.

Unlike case-based reasoning, analogy is inter-domains. Someone could be searching for the "Rolls-Royce of houses" meaning a house that would be among houses what a Rolls-Royce is among cars. Someone else could have said "I found a house that is a Rolls Royce." Since the two domains are different, the resemblance relations are fuzzy and the features and criteria are hard to be aligned and matched. Contrary to case-based reasoning, if a Rolls-Royce can serve as a source for a large villa, a large villa will hardly serve as a source for a car. The main reason (unlike case-based reasoning) the source is not a particular thing, but a category of things that serves to make the target inherit the properties of the category. If a Rolls Royce is luxury, opulence, very expensive, distinctive, solid, well known, then the house that is in the category of Rolls Royce will have these features. It will inherit these properties as being an attributive category just as a particular living being will inherit the properties of being a mammal [5]. In summary, we argue that analogy and metaphor are two faces of the same coin.

4   Models of Solving Analogies and Metaphors for Fuzzy Inference Making

There are many cognitive models of analogy. See [29] for a review. As for other kinds of cognitive processes, there are two different approaches of modeling and simulating the resolution of analogies. The first approach is based on a bottom-up decentralized process such as the model named AMBR that stands for "Associative Memory-Based Reasoning." It goes from local to global, blending episodic-contextual and long term memories. For solving analogies, memory and reasoning are highly integrated in Neural Nets and high level features are built starting from the local level. There is an initial distribution of activation resulting from previously solved problems as sources that can prime the relevant features of the target. The functioning of the model can be seen as a collection of basic units that are domain specific cognitive agents that collaborate according to the declarative (what) and procedural (how/why) knowledge they encapsulate. Thus, as agents (e.g. "railway" and "sea") that entail what (support of train or boat) and how/why (for moving) collaborate to produce analogies such as "The railway is to the train as the sea is to the boat." Similarly, "Juliet" and "sun" entail what (important) and how/why (for living) collaborate to produce metaphors such as "Juliet is my sun."

The second approach is somewhat based on a top-down centralized process such the one used in the model named SMT that stands for "Structure Mapping Theory" [21, 30, 31]. In SMT, the similarity between the target and the source is evaluated by computing commonalities and differences. The former provides generalization, abstraction and schemas while the latter provides alignable differences, having "some expression in the base and some corresponding but different expression in the target." The computation of commonalities and differences strengthens the structural alignment of features that guide the analogical process.

These two approaches of analogy are based on the similarity computation of T and S; the comparison being symmetrical and reversible: T can be compared to S and S can be compared to T. When two objects, situations or domains are comparable, either one or the other can serve as a source or as target. However, this is a particular case of analogy. Most source-target analogies are oriented: a target can imply a source, while the reverse is not true.

We assert that there is a clear distinction between physical world objects and the categories humans use to represent them, in order to think, talk and communicate about them [32]. Categories as sources for understanding can be used literally as in deduction and induction, but also unliterally in analogies and metaphors in a fuzzy way. As [33], we maintain that in both cases, categorizing a target as a source type yields unseen features. If someone reads that in a fictitious country "Xs are birds," then one can infer not only that "Xs are animals," but also that "Xs fly." As semantic relations, analogies and metaphors are based on categorization. They activate a category and its attached features. They also activate the super-ordinate categories and their respective attached features.

Our categorization approach of analogies is the one that has already been proposed for metaphors understanding [5, 18, 34] in which a source (e.g. sun) is a cognitive vehicle to transfer meanings to the target topic (e.g. Juliet) with the notion that vehicles in metaphors are attributive categories.

In the past, most of the cognitive models of metaphor understanding have adopted the approach according to which metaphor is an implicit comparison: understanding a metaphor "X (topic) is Y (vehicle)" consists in converting it into a simile "X (the topic) is like Y (the vehicle)". This comparison-based model of metaphor understanding is a mechanism of property matching. This is the reason these models are confronted with the problem of measuring the similarity of properties as well as with the problem of calculating the distance between properties, which makes a simile literal or unliteral—metaphoric.

More recently, an alternative attributive categorization based approach is the Glucksberg's class inclusion model: a metaphoric statement of the type "X is Y" is solved by looking for the category, represented by the term Y, which furnishes source properties that are potentially relevant for the target topic X.

The general hypothesis is that metaphor understanding consists of including the topic in the category of the source-vehicle and attributing to it the properties of that category that are compatible with what is already known about the topic. We assume that interpretation is constructed on-line and that knowledge about the topic intervenes at an early stage in processing by constraining the selection of features.

BBM [10] noticed that crisp sets that represent precise and certain descriptions of objects might be regarded as particular cases of fuzzy sets. Similarly, we argue that deduction, induction, abduction, induction, analogy and metaphor are particular cases of metaphor, from the less fuzzy-certain, to the more fuzzy uncertain, analogy a particular case of metaphor. According to BBM [11], gradual reasoning can be obtained by using linguistic modifiers such as in [35, 36], the link between gradual reasoning and analogical reasoning corresponding to the utilization of a relationship between variations of X and variations of Y expressed in gradual knowledge to infer a value of Y from a given value of X. Thus BBM and collaborators introduced a general framework that represents analogy, on the basis of a link between variables and measures of comparison between values of variables. This analogical scheme is a common description of several forms of reasoning used in fuzzy control or in the management of knowledge-based systems, such as deductive reasoning, inductive reasoning or prototypical reasoning and gradual reasoning.

A general model for the simulation of those modes of inference-making is a model based on a fuzzy semantic network [37]. Making hierarchies of categories of the semantic network with Galois Lattices [38] allows partonomy, which is the decomposition of an object into its physical parts (the what), and meronomy, which is the decomposition of a category description into its cognitive parts (the how and why as well as conceptual features). Unlike classical Galois Lattices, the inheritance of properties and membership link (e.g. "is a" for category; "is a kind of" for subordinate category to super-ordinate category), can be interval-valued for fuzzy inclusion all along the path from instances to subordinate categories, then to the highest general super-ordinate categories. Another fuzzy measure is the extent a given feature can possibly be the attached feature of a given category. Within the lineage of categories, it is important for the structure of a category to distinguish among levels of categorization [39]: among subordinate (siamese), basic level (cat) and super-ordinate (animal) categories. A distinction that can be made with the partonomy and meronomy decomposition of descriptions; allowing gradual evaluation of concreteness of categories as well as of the domain of comparisons.

These fuzzy and categorical approaches of analogy differ from those that are based on similarity computed from features comparisons. These are two different cognitive approaches since psychological studies and in cognitive science show that similarity does not match categorization [20]. For instance, the similarity score (Russia, Cuba) is 7 and (Cuba, Jamaica) is 8, similarity (Russia, Jamaica) should be around 7, but the similarity score is 1! People categorize things differently from the way they evaluate things to be very similar: similar objects can belong to different categories while dissimilar objects can belong to the same category.

BBM [11] proposed a fuzzy prototype-based reasoning for making and solving analogies. A fuzzy prototype of a category enables one to generate typicality and the set of relevant objects and therefore can be used for matching source to target, as Tverski's proposal. The degree of typicality depends on both the resemblance to other objects in the same category and on the dissimilarity to objects in other categories. Thus, the analogical question at hand is: "does the target gradually satisfy the prototype of the source category?" These are solutions for the different modes of reasoning, including analogy and metaphor.

5 Discussion

For cognitive purposes, objects are psychologically grouped in categories. Once a category exists, it has an extension that includes all the instances of the category (even innumerable) and an intension that includes the properties (even innumerable) shared by the objects (e.g. unseen). Also for cognitive purposes, categories entail categories as well as forms of categorical hierarchies. The importance of categorization can be noticed from the following points of view:

- Reasoning: As categorization corresponds often to an abductive process. When putting an object in an existing category, we provide it the "rule" or "set" of properties of the category; and because the "is-a-kind-of" relation entails modus ponens (Socrate is mortal because Socrate is a kind of person that is a kind of mortal), but not modus tollens (which is not based on categories; i.e. things that would be a non-person that are non-mortal);
- Comprehension by inference: Since two things, or two categories, are put together, common properties of the super-ordinate category, act as a filter, indicating the "what-is-about" in terms of structure, functionality and usability. For instance, a piano and a guitar are put in "music and band playing music," while a piano and a fridge are put in "large heavy objects" and "how to carry large heavy objects"; and
- Comprehension of the world structure: Since a category in a hierarchy of categories factorize different kinds of properties and provide the causal links between procedure-function and structure because the "how-to-use" the object as well as the "in-order-to" will be based on other features of the object, such as structural properties. For instance, notice that in folk taxonomy "to have wings" and "flying" are properties of birds.

Analogies and metaphors are usual modes of thinking and reasoning although based on false categorization: "electricity is like water," or "this lawyer is really a shark." Thus, there is a powerful use of approximation and imprecision by the brain using analogies and metaphors through fuzzy inference-making.

In their prominent paper on the fuzzy approach to analogical reasoning, BBM and Valverde [12] address the problem of the representation of resemblances involved in analogical reasoning and use fuzzy relations to compare situations. As fuzziness entails the diverse forms of reasoning, from true literal sentences to false unliteral sentences, it is a powerful computation mode for human thinking and reasoning that is mainly metaphor and analogy-based reasoning. In addition, analogical and metaphorical sentences often include modifiers that BBM and Marsala put as the core of interpretable Fuzzy Systems [16]. For instance, when describing electricity, a common analogy is a water tank, where charge stands for the water amount, voltage for the water pressure, and current for the water flow. They are said to be equivalent, but electricity is not used to explain water flow. There are even situations where the water analogy is rather misleading. Electricity is like water but they cannot be mixed. Water is largely used to produce electricity, not the reverse. One might think that a metaphor such as "this is truly a gem" means a true literal sentence although this is metaphorical "image-based language" that strengthens the metaphor.

According to BBM [12], we use fuzzy relations to compare situations that can be used to model a natural analogy: resemblance relations can be used to define a kind of analogical scheme compatible with approximate reasoning in fuzzy logic, with measures of satisfiability, resemblance and inclusion. These fuzzy relations can be regarded as measures of a categorization process devoted to analogy and metaphor with the purpose of transmitting knowledge from the source to the target.

References

1. Zibetti, E., Tijus, C.: Understanding actions: contextual dimensions and heuristics. In International and Interdisciplinary Conference on Modeling and Using Context, pp. 542–555. Springer, Berlin, Heidelberg (2005)
2. Hard, B.M., Meyer, M., Baldwin, D.: Attention reorganizes as structure is detected in dynamic action. Memory & Cognition **47**(1), 17–32 (2018)
3. Picard, J.: Les trois modes du raisonnement analogique. Revue Philosophique **104**, 242–282 (1927)
4. Goblot, E.: Traité de logique. A. Colin (1920)
5. Glucksberg, S., McGlone, M.S., Manfredi, D.: Property attribution in metaphor comprehension. J. Mem. Lang. **36**(1), 50–67 (1997)
6. Lakoff, G., Johnson, M.: Metaphors we live by. University of Chicago press (2008)
7. Bouchon-Meunier, B., Ramdani, M., Valverde, L.: Fuzzy logic, inductive and analogical reasoning. In: International Workshop on Fuzzy Logic in Artificial Intelligence (pp. 38–50) Springer, Berlin, Heidelberg (1993)
8. Bouchon-Meunier, B., Valverde, L.: Analogy relations and inference. In: Second IEEE International Conference on Fuzzy Systems, pp. 1140–1144 (1993)
9. Bouchon-Meunier, B., Valverde, L.: A resemblance approach to analogical reasoning functions. In: International Workshop on Fuzzy Logic in Artificial Intelligence, pp. 266–272 Springer, Berlin, Heidelberg (1995)
10. Bouchon-Meunier, B., Rifqi, M., Bothorel, S.: Towards general measures of comparison of objects. Fuzzy Sets Syst. **84**(2), 143–153 (1996)
11. Bouchon-Meunier, B., Delechamp, J., Marsala, C., Rifqi, M.: Several forms of fuzzy analogical reasoning. In: Proceedings of the Sixth IEEE International Conference on Fuzzy Systems, vol. 1, pp. 45–50 (1997)
12. Bouchon-Meunier, B., Valverde, L.: A fuzzy approach to analogical reasoning. Soft. Comput. **3**(3), 141–147 (1999)
13. Bouchon-Meunier, B., Delechamp, J., Marsala, C., Rifqi, M.: Analogy as a basis of various forms of approximate reasoning. In: Uncertainty in Intelligent and Information Systems, pp. 70–79 (2000)
14. Bouchon-Meunier, B.: Une approche floue du raisonnement par analogie. In Tijus, C. (ed.) Métaphores et Analogies. Collection Traité de Sciences Cognitives, Hermes (2003)
15. Bouchon-Meunier, B., Mesiar, R., Marsala, C., Rifqi, M.: Compositional rule of inference as an analogical scheme. Fuzzy Sets and Syst. **138**(1), 53–65 (2003)

16. Bouchon-Meunier, B., Marsala, C.: Fuzzy modifiers at the core of interpretable fuzzy systems. In: Fifty Years of Fuzzy Logic and its Applications, pp. 51–63. Springer, Cham (2015)
17. Grice, H.P.: Logic and Conversation, 41–58 (1975)
18. Glucksberg, S.: The psycholinguistics of metaphor. Trends in Cogn Sci **7**(2), 92–96 (2003)
19. Gombrich, E.H.: Mediations on a hobby horse. In: Meditations on a Hobby Horse and Other Essays on the Theory of Art. L.L. Whyte, London (1963)
20. Tversky, A.: Features of similarity. Psychol. Rev. **84**, 327–352 (1977)
21. Gentner, D., Markman, A.B.: Structure mapping in analogy and similarity. Am. Psychol. **52**(1), 45–56 (1997)
22. Berthoz, A.: Simplexity: Simplifying Principles for a Complex World. Yale University Press, USA (2012)
23. Raccah, P.Y.: Linguistic argumentation as a shortcut for the empirical study of argumentative strategies. In: Reflections on Theoretical Issues in Argumentation Theory, pp. 279–293. Springer, Cham (2015)
24. Bassok, M., Holyoak, K.J.: Interdomain transfer between isomorphic topics in algebra and physics. J. Exp. Psychol. Learn. Mem. Cogn. **15**(1), 153–166 (1999)
25. Kotovsky, K., Hayes, J.R., Simon, H.A.: Why are some problems hard? Evidence from Tower of Hanoi. Cogn. Psychol. **17**(2), 248–294 (1985)
26. Megalakaki, O., Tijus, C., Baiche, R., Poitrenaud, S.: The effect of semantics on problem solving is to reduce relational complexity. Think. Reason. **18**(2), 159–182 (2012)
27. Zhang, J.: The nature of external representations in problem solving. Cogn. Sci. **21**(2), 179–217 (1997)
28. Nguyen-Xuan, A., Tijus, C.: Rules discovery: transfer and generalization. In: IEEE International Conference on Research, Innovation and Vision for the Future, RIVF, pp. 9–16 (2008)
29. Kokinov, B., French, R.M.: Computational models of analogy making. Encycloped. Cogn. Sci. **1**, 113–118 (2003)
30. Forbus, K.D., Ferguson, R.W., Lovett, A., Gentner, D.: Extending SME to handle large-scale cognitive modeling. Cogn. Sci. **41**(5), 1152–1201 (2017)
31. Lovett, A., Forbus, K.: Modeling visual problem solving as analogical reasoning. Psychol. Rev. **124**(1), 60–90 (2017)
32. Tijus, C., Poitrenaud, S., Chene, D.: Similarity and categorization: taxonomic and meronomic parts of similes, In: Proceedings of the 6th European Congress on System Sciences, vol. 38 (2005)
33. Anderson, J.R.: The adaptive nature of human categorization. Psychol. Rev. **98**(3), 409–429 (1991)
34. Pudelko, B., Hamilton, E., Legros, D., Tijus, C.: How context contributes to metaphor understanding. In: International and Interdisciplinary Conference on Modeling and Using Context, pp. 511–514. Springer, Berlin, Heidelberg (1999)

The Role of the Context in Decision and Optimization Problems

Maria T. Lamata, David A. Pelta, and José Luis Verdegay

Abstract To build Intelligent Systems that act in daily life like people do, it is very important to know in depth the mechanisms that govern the decision processes that human beings follow. The context in which a decision process is developed is a key aspect that needs to be known in depth. In this paper, this aspect is defined and studied. Accordingly the definition of General Decision Problem is modified and, by means of a simple example, it is shown how the solutions of decision and optimization problems can vary depending on the context (To Professor Bernadette Bouchon-Meunier, with deep admiration for her scientific and academic work. Many thanks Bernadette for keeping us among your friends for so long.).

1 Introduction

Huawei [4] presented a study on the similarities between the human brain and Artificial Intelligence, i.e., Intelligent Systems, which reveals that the average European is unaware of 99.74% of the actual decisions they make every day, showing how hard our brain works without us having to consciously engage it.

It is commonly accepted that the human brain makes approximately 35,000 decisions a day [3]. However, the new research, polling 10,000 Europeans, reveals that we are aware of just 0.26% of these decisions with respondents on average believing they make only 92 decisions per day.

M. T. Lamata · D. A. Pelta · J. L. Verdegay (✉)
Universidad de Granada, E.T.S. de Ingenierías Informática y de Telecomunicación, Calle Periodista Daniel Saucedo Aranda, s/n, 18014 Granada, Spain
e-mail: verdegay@decsai.ugr.es

M. T. Lamata
e-mail: mtl@decsai.ugr.es

D. A. Pelta
e-mail: dpelta@decsai.ugr.es

M.-J. Lesot and C. Marsala (eds.), *Fuzzy Approaches for Soft Computing and Approximate Reasoning: Theories and Applications*, Studies in Fuzziness and Soft Computing 394, https://doi.org/10.1007/978-3-030-54341-9_7

Therefore, with regard to the design of Intelligent Systems, the in-depth knowledge of the mechanisms that drive our decision-making processes becomes fundamental. This, which is something commonly performed by human beings, is not easy to model. The way each person decides is different and also, depending on the circumstances that are given, such decision may change.

Thinking about building an Intelligent System, what we usually do is to provide it with "rational" decision mechanisms that, however flexible they are in their theoretical-practical approaches, are usually very rigids, and therefore they are far from the human beings deciding way. If we train such a system to act in a certain way, it would be difficult to modify it when some changes in its context are detected. However this is very well done by persons.

More than 20 years ago H. Simon stated in [12] an absolutely brilliant thought: "*Human beings, viewed as behaving systems, are quite simple. The apparent complexity of our behavior over time is largely a reflection of the complexity of the environment in which we find ourselves.*" This phrase marks the starting point of this paper, since it is evident that the behavior of human beings also depends on the environment in which they move. Therefore decision-making is directly influenced by the environment, in which we include those small changes that, although imperceptible, modify our rules of behavior. The key is therefore the "environment" in which decisions are made.

In the field of Decision Theory, the Environment is not that context to which we are referring: the Environment is a concept associated to the probabilistic nature of the information available when this information is incomplete. Hence it seems necessary to reconsider the definition of General Decision Problem as in [5, 6] to firstly make patent the type of information available (complete or incomplete) and secondly to include the referred "context", which in all the sequel of this paper will be called Framework for distinguishing it from the Environment.

What we are seeking in this article is to show with a simple example how the Framework in which decisions are made has a decisive influence on the solution of the problem we consider. In order to do so, the following section addresses the definition of a General Decision problem. Then we describe different decision Frameworks that we can consider and the optimization problems that may arise associated to them. Finally, an example illustrates how the solutions vary depending on the Frameworks.

2   The General Decision Problem

As it is well known [5] the approach of a decision problem requires knowing the following essential elements,

- One decision-maker, which can be either an individual or a group,
- a set of actions on which the decision maker can choose,
- a set, called the Environment, which is constituted by the situations (states) that the decision maker can encounter when choosing, and cannot control,

- a set of consequences, associated with each action and each state,
- the criterion that sorts the consequences,
- the nature of the information available,
- the duration of the process, and
- the (social) framework in which the decision-making process takes place.

Thus, if we assume in what follows: (a) a single decision maker, and hence we do not consider Group Decision Making (GDM) problems, (b) single criterion, thus avoiding the Multi-criteria Decision Making (MCDM) problems, and (c) the duration of the process is only one step, then a decision problem is described by a sextet $(X, I, E, C, \leq, K)$ that includes the set X of possible actions for the decision maker; the available information I; the environment E; the set C of the consequences of actions; the criterion $\leq$ that sorts the consequences; and the framework K, which is the context in which the decision maker decides.

On two of these elements, the available information I and the framework K, we have to specify some points that will help us to better model the problem to be solved.

(a) Regarding the information available, it is always supposed to be of a probabilistic nature (due to the presence of uncertainty). However, the use of probabilities may not properly fit the nature of the information available. What usually assumed is that we do not know exactly what would happen if a particular course of action will be adopted [7]. But, of course, this lack of information may have different characteristics from the probabilistic ones. In this sense, the Smithson taxonomy [12] can be particularly useful to recognize the appropriate tools to properly model the problem.

(b) Secondly, regarding the Framework, as we have mentioned above, the action that the decision maker ultimately chooses as optimal can be conditioned by the Framework K in which the problem is developed. To be concrete, a Framework, regardless of the nature of the information available, is defined as a set of rules, often established in the form of logical predicates, that establish the qualitative characteristics that the available decisions must have. As it may be evident, provided some framework dependent problem its solution may vary according to the framework.

Several frameworks can be considered. Among others, Ethical, Concurrence, Adversarial (in the presence of adversaries), Crisis (catastrophes), Stress, Sustainability, Dynamic, Corporate Social Responsibility (CSR) or Induced. Here, and for the sake of illustration of its influence in the final solution of the problems, we only consider four frameworks: Ethical, CSR, Sustainability and Induced, which in the following are slightly described.

If the final decisions have to be made by taking into account some "code", what is typical in decision processes that are developed in very specific and professional contexts (legal, military, medical, etc.) the framework is defined as an Ethical one. The Ethical Framework is usually defined by a set of "good practices" to which the decision makers must conform.

A Sustainability Framework is considered when we can associate the problem to what is understood by sustainable decision in a specific ecosystem, that is, a decision that must meet the expectations of the moment when it is taken, and at the same time not compromise the choices that may be made about the problem in the future [2]. This framework, although generally associated with environmental issues, obviously it is not limited to that context.

The Corporate Social Responsibility (CSR) is an economical concept to conduct companies based on the management of the impacts that their activity generates on their customers, employees, shareholders, local communities, the environment and on society in general [10]. In this way a CSR Framework is understood as an action way through which the decision maker can voluntarily integrate social and environmental dimensions into the consequences of his actions [5].

In turn, if some psychological (nudge) factors are considered [13] we could assume an Induced Framework, as the deliberated design of a framework to make decisions can help the decision maker to take better decisions. The previous personalization (the so called induction) of the Framework in which we have to make decisions is usually a very convenient mechanism for effective decision making.

Let $(X, I, E, C, \leq, K)$ be a generic decision making problem as defined above. With regard to solve this problem what we are looking for it is to make the best decision among the possible ones, which in turn, means that under every decision problem, there is an underlying optimization problem.

If I and K are concreted by the decision maker, this optimization problem may be represented as a tuple $(X^{I_k}, E^{I_k}, z^{I_k}, \leq^{I_k})$, where I_k stands for a given type of information (I) and a frame of behaviour (K) and the tuple, as a whole, represents the version of the considered decision-making problem when the type of information and the context have been concreted.

Then the problem becomes to find that alternative $x^* \in X^{I_k}$ such that

$$z^{I_k}(x^*) = Max_{\leq}{}^{f_k}\{z^{I_k}(x) : x \in X^{I_k \times}; z^{I_k} : X^{I_k} \times E^{I_k} \to U^{I_k}\} \tag{1}$$

where z^{I_k} is the so-called objective function, that gives the reward associated to each alternative, for each state of nature that is considered. Said reward can be defined by means of numerical variables, linguistic ones, qualitative ones, etc. and, in any case, sorted by the criterion $\leq^{I_k}$ that takes part into the problem.

As said above, from the point of view of the nature that the information have, two main cases can be distinguished from (1), according to the information available is complete or incomplete:

(a) If the available information is complete, the results are valued in the real line, $X \subset R^n$ and we know exactly what the state of nature is, the problem (1) becomes

$$z_K(x*) = Max_K\{z_K(x) : x \in X_K; f_K : X_K \times E_K \to U_K\}$$

which can be easily formulated as a General Optimisation problem (GOP) in the following terms:

$$\text{Maximize} \quad z(x)$$
$$\text{Subject to:} \quad g_i(x) \leq 0, \, for \, i \in M = \{1, \, \ldots, \, m\}$$
$$x \in K$$

where z and $g_i(x)$, $i \in M$, are functions that take values in the real line and K is a set of rules or constraints that define the Framework in which the problem is to be developed.

(b) If the information is incomplete, and particularly of a fuzzy nature, and the set of alternatives and the results are also fuzzy, then the problem (1) can be posed as

$$z^{f_k}\left(x^*\right) = Max_{\leq}{}^{f_k}\{z^{f_k}(x) : x \in X^{f_k}; z^{f_k} : X^{f_k} \times E^{f_k} \rightarrow U^{f_k}\}$$

where the superscript "f" stands for fuzzy. As in the case of complete information, this problem is a Fuzzy Mathematical Programming problem that can adopt different versions. Particularly it can be formulated as,

$$Maximize : z(x)^f$$

$$Subject \, to : g_i(x)^f \leq^f 0, \, for \, i \in M \tag{2}$$

$$x \in K$$

where the objective function, the set of constraints or the coefficients that take part into the problem, one by one, partially, or all together may have a fuzzy nature [14, 15].

If all the elements in (2) are convex, one has a Fuzzy Convex Programming (FCP) problem. Particular relevant FCP problems are Fuzzy Linear Programming (FLP) ones, Fuzzy Quadratic Programming (FQP) ones or Fuzzy Geometrical Programming (FGP) ones. Additionally, and in turn, depending on the context to be considered, different specific models can arise.

Thus, for instance, if both objective and constraints are defined by linear functions and the Framework K is defined, for the sake of illustration, as "non negative solutions", we obtain the model corresponding to the well known FLP problem:

$$Maximize \, z = c^f x$$

$$Subject\ to: A^f x \leq^f b^f \ (3)$$

$$x \geq 0$$

where c^f, b^f are vectors of n and m fuzzy numbers, respectively, A^f is an $n \times m$ matrix of fuzzy numbers and $\leq^f$ is a comparison relation among fuzzy numbers [1, 8, 9, 14, 15].

The influence that Frameworks have in the solution of decision making problems, and hence in its corresponding optimization problems, will be illustrated in the following section.

3   An Illustrative Example

Let consider an easy example to show how the solution may vary according to the framework which we can consider: Let us suppose two training centers that can grant to the people who follow their courses diplomas of three levels: technical, graduate and master. The first center can issue 6 master's degrees, 2 graduate degrees and 4 technical degrees per year. The second center can issue 2 master's degrees, 2 graduate degrees and 12 technical degrees per year. The annual costs of these centers are 200 K euros and 160 K euros respectively. It is known that in a time horizon of 6 years, 12 people with a master's degree, 8 with a graduate level and 24 with a technical level will be needed. How many years should each Center operate to meet the needs at the lowest possible cost?

The data of the problem can be easily summarized in the following Table 1.

Thus the problem can be directly formulated as,

$$Minimize\quad 200x_1 + 160x_2$$
$$Subject\ to:$$
$$6x_1 + 2x_2 \geq 12$$
$$2x_1 + 2x_2 \geq 8$$
$$4x_1 + 12x_2 \geq 24$$
$$6 \geq x_1, x_2 \geq 0$$

Table 1 Numerical data

	Master	Graduate	Technical	Cost
Center 1	6	2	4	200
Center 2	2	2	12	160
Needs	12	8	24	

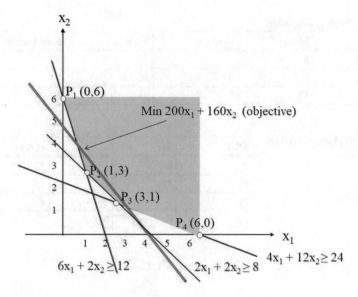

Fig. 1 Graphical representation of the problem

Table 2 Values of the objective function

Induced	x_1	x_2	Objective
P_1	0	6	960
P_2	1	3	680
P_3	3	1	760
P_4	6	0	1200

Graphically the problem, feasible solutions set and objective function, may be represented as in Fig. 1.

The feasible region of this problem (in grey in Fig. 1) has four extreme points P_i $(i = 1, 2, 3, 4)$. As it is known, at some of these extreme points the objective function will reach its optimum value.

The values that take the two variables that take part into the problem in each of those extreme points (possible solutions) and the corresponding value that reaches the objective function are shown in the following Table 2.

3.1 Induced Framework

Let us refer firstly to an Induced Framework. In this case we consider that the "induction" is designed to choose the solution that provides a minimum cost. Then, in terms of cost,

Table 3 Surplus of diplomas

	Master	Graduate	Technical	Final Excess
Center 1	6	2	4	
Center 2	6	6	36	
Needs	12	8	24 (+16)	16

Table 4 Diplomas issued for each solution

Ethic	x_1	x_2	Excess
P_1	0	6	52
P_2	1	3	16
P_3	3	1	8
P_4	6	0	28

$$P_2 \leq P_3 \leq P_1 \leq P_4$$

the best (induced) solution is that one provided by the point P_2.

3.2 Ethical Framework

Let us suppose now that we are working in an Ethical Framework. The best solution (P_2 from the objective value point view) supposes to act $x_1 = 1$ year in the first Center and $x_2 = 3$ years in the second (Table 3).

But in such a Framework, this solution could not be applicable because it would leave 16 people unemployed, something that the "ethics" of the organization that is solving the problem does not consider admissible. However, from this ethical view, which considers reducing the number of unemployed people to the maximum, the solution should be the one given by P_3, since it produces the least number of unemployed people: 8. Table 4 summarizes these data,

3.3 Sustainability Framework

Let us consider a Sustainability Framework. It seems obvious that what will interest us most is having a valid solution for the longest number of years. The solution would be the one provided by P_1 or P_4 since any of them keeps the operational program the same number of years: 6. But in this case, having a lower cost P_1 than P_4 the solution would be the one that provides the point P_1 (Table 5)

Table 5 Number of years that the program is operative

Sustainability	x_1	x_2	Duration
P_1	0	6	6
P_2	1	3	4
P_3	3	1	4
P_4	6	0	6

Table 6 Number of master level diplomas

CSR	x_1	x_2	Master
P_1	0	6	12
P_2	1	3	12
P_3	3	1	20
P_4	6	0	38

3.4 Corporate Social Responsibility Framework

In a Framework of CSR, and for the reasons of illustration of this example, we will consider that the index that measures said CSR is the number of master's degree diplomas that can be issued, independently of other indexes (costs, duration, etc.). Thus, one solution is better than another if the number of master's degrees obtained with one solution is greater than that of another. Table 6 provides the data for the different solutions, showing that in this case it would be point **P₄** that gave us the optimal solution for our problem.

Therefore, it becomes evident that the solution of a problem may vary according to the Framework,

	Induced	Ethical	Sustainability	CSR
P_1			●	
P_2	●			
P_3		●		
P_4				●

Therefore, as the example has illustrated, the role of the Framework is such that may change the solution that conventional and classical solution methods provide to the problems.

4 Conclusions

The introduction of the Framework as a new element in the formulation of a General Decision Making problem, has proved essential for the correct formulation of the subsequent decision process and its corresponding optimization problem for choosing the best alternative.

Although there is a wide variety of different Frameworks, in this paper, for the sake of illustration of the relevance of the concept, we have considered only four: induced, ethical, sustainability and Corporate Social Responsibility, which in turn have been briefly described. Regarding them, through the resolution of a simple example of linear programming, it has been shown how the solution of a given problem can vary as the frameworks vary.

The extension of this new model to the case in which the data taking part into the problem have a fuzzy nature will be approached in a forthcoming paper.

Acknowledgements Research supported by the projects TIN2014-55024-P and TIN2017-86647-P (MINECO/AEI/FEDER, UE).

References

1. Bellman, R.E., Zadeh, L.A.: Decision-making in a fuzzy environment. Man. Sci. **1**, B-141-B-164 (1970)
2. Our Common Future: Brundtland Report 20 March 1987. United Nations
3. Farber N.: Decision-Making Made Ridiculously Simple (2010). https://www.psychologytoday.com/blog/the-blame-game/201607/decision-making-made-ridiculously-simple
4. https://www.huawei.com/es/press-events/news
5. Lamata, M.T., Pelta, D.A., Verdegay, J.L.: Optimisation problems as decision problems: the case of fuzzy optimisation problems. Inf Sci 460–461, 377–388 (2018)
6. Lamata, M.T., Verdegay, J.L.: On new frameworks for decision making and optimization. In: Gil, E., Gil, E., Gil, J., Gil, M.Á. (eds) The Mathematics of the Uncertain: A Tribute to Pedro Gil. Studies in Systems, Decision and Control 142, 629–641. Springer (2018)
7. Lindley, D.V.: Making Decisions. John Wiley & Sons Ltd (1971)
8. Lodwick, W., Thipwiwatpotjana, P.: Flexible and generalized uncertainty optimization. In: Theory and Methods. Studies in Computational Intelligence. Springer (2017)
9. Luhandjula, M.K.: Fuzzy optimization: milestones and perspectives. Fuzzy Sets Syst. **274**, 4–11 (2015)
10. Observatory of CSR: https://observatoriorsc.org/la-rsc-que-es/
11. Simon, H.: The Sciences of the Artificial. The MIT Press (1996)
12. Smithson, M.: Ignorance and Uncertainty: Emerging Paradigms. Springer Verlag, New York (1989)
13. Thaler, R., Sunstein, C.: Nudge, Improving Decisions About Health, Wealth and Happiness. Yale University Press (2008)
14. Verdegay, J.L.: Fuzzy mathematical programming. In: Gupta, M.M., Sánchez, E. (eds.) Fuzzy Information and Decision Processes, pp. 231–237. North Holland (1982)
15. Verdegay, J.L.: Progress on fuzzy mathematical programming: a personal perspective. Fuzzy Sets Syst. **281**, 219–226 (2015)

Decision Rules Under Vague and Uncertain Information

Giulianella Coletti

Abstract A decision processes in presence of uncertainty and fuzziness can be performed by using the interpretation of a fuzzy set as a pair whose elements are a suitable crisp event E_φ of the kind "You claim that the variable X has the property φ" and an assessment consistent with an uncertainty conditional measure on the conditional events $E_\varphi|\{X = x\}$. The decision framework is based on uncertain fuzzy IF-THEN rules performed through the concept of degree of implication between two "fuzzy events", expressed by a conditional uncertainty measure.

1 Introduction

Fuzzy decision making is a broad field where the classic decision process are generalized for handling processes in which (as specified in the pioneering article by Bellman and Zadeh [1]) "the goals and/or the constraints, but not necessarily the system under control, are fuzzy, that is the goals and/or the constraints constitute classes of alternatives whose boundaries are not sharply defined". In this paper only a particular aspect of decision making is considered: the decision processes using rules (which are functions mapping an observation to an appropriate action) for making deductions or choices.

Decision rules had an important role in statistics and economics, and are closely related to the concept of a strategy in game theory. Nevertheless this aspect of decision making has gained great importance with the emergence of expert systems and now are used in many fields, for instance to perform lexical analysis to compile or interpret computer programs, or in natural language processing, but mainly in specific ambits of in A.I.

G. Coletti (✉)
Department of Mathematics and Computer Sciences, via Vanvitelli, 1, 06123 Perugia, Italy
e-mail: giulianella.coletti@unipg.it

M.-J. Lesot and C. Marsala (eds.), *Fuzzy Approaches for Soft Computing and Approximate Reasoning: Theories and Applications*, Studies in Fuzziness and Soft Computing 394,
https://doi.org/10.1007/978-3-030-54341-9_8

In fact, most decisions, that any person (in particular a field expert) makes, are in fact logical decisions: he/she considers the available information and makes a decision based on it, by using a reasoning which generalizes the classic "strict" modus ponens.

In order to automate this reasoning it is necessary to define a set of rules coherent with a framework of reference. The classic general starting point is formed by a set of rules that have one input and one output, with this form: $R = $ "if x is φ_i then y is ψ_i" ($i = 1, ..., n$), where x is a value taken by a variable X, y a value taken by a variable Y and φ_i and ψ_i are "properties" of variable X and Y, respectively.

In presence of not complete information (and so in presence of uncertainty), the above statements need to be equipped with an evaluation of degrees of belief, expressed through some uncertainty measure m. One of the formalization of the above pairs (rule, degree of belief) is obtained through a suitable conditional measure $m(.|.)$, i.e. by performing an assessment on the conditional events ($\{y$ is $\psi_i\}|\{x$ is $\varphi_i\}$), consistent with a conditional measure of reference. To make this, it is necessary to refer to a concept of conditional uncertainty measure able to regard both the events as variables, having the same status (that is both can be true or false), but a different role (the second one having the role of hypothesis).

However also the elements of the rules (events) itself can be affect by imprecision, ambiguity or vagueness, for instance when either φ_i or/and ψ_i are expressed in natural language. As it is well known, to manage this situation, the most convincing and well founded theory is the fuzzy set theory [42], which considers a scale of membership degrees not restricted to 0 and 1, but taking values in [0, 1].

Now a problem arises when we want to consistently model the uncertainty of the implication and the vagueness of the statements forming either the input and the output. The problem is obviously due to the fact that all the uncertainty measures have as domain Boolean events, that is facts expressed by a Boolean sentence, which can only be true or false (i.e. taking only 0 and 1 as truth values). Do to the Birkhoff's representation theorem [2], it is possible to represent by subsets of a suitable set called sample space or set of states of world.

Then two possible lines can be followed: defining conditional uncertainty measures in objects generalizing pairs of Boolean events (see [35, 37, 39, 40, 43, 44], or giving an interpretation of fuzzy sets and fuzzy events in therms of coherent conditional measures on suitable (Boolean) conditional events. In a series of works our research group outlined the goals and the limits of this second approach.

The basic concepts for this proposal are: the concept of event as any fact represented by a Boolean sentence, which removes any fence among different sources of information; the concept of conditional event as an ordered pair of events and finally the concept of conditional uncertainty measure directly defined as a function of two variables (events) satisfying a set of axioms. In this context the main tool is the notion of coherent assessment and coherent extension, which permit (thanks to their characterizations) to test the consistency of the initial data base and to make inference to new entities.

The paper starts, presenting a series of concepts and results obtained, during the last 20 years, together with B. Bouchon-Meunier, C. Marsala, D. Petturiti, R. Scozzafava and B. Vantaggi, relative to the concepts of coherent conditional probability and T-conditional possibility (with T any continuous t-norm). Only concepts and results are selected useful to better understand the representation of a fuzzy set as a pair formed by a Boolean event of the kind "You claim that the variable X has the property γ" and a function of $\mu(x) = \varphi(E_\gamma | \{X = x\})$, where φ is either a conditional probability or a T-conditional possibility. Then notions and results related to coherent conditional probability and T-conditional possibility provide of this interpretation; the semantic is very clear per se: for every possible value x of a variable X, $\mu(x)$ is the measure of how many one thinks that a generic person (named You) can claim that X has property γ under the hypothesis that $X = x$. The same interpretation has been given also in therms of coherent conditional plausibility [30] or convex capacity [27], but the limits of the two main interpretations (the probabilistic and the possibilistic ones) are not overcome by relaxing the properties of the measure, moreover the computational difficulty increases. Nevertheless they are useful when the uncertainty on the values taken by the variables is necessarily treated by these measures.

These interpretations of fuzzy sets allowed to propose a solution to some problems where uncertainty and fuzziness are jointly present, without introducing new concepts (see for instance [9, 12, 14, 20, 21, 26, 28, 29]), moreover some notions of fuzzy set theory have been reinterpreted, some time generalized (see for instance [23, 41]) in a natual way. As an example the management of the notion of fuzzy decision rule in presence of uncertainty is presented here, in both probabilistic and possibilistic uncertainty frameworks . The interpretation of the rules is a conditional probability or T-conditional possibility $\varphi(E_\psi | E_\gamma)$ and the inference is ruled by coherence, possibly using entailment rules of the probabilistic or possibilistic reasoning [15, 23, 34]. The most famous rules of interpretation and inference, due to Mamdani [36] are obtained as as particular case of cf conditional possibility when one is in presence of an uninformative possibility distribution on the values of the variables.

2 Coherent Conditional Decomposable Measures

In the literature, conditional measures are usually introduced as a derived notion of the unconditional ones, defined as solution of an equation involving the joint measure and its marginals. This is a restrictive and misleading vision of conditioning, which only emphasizes that, given a conditioning event, the conditional measure satisfies the same rules of the unconditional one, and so in fact corresponding trivially to just a modification of the "sample space".

It is instead essential to regard also the conditioning event as a "variable", i.e. the "status" of it can not be just that of something representing a given fact, but that of an uncertain event for which the knowledge of its truth value is not required. In other words, even if beliefs may come from various sources, they can be treated in the same way, since the relevant conditioning events (including both statistical data and any perception-based information) can always be considered as being assumed propositions. This is at basis of the "reasoning under hypotheses" and also of the aim of this proposal. Moreover, by a syntactic point of view, the classical approach leads to a partial definition, indeed, for some pair of events the solution of the equation (the conditional measure) can either not exist or not be unique.

So it is preferable to define conditional measures in an axiomatic way, directly as a function defined on a suitable set of conditional events, satisfying a set of rules (axioms). For simplicity we refer only to the family of decomposable measures.

2.1 Definitions and Main Results

Let us consider a pair of commutative, associative, increasing binary operations $(\oplus, \odot)$ from $[0, 1]^2$ to $\mathscr{R}_0^+$, having 0 and 1 as neutral element, respectively, and with $\odot$ continuous and distributive over $\oplus$. A real function φ defined on an arbitrary set of conditional events $\mathscr{G} = \{E_i|H_i\}_{i \in I}$ is a *coherent conditional assessment* with respect to an $(\oplus, \odot)$-decomposable conditional measure [18], if and only if it is the restriction of a $(\oplus, \odot)$-*decomposable conditional measure* $\varphi' : \mathscr{B} \times \mathscr{H} \to [0, 1]$ (i.e. $\varphi'_{|\mathscr{G}} = \varphi$ where $\mathscr{B}$ is a Boolean algebra, $\mathscr{B}^0 = (\mathscr{B} \setminus \{\emptyset\})$ and $\mathscr{H} \subseteq \mathscr{B}^0$ is an additive set (i.e., it is closed under finite logical sums),

(C1) $\varphi'(E|H) = \varphi'(E \wedge H|H)$, for every $E \in \mathscr{B}$ and $H \in \mathscr{H}$;

(C2) $\varphi'(\cdot|H)$ is a $\oplus$-decomposable measure on $\mathscr{B}$, for any $H \in \mathscr{H}$;
   (i.e. $\varphi'(\Omega) = 1$, $\varphi'(\emptyset) = 0$ and for every $E, F \in \mathscr{B}, H \in \mathscr{H}$ with $E \wedge F \wedge H = \emptyset \; \varphi'(E \vee F|H) = \varphi'(E|H) \oplus \varphi'(F|H)$

(C3) $\varphi'(E \wedge F|H) = \varphi'(E|H) \odot \varphi'(F|E \wedge H)$, for every $H, E \wedge H \in \mathscr{H}, E, F \in \mathscr{B}$.

In particular, for $\oplus$ and $\odot$ equal to sum and product, respectively, we obtain conditional probability introduced many years ago in [31], for $\oplus$ equal to max and $\odot$ equal min one obtains a conditional possibility, introduced in [5, 6]. If the min is replaced by any other t-norm T, one has T-conditional possibility introduced in [7] and deeply studied in [10, 25].

By using characterization theorems of coherent assessment (see for instance [8, 17, 19] for probability and [7, 10, 11, 25]) for possibility, it is easy to prove coherence for particular assessments, which are basic for the aim of this proposal.

By referring to an event (evidence) E and a partition $\mathscr{L}$ we call likelihood function any $f : \{E\} \times \mathscr{L} \to [0, 1]$ satisfying

(L1) $f(E|H_i) = 0$ if $E \wedge H_i = \emptyset$ and $f(E|H_i) = 1$ if $H_i \subseteq E$

Remark 8.1 The name of likelihood is usually associate to experiments or, in general, to statistical data, but this is one of the semantic point of views, not the unique. It can be assessed as a "subjective" evaluation of the different degrees of belief of a person on an event E, when he/she consider it under different hypotheses. From a syntactic point of view, it is a conditional measure regarded as function of the conditioning event. So its properties come from the axioms and the concept of coherence.

Theorem 8.1 *Let $\mathscr{L} = \{H_i\}_{i\in I}$ be an arbitrary partition of Ω and E an event. For every likelihood function $f : \{E\} \times \mathscr{L} \to [0, 1]$ the following statements hold:*

(i) *f is a coherent conditional probability;*
(ii) *f is a coherent T-conditional possibility (for every continuous t-norm T).*

Remark 8.2 The previous result points out that "syntactically" a probabilistic likelihood function is indistinguishable from a possibilistic likelihood function and both are completely free to assume any value in $[0, 1]$, a part the trivial condition $(L1)$. Obviously when one considers the extension of a likelihood function on the elements of the additive set generated by $\mathscr{L}$, the rules of conditional probability and conditional possibility drive to functions with different properties (see [13, 22]. We only stress that, as a function of the conditioning event, a conditional probability can not be a probability and a conditional possibility can not be a possibility.

Suppose now to have a finite set of different likelihood functions $f_i : \{E_i\} \times \mathscr{L} \to [0, 1]$ referred the same partition $\mathscr{L}$ but to different events E_i belonging to a set $\mathscr{C}$. Our aim only requires to refer to special classes of events E_i, that is those containing events *almost logically independent with respect to the partition $\mathscr{L}$* , i.e. satisfying the following conditions:

(i) The events in $\mathscr{C}$ are logically independent, i.e.,
$\bigwedge_{j=1}^{m} E_j^* \neq \emptyset$;
(ii) For every $H_i \in \mathscr{L}$, $\bigwedge_{j=1}^{m} E_j^* \wedge H_i = \emptyset \Longrightarrow E_j^* \wedge H_i = \emptyset$ for some $j = 1, \ldots, m$.

where E_j^* indicates either E_j or E_j^c.
As proved in [14] the following result holds:

Theorem 8.2 *Let $\mathscr{L} = \{H_i\}_{i\in I}$ be an arbitrary partition of Ω, $\mathrm{e} = \{E_j\}_{j=1,\ldots,m}$ a set of almost $\mathscr{L}$-logically independent events, and $\mathscr{F} = \{f_j\}_{j=1,\ldots,m}$ a set of likelihood functions, where f_j is defined on $\{E_j\} \times \mathscr{L}$. Let v and π be a finitely additive probability and a finitely maxitive possibility on $\mathscr{A}_{\mathscr{L}}$ with $\langle \mathscr{L} \rangle \subseteq \mathscr{A}_{\mathscr{L}} \subseteq \langle \mathscr{L} \rangle^*$, respectively. The following statements hold:*

(i) *The assessment $\{\mathscr{F}, v\}$ is a coherent conditional probability;*
(ii) *The assessment $\{\mathscr{F}, \pi\}$ is a coherent T-conditional possibility (for every continuous t-norm T).*

2.2 Coherent Extensions

For coherent conditional probability assessments the fundamental theorem, essentially due to de Finetti [32] holds. An analogous result for coherent conditional possibilities has been proved in [25] in the finite case and in [10] in the general case.

Theorem 8.3 *Let $\mathcal{G}'$ be an arbitrary set of conditional events with $\mathcal{G} \subset \mathcal{G}'$ and P a real function on $\mathcal{G}$. Then, there exists a (non-necessarily unique) coherent conditional probability P' extending P on $\mathcal{G}'$ if and only if P is a coherent conditional probability on $\mathcal{G}$. Moreover, if $\mathcal{G}' = \mathcal{G} \cup \{E|H\}$, then the coherent values for the probability (possibility) of $E|H$ form a closed interval $[\varphi_*, \varphi^*]$ with $\varphi_* \leq \varphi^*$*

In general, starting from a coherent conditional probability or possibility φ on $\mathcal{G} = \{E_i|H_i\}_{i \in I}$ and extending it to a superset $\mathcal{G}' = \mathcal{G} \cup \{E_j|H_j\}_{j \in J}$, in the case J is finite, we could require to proceed step-wise, choosing at each step a value in the corresponding extension interval and taking it into account in the next step. Obviously the final coherent conditional probability or possibility φ^* on $\mathcal{G}'$ depends on the choices made during the process and, also in the case where one fixes a rule for choosing, for instance the minimum (or the maximum) value in the interval, the final assessment will depend on the chosen order for the extension on the $E_j|H_j$'s.

Then becomes interesting to discover if, for some specific class of conditional events, it is possible to find a choice function able to provide a coherent extension independent of the order.

The following Theorem 8.4 goes in this direction and highlights the minimum as a particularly effective t-norm.

Indicate by T_L the Łukasiewicz t-norm, by T_M the Gödel t-norm (min), and by T_p the product, we recall the next theorem, proved in [30] for T_M and in [16] for T_p, where the problem is deeply studied for all the Frak's t-norms.

Theorem 8.4 *Let $\mathcal{L}$ be a partition and $\mathcal{F} = \{E_j\}_{j=1,\dots,m}$ a finite set of events almost logically independent with respect to $\mathcal{L}$. Given a set of likelihood functions $f_i :$ $\{E_i\} \times \mathcal{L} \to [0,1]$ $(i = 1, \dots, n)$ consider the set $\mathcal{A}$ of events $A = \bigwedge E_i$ obtained as intersection of the elements of any subset of $\mathcal{F}$. The assessment $\{f(A|H_j), : A \in \mathcal{A}, H_j \in \mathcal{L}\}$, with $T_\odot$ equal to either T_M or T_p*

$$f(A|H_j) = T_\odot \{f_i(E_i|H_j) : A \subseteq E_i\}$$

is both a coherent conditional probability and a coherent conditional possibility.

A similar result does not hold for T_L, as proved in [30].

3 An Interpretation of Fuzzy Sets in Terms of Likelihood Function

Let X be a (non-necessarily numerical) variable, with range $\mathcal{C}_X$, and, for any $x \in \mathcal{C}_X$, let us indicate by x the event $\{X = x\}$, which are the elements of partition $\mathcal{L}$. Let γ be any *property* related to the variable X and let us refer to the state of information of a real (or fictitious) person that will be denoted by "You". Moreover, consider the Boolean event: $E_\gamma =$ "You claim that X is γ".

An uncertainty assessment (either a coherent conditional probability or a coherent conditional possibility) $\{\varphi(E_\gamma|x)\}_{x\in\mathscr{C}_X}$ measures the degree of belief of You in E_φ, when X assumes the different values of its range.

It is important at this moment to underline the total freedom of choice of the function $\varphi(E_\gamma|\cdot)$ as shown by Theorem 8.1. Then $\varphi(E_\gamma|\cdot)$ comes out to be a natural interpretation of the membership function $\mu_\varphi(\cdot)$, according to [20] or [19, 21] for probabilistic likelihood and [14] for possibilistic likelihood.

Definition 8.1 For any variable X with range $\mathscr{C}_X$ and a related property γ, a **fuzzy subset** E_γ^* of $\mathscr{C}_X$ is any pair

$$E_\gamma^* = (E_\gamma, \mu_\gamma),$$

with $\mu_\gamma(x) = \varphi(E_\gamma|x)$ for every $x \in \mathscr{C}_X$.

By this interpretation the operations between fuzzy subsets (union, intersection and complementation) can and must be obtained directly by using the rules of coherent conditional probability and possibility, taking into account the almost logical independence between E_γ and E_ψ w.r.t. the partition $\mathscr{L} = \{(X = x)\}_{x\in\mathscr{C}_X}$ (or w.r.t. X for short). For this aim, given two fuzzy subsets E_γ^*, E_ψ^*, with the events E_γ, E_ψ almost logically independent w.r.t. X, let us define:

$$E_{\gamma\wedge\psi} = E_\gamma \wedge E_\psi; \quad E_{\gamma\vee\psi} = E_\gamma \vee E_\psi.$$

In the probabilistic setting, for any given x in the range of X, the assessment $P(E_\gamma \wedge E_\psi|x)$ is coherent if and only if it holds

$$T_L(P(E_\gamma|x), P(E_\psi|x)) \leq P(E_\gamma \wedge E_\psi|x) \leq T_M(P(E_\gamma|x), P(E_\psi|x)).$$

From probability properties, we must also have

$$P(E_\gamma \vee E_\psi|x) = P(E_\gamma|x) + P(E_\psi|x) - P(E_\gamma \wedge E_\psi|x).$$

Concerning possibilistic setting, for any given x in the range of X, the assessment $\Pi(E_\gamma \wedge E_\psi|x)$ is coherent if and only if it holds

$$0 \leq \Pi(E_\gamma \wedge E_\psi|x) \leq T_M(\Pi(E_\gamma|x), \Pi(E_\psi|x))$$

$$\Pi(E_\gamma \vee E_\psi|x) = \max(\Pi(E_\gamma|x), \Pi(E_\psi|x)).$$

Finally recall that in this context the complement of a fuzzy subset is defined as

$$(E_\gamma^*)' = (E_{\neg\gamma}, \mu_{\neg\gamma}) = (E_{\neg\gamma}, 1 - \mu_\gamma).$$

Obviously $E_{\neg\gamma} \neq (E_\gamma)^c$, moreover the events "You claim X is γ" and "You claim X is $\neg\gamma$" are logically independent: we can claim both "X is γ" and "X is $\neg\gamma$", or only one of them or finally neither of them. and it is natural to consider them almost logically independent. Then, while $E_\gamma \vee (E_\gamma)^c = \Omega$, one has $E_\gamma \vee E_{\neg\gamma} \subset \Omega$ and so $(E_\gamma^*)'$ is generally only a fuzzy subset of the universe $\mathscr{C}_X$, for which in probabilistic and possibilistic setting one has respectively

$$\mu_{\gamma \vee \neg\gamma}(x) = \mu_\gamma(x) + \mu_{\neg\gamma}(x) - \mu_{\gamma \wedge \neg\gamma}(x); \qquad \mu_{\gamma \vee \neg\gamma}(x) = \max(\mu_\gamma(x), \mu_{\neg\gamma}(x)).$$

The two equations coincide when the intersection is computed through T_M.

A similar discussion holds even for E_γ and E_ψ, where ψ is the superlative of γ.

3.1 Conditional Probability and Possibility of Fuzzy Events

For simplicity from now only finite frameworks will be considered, for results about the infinite settings see for instance[13, 14, 29, 30].

In this context, a fuzzy event, as introduced by Zadeh, is actually the Boolean event E_γ = "You claim that X is γ", which, however, is endowed with information expressed by a likelihood defined on $x \in \mathscr{C}_X$.

Now let $\mathbf{X} = (X_1, \ldots, X_m)$ be a vector with range $\mathscr{C}_{\mathbf{X}}$, where each component X_i has range $\mathscr{C}_{X_i}$. Consider a finite family of fuzzy subsets $E_{\gamma_i}^* = \{E_{\gamma_i}, \mu_{\gamma_i}\}$, with $i = 1, \ldots, m$, related to the (possibly coincident) components X_i of $\mathbf{X}$, where the events $\{E_{\gamma_i}\}_{i=1,\ldots,m}$ are assumed to be almost $\mathscr{C}_{\mathbf{X}}$-logically independent.

For every joint probability distribution P or possibility distribution Π the global assessment $\{\mu_{\gamma_i}, \pi\}_{i=1,\ldots,m}$ is a coherent T-conditional possibility by Theorem 8.2 and so coherently extendible to E_{γ_i}.

It is easy to see that, indicate by P_i and Π_i the marginal distributions on X_i the only coherent value are: for probabilistic interpretation,

$$P(E_{\gamma_i}) = \sum_{x \in \mathscr{C}_{X_i}} \mu_{\gamma_i}(x), P_i(x)), \tag{1}$$

(which formally coincides with Zadeh's definition of probability of a "fuzzy event" [43]) and, for possibilistic interpretation,

$$\Pi(E_{\gamma_i}) = \max_{x \in \mathscr{C}_X} T(\mu_{\gamma_i}(x), \pi(x)). \tag{2}$$

(which, for $T = T_M$ exactly corresponds to the possibility of a "fuzzy event" introduced by Zadeh in [44]).

These coherent (conditional) probability and possibility P can be furthermore extended to the conditional events $A|B$ where A, B are events of the algebra $\mathscr{B}$, with $B \neq \emptyset$. This extension is not unique in general, but, if we choose T_M or T_p as rule for computing intersection of $E_{\gamma_i}'s$, from Theorem 8.4 we have the coherent extensions $P_\odot$ and $\Pi_\odot$ (with $\odot$ either T_M or T_p) on $\mathscr{A}$.

Then for the events $A = E_{\gamma_i}$ and $B = E_{\gamma_j}$ $(i \neq j)$, in the probabilistic setting one necessarily has, in the case where $P_\odot(E_{\gamma_j}) > 0$ the following coherent $T_\odot$-extension.

$$P_\odot(E_{\gamma_i}|E_{\gamma_j}) = \frac{\displaystyle\sum_{(x_i,x_j)\in\mathscr{C}_{(x_i,x_j)}} (\mu_{\gamma_i} \odot \mu_{\gamma_j})(x_i, x_j) P_{ij}(x_i, x_j)}{\displaystyle\sum_{x\in\mathscr{C}_X} \mu_{\gamma_j}(x) P_j(x)}. \tag{3}$$

In the case where $P_\odot(E_{\gamma_j}) = 0$ for some event E_{φ_j}, to have a unique extension to the events $E_{\varphi_i}|E_{\varphi_j}$ we need also the conditional probability $P(A_x|B)$, where B is the logical sum of the events x such that $P(E_{\gamma_j}|x) = 0$ (see [23]).

In the possibilistic setting one has for $A = E_{\varphi_i}$ and $B = E_{\varphi_j}$ the coherent extension $\Pi(E_{\varphi_i}|E_{\varphi_j})$ and for $i \neq j$ it is a solution of the equation

$$\Pi_\odot(E_{\gamma_i} \wedge E_{\gamma_j}) = T(x, \Pi(E_{\gamma_j})). \tag{4}$$

where $\Pi(E_{\gamma_i})$ is defined in Eq. (2) and

$$\Pi_\odot(E_{\gamma_i} \wedge E_{\gamma_j}) = \max_{(x_i,x_j)\in\mathscr{C}_{(x_i,x_j)}} T((\mu_{\gamma_i} \odot \mu_{\gamma_j})(x_i, x_j), \Pi_{i,j}(x_i, x_j)), \tag{5}$$

Remark 8.3 The coherent values $P_\odot(E_{\varphi_i}|E_{\varphi_j})$ and $\Pi_\odot(E_{\varphi_i}|E_{\varphi_j})$ are computed by Eqs. (3) and (4), when the events E_{φ_i} and E_{φ_j} are almost $\mathscr{C}_X$-logically independent. When E_{φ_i} and E_{φ_j} are logically dependent, the values must be chosen in the coherence intervals. For instance if $E_{\varphi_i} = E_{\varphi_j}$ the only coherent assessment is 1.

4 Probabilistic and Possibilistic Fuzzy IF-THEN Rules

As it is well known a relevant problem in literature is the managing of fuzzy IF-THEN rules based systems. Many scientist works in methods to learn fuzzy decision rules (for a review see [4]); this is not the argument of this section, which refers to system in which a person, usually a field expert, crafts or curates rule sets.

They are essentially composed by:

- A collection of rules of the form "IF A THEN B" with a given degree of belief", where either premise A and consequence B of the rule can contain a fuzzy instance, representing a specific type of knowledge base;
- An interpretation rule of the implication;
- An inference method.

Moreover, for expressing the degree of belief, it is necessary to chose a measure of uncertainty.

But immediately a problem occurs: where this uncertainty measure is defined? Makes some sense (from a mathematical point of view) to attribute a probability or a possibility to an implication (to be interpreted) of fuzzy sets?

By using the proposal and the results summarized in the sections before, we proposed an interpretation of rules X IS γ THEN YS IS ψ, with degree of belief k in therms of conditional measure of the conditional events $E_\psi | E_\gamma$, where either γ and ψ can be fuzzy.

To better explain this interpretation we start from the following easy example:

Example 8.1 From an urn containing balls of different colours (Red, Black, Green, Yellow) and diameters $(d_1, ..., d_6)$. The composition of the urn and so the joint probability $P(C = c, D = d)$, where C is a variable ranging in the different colours and D that ranging on different diameters. Let us consider the following rules (without fuzzy properties):

"IF $C = R$ THEN the gain is 2000 times the diameter"
"IF $C = B$ is black THEN the gain is 100 times the diameter"
"IF $C = G$ is red THEN the gain is 1000 times the diameter"
"IF $C = Y$ is red THEN the gain is 500 times the diameter"

If one must choose among the above rules, it is necessary to transform the rules in others involving directly the diameter of the kind "IF $C = R$ THEN $D = d_i$" and associate to them a probabilistic degree of belief. Then the most natural way to perform them is through conditional probability, i.e. by computing $P(D = d_i | C = c_j)$ $(i = 1, ..., 6)$ $(c_j = R, B, G, Y)$, where both conditioning and conditioned events vary.

Since we hypothesised to know the composition of the urn and so the joint percentage, the conditional probabilities above are univocally assessed. In more general situations, when the probability distribution is not completely available or the membership function is defined on a different partition (even if related to that of the space), the unicity of the conditional probability is not guaranteed, so the lower and upper envelope of a class of such probabilities must be computed.

This is one of the reasons for using probability with possibility in fuzzy IF-THEN rules depends on the prior information. More precisely, if membership functions are assessed in a space where no probability is available, while a probabilistic information is present on a different space, the best alternative is to extend the available "prior" probabilistic information to the space where the membership is defined. In the case that the two partition are under suitable constraints, the upper envelope of the class turns out to be a possibility measure [24].

Example 8.2 (Example 1 continues) Consider now a similar problem, but with fuzzy statements in the input and in the output such as:

"IF C is Dark THEN the gain is 2000 times a small diameter"
"If C is Dark THEN the gain is 1000 times a large diameter"
"If C is Light THEN the gain is 400 times a small diameter"
"If C is Light THEN the gain is 800 times a large diameter".

The reasoning does not change, so, by using the above interpretation and, in particular, Eq. 3, one can compute $P(E_s|E_D)$, $P(E_s|E_L)$, $P(E_l|E_D)$, $P(E_l|E_L)$, where E_s and E_D stand, respectively, for "You claim that D is small" and "You claim that C is Dark", equipped with the relevant $\mu_s(c_j) = P(E_s|c_j)$ and $\mu_D(c_i) = P(E_D|d_i)$. Similarly for the other conditional events. If the available information on $\mathscr{C}_{X,Y}$ is represented by a joint possibility distribution, then, thanks to Theorems 8.2 and 8.3, we can extend to any other conditional event.

Recall that, since the events E_{γ_i} are almost logically independent with respect to the joint partition, from Theorem 8.4, one of the coherent extension on the events can be obtained by computing all the intersections of E_{γ_i} by either T_M or T_p, and then extending this coherent assessment by rules of coherent probability or possibility.

Remark 8.4 Mamdani interpretation formula coincides with the conditional possibility $\Pi_\odot(E_\psi|E_\gamma)$ in the trivial case where the joint possibility distribution is the uninformative one (i.e. $\Pi(x, y) = 1$ for every $(x, y) \in \mathscr{C}_{X,Y}$). This point of view conserves the sense of direction to the rule, i.e. highlights which is the antecedent and the consequent, which is impossible to identify directly from the Mamdani interpretation rule (for similar arguments see [3] and [33], where a systematic attempt to semantic aspect of fuzzy rules in the framework of possibility theory is made). Moreover the inference rule for a new conditional event $(E_{\psi'}|E_{\gamma'})$ is one of the coherent (possibilistic) likelihood, obtained by imposing the constraint $\Pi_\odot(E_\psi|E_\gamma) = \Pi_\odot(E_{\psi'}|E_{\gamma'})$.

Remark 8.5 In this paper no emphasis was placed on possible events of zero probability or possibility, that can be dealt with without problems in this framework. In the decision-making rules it is obvious that to take into account unexpected hypotheses could be of great advantage.

5 Conclusions

The paper performs a decision processes involving uncertainty and fuzziness by using the interpretation of a fuzzy set as a crisp event E_φ of the kind "You claim that the variable X has the property φ" and a coherent conditional probability or possibility assessment.

In order to apply this interpretation to a dynamic process of decision, it is necessary to study which concepts, elaborated in the fuzzy set theory, can be reinterpreted in either probabilistic and possibilistic ambit, maintaining coherence with the framework of reference.

References

1. Bellman, R.E., Zadeh, L.A.: Decision making in a fuzzy environment. Manag. Sci. **17**(4), 141–164 (1970)
2. Birkhoff, G.: Rings of sets. Duke Math. J. **3**(3), 443–454 (1937)
3. Bouchon-Meunier, B.: La Logique Floue. Presses Universitaires de France, Paris (1993)
4. Bouchon-Meunier, B., Marsala, C.: Learning fuzzy decision rules. In: Bezdek J.C., Dubois D., Prade H. (eds) Fuzzy Sets in Approximate Reasoning and Information Systems. The Handbooks of Fuzzy Sets Series, vol. 5, pp. 279–304. Springer (1999)
5. Bouchon-Meunier, B., Coletti G., Marsala C.: Possibilistic conditional events. In: Proceedings of the 8th International Conference on Information Processing and Management of Uncertainty in Knowledge-Based Systems, pp. 1561–1566. Madrid, Spain (2000)
6. Bouchon-Meunier, B., Coletti, G., Marsala, C.: Conditional possibility and necessity. In: Bouchon-Meunier, B., et al. (ed.) Technologies for Constructing Intelligent Systems 2. Studies in Fuzziness and Soft Computing, vol. 90, pp. 59–71. Physica, Heidelberg (2002)
7. Bouchon-Meunier, B., Coletti, G., Marsala, C.: Independence and possibilistic conditioning. Ann. Math. Artif. Intell. **35**, 107–123 (2004)
8. Coletti, G.: Coherent numerical and ordinal probabilistic assessments. IEEE Trans. Syst. Man. Cybernet. **24**, 1747–1754 (1994)
9. Coletti, G., Gervasi, O., Tasso, S., Vantaggi, B.: Generalized Bayesian inference in a fuzzy context: From theory to a virtual reality application. Comput. Stat. Data Anal. **56**(4), 967–980 (2012)
10. Coletti, G., Petturiti, D.: Finitely maxitive T-conditional possibility theory: coherence and extension. Int. J. Approx. Reason. **71**, 64–88 (2016)
11. Coletti, G., Petturiti, D.: Finitely maxitive conditional possibilities, Bayesian-like inference, disintegrability and conglomerability. Fuzzy Sets Systems **284**, 31–55 (2016)
12. Coletti, G., Petturiti, D., Vantaggi, B.: Likelihood in a possibilistic and probabilistic context: a comparison. In: Borgelt C. et al. (eds) Combining Soft Computing and Statistical Methods in Data Analysis. Advances in Intelligent and Soft Computing, vol. 77,pp. 89–96. Springer (2010)
13. Coletti, G., Petturiti, D., Vantaggi, B.: Possibilistic and probabilistic likelihood functions and their extensions: common features and specific characteristics. Fuzzy Sets Systems **250**, 25–51 (2014)
14. Coletti, G., Petturiti, D., Vantaggi, B.: Fuzzy memberships as likelihood functions in a possibilistic framework. Int. J. Approx Reason **88**, 547–566 (2017)
15. Coletti, G., Petturiti, D., Vantaggi, B.: Interval-based possibilistic logic in a coherent setting. In: Benferhat, S., Tabia, K., Ali, M. (eds.), Advances in Artificial Intelligence: From Theory to Practice, vol. 10351, pp. 75–84. LNAI, Springer (2017)
16. Coletti, G., Petturiti, D., Vantaggi, B.: Preferences on fuzzy lotteries (2018)
17. Coletti, G., Scozzafava, R.: Characterization of coherent conditional probabilities as a tool for their assessment and extension. Int. J. Uncert. Fuzz. Knowl. Based Syst. **4**, 103–127 (1996)
18. Coletti, G., Scozzafava, R.: From conditional events to conditional measures: a new axiomatic approach. Ann. Math. Artif. Intell. **32**, 373–392 (2001)
19. Coletti, G., Scozzafava, R.: Probabilistic Logic in a Coherent Setting. Kluwer, Dordrecht (2002)
20. Coletti, G., Scozzafava, R.: Conditional probability, fuzzy sets, and possibility: a unifying view. Fuzzy Sets Systems **144**, 227–249 (2004)
21. Coletti, G., Scozzafava, R.: Conditional probability and fuzzy information. Comput. Stat. Data Anal. **51**, 115–132 (2006)
22. Coletti, G., Scozzafava, R., Vantaggi, B.: Integrated likelihood in a finitely additive setting. In: Sossai, C., Chemello, G. (eds.) ECSQARU 2009, LNAI 5590, pp. 554–565. Springer, Berlin, Heidelberg (2009)
23. Coletti, G., Scozzafava, R., Vantaggi, B.: Coherent conditional probability, fuzzy inclusion and default rules. In: Yager, R. et al. (eds.) Soft Computing: State of the Art Theory and Novel Applications. Studies in Fuzziness and Soft Computing, vol. 291, pp. 193–208. Springer (2013)

24. Coletti, G., Scozzafava, R., Vantaggi, B.: Inferential processes leading to possibility and necessity. Inform. Sci. **245**, 132–145 (2013)
25. Coletti, G., Vantaggi, B.: T-conditional possibilities: coherence and inference. Fuzzy Sets Syst. **160**(3), 306–324 (2009)
26. Coletti, G., Vantaggi, B.: Hybrid models: probabilistic and fuzzy information. In: Kruse, R. et al. (eds.) Synergies of Soft Computing and Statistics for Intelligent Data Analysis, Advances in Intelligent Systems and Computin, vol. 190, pp. 389–398 (2012)
27. Coletti, G., Vantaggi, B.: Conditional non-additive measures and fuzzy sets. In: Proceedings of 8th International Symposium on Imprecise Probability: Theories and Applications. (2013)
28. Coletti, G., Vantaggi, B.: Probabilistic reasoning in a fuzzy context. In: Zadeh, L. et al. (eds.) Recent Developments and New Directions in Soft Computing. Studies in Fuzziness and Soft Computing, vol. 317, pp. 97–115. Springer, Cham (2014)
29. Coletti, G., Vantaggi, B.: Knowledge processing in decisions under fuzziness and uncertainty. Proc. IFSA-SCIS **2017**, 1–6 (2017)
30. Coletti, G., Vantaggi, B.: Coherent conditional plausibility: a tool for handling fuzziness and uncertainty under partial information. In: Collan, M., Kacprzyk, J. (eds.) Soft Computing Applications for Group Decision-making and Consensus Modeling. Studies in Fuzziness and Soft Computing, vol. 357. Springer, Cham (2018)
31. de Finetti, B.: Sul significato soggettivo della probabilità. Fund. Math. **17**, 298–329 (1931)
32. de Finetti, B.: Teoria della probabilitá vol. I, II. Einaudi, Torino (1970) (Engl. Transl. Theory of probability. Wiley, London (1974))
33. Dubois, D., Prade, H.: What are fuzzy rules and how to use them. Fuzzy Sets Syst. **84**(2), 169–185 (1996)
34. Gilio, A.: Probabilistic reasoning under coherence in system P. Ann. Math. Artif. Intell. **34**, 5–34 (2002)
35. Godo, L., Marchioni, E.: Coherent conditional probability in a fuzzy logic setting. Log. J. IGPL **14**(3), 457–481 (2006)
36. Mamdani, E.H., Assilian, S.: An experiment in linguistic synthesis with a fuzzy logic controller. Int. J. Man-Machine Stud. **7**(1), 1–13 (1975)
37. Montagna, F.: A notion of coherence for books on conditional events in many-valued logic. J. Log. Comput. **44**(3), 563–593 (2012)
38. Marchioni, E., Godo, L.: A logic for reasoning about coherent conditional probability: a modal fuzzy logic approach. In: Alferes, J.J., Leite J. (eds.) Logics in Artificial Intelligence. JELIA 2004, vol. 3229, pp. 213–225. LNCS, Springer, Berlin, Heidelberg (2004)
39. Mundici, D.: Faithful and invariant conditional probability in Łukasiewicz logic. In: Makinson, D., Malinowski, J., Wansing H. (eds.) Proceedings of Conference Trends in Logic IV, Torun, Poland, 2006, (Trends in Logic, Vol. 28), pp. 213–232. Springer, New York (2008)
40. Mundici, D.: Conditionals and independence in many-valued logics. In: Sossai, C., Chemello, G. (eds.) Proceedings of ECSQARU 09, LNAI, vol. 5590, pp. 16–21 (2009)
41. Scozzafava, R., Vantaggi, B.: Fuzzy inclusion and similarity through coherent conditional probability. Fuzzy Sets Syst. **160**, 292–305 (2009)
42. Zadeh, L.A.: Fuzzy sets. Inform. Control **8**, 338–353 (1965)
43. Zadeh, L.A.: Probability measures of fuzzy events. J. Math. Anal. App. **23**(2), 421–427 (1968)
44. Zadeh, L.A.: Fuzzy sets as a basis for a theory of possibility. Fuzzy Sets Syst. **100**, 9–34 (1999)

Abstract Models for Systems Identification

Dan A. Ralescu and Anca L. Ralescu

Abstract We consider the problem of system identification (or minimal realization), for various classes of systems, for which we define some abstract versions of the minimal realization process. The models we discuss are used for the study of subsets (or, subcategories) in which a unique minimal realization exists.

1 Introduction

Systems identification is an old problem with a well established history. Important, both from mathematical and practical points of view, this problem is stated as follows: in the class of all systems with a given behavior, look for a system whose characteristics are "best". Of course, the terms "characteristics" and "best" have to be specified. Most often, we start with a behavior and then try to identify a system which has a certain "optimality" of structure.

An early work on the philosophical point of view and history of these ideas can be found in Gaines [2]. Zadeh [11] defined system identification as the "determination on the basis of input and output, of a system within a specified class of systems, to which the system under test is *equivalent*". In an early stage, linear systems were the main object of investigation, and Kalman [6] gave a uniqueness result for minimal realizations. By "minimal" we mean this "optimality" of structure mentioned above; this term will be defined precisely later in this paper. The first general categorical minimal realization result is due to Goguen [3], whose result applies to various classes of systems, including linear systems, automata, fuzzy systems. Goguen [4] also first

D. A. Ralescu
Department of Mathematical Sciences, University of Cincinnati,
ML 0025, Cincinnati, OH 45221–0025, USA
e-mail: Dan.Ralescu@uc.edu

A. L. Ralescu (✉)
EECS Department, University of Cincinnati, ML 0030, Cincinnati, OH 45221–0030, USA
e-mail: Anca.Ralescu@uc.edu

© The Editor(s) (if applicable) and The Author(s), under exclusive license   99
to Springer Nature Switzerland AG 2021
M.-J. Lesot and C. Marsala (eds.), *Fuzzy Approaches for Soft Computing and Approximate Reasoning: Theories and Applications*, Studies in Fuzziness and Soft Computing 394,
https://doi.org/10.1007/978-3-030-54341-9_9

suggested that minimal realization should be viewed as adjoint functors. Negoita, Ralescu and Ratiu [9] proved that an equivalence exists between the category of reachable systems with a given behavior and some category of equivalence relations on the input space (see Ralescu [10]). In this way, the minimal realization corresponds to the Nerode equivalence (a well known result). The new fact is that this equivalence is the supremum of all other relations in that category.

For more details on the prerequisites, the reader is referred to Arbib and Zeiger [1] (for minimal realizations) and Mac Lane [7] (for category theory).

The term "minimal", for example, denotes some "optimality of structure", of the system under consideration. This problem was put first for linear systems, and then generalized for deterministic dynamic systems (see [2]). Later on, the categorical approach to systems theory permitted to include a broad class of systems, as probabilistic or fuzzy systems (see [1]) Some minimal realization theorems were proved, such that in some special conditions the minimal realization exists (see [7]).

From this point on, the paper is organized as follows: In Sect. 2 we recall the minimal realization problem and some of its properties. Section 3.1 develops the first model, which is based on equivalence relations. The sets which support minimal realization are in connection with the systems of representants of an equivalence relation. This model corresponds to the external behavior point of view. In Sect. 3.2 we give a model based on ordering relations. This version is qualified as "minimal". Both models in Sects. 3.1 and 3.2 include systems identification. This is not however the only example; one can find others, such as the "integer part" function and the "congruence modulo n". The most powerful model seems to be the categorical model which is introduced in Sect. 3.3. The admissible subcategories are those for which the inclusion functor admits an adjoint. Some conclusions and further developments of the subject are discussed in the last section.

2   Background of the Minimal Realization Problem

We briefly sketch the minimal realization problem for dynamic deterministic systems. Throughout the paper, such a system will be a 6-tuple $\mathscr{S} = (X, U, Y, \delta, \beta, x_0)$ where X, U, Y are arbitrary sets called, respectively, the *state-space, input-space* and *output-space*. When the sets are finite, S is called an *automaton*. The dynamics δ is a function $\delta : X \times U \to X$, and the *output function* is $\beta : X \to Y$. The system $\mathscr{S}$ is "initialized", and $x_0 \in X$ is called its *initial state*. We can build the *category of systems*, denoted by Sys, whose objects are systems $\mathscr{S}$ as above. A morphism between two systems $\mathscr{S} = (X, U, Y, \delta, \beta, x_0)$ and $\mathscr{S}' = (X', U', Y', \delta', \beta', x_0')$ is a triple $\mathscr{F} = (f, g, h)$, where $f : X \longrightarrow X', g : U \longrightarrow U', h : Y \longrightarrow Y'$ are functions, such that the diagrams

$$X \times U \xrightarrow{\delta} X \xrightarrow{\beta} Y$$
$$f \times g \downarrow \qquad f \downarrow \qquad \downarrow h$$
$$X' \times U' \xrightarrow{\delta'} X' \xrightarrow{\beta'} Y'$$

are commutative, and $f(x_0) = x_0'$. Briefly speaking, morphisms between systems must preserve the initial states and commute with dynamics and output maps. The dynamics δ of a system $\mathscr{S}$ is extended to an action of the free monoid U^* on the state space X. This extension is defined recursively by:

1. $\delta(x, \Lambda) = x, \forall x \in X$
2. $\delta(x, \theta\theta') = \delta(\delta(x, \theta), \theta'), \forall x \in X, \theta, \theta' \in U^*$

Thus, the extended dynamics becomes a function from $X \times U^*$ and we may define the *reachability map*

$$\delta_{x_0} : U^* \longrightarrow X, \delta_{x_0}(\theta) = \delta(x_0, \theta)$$

and the *behavior map* from initial state x_0 :

$$f_{x_0} : U^* \longrightarrow Y, f_{x_0} = \beta \circ \delta_{x_0}$$

The behavior of f_{x_0} depends, of course, on the initial state $x_0 \in X$. In general, we may define a family of behavior maps $(f_x)_{x \in X}$, for any state $x \in X$ (thought as the initial state). A system $\mathscr{S}$ is called *reachable* (from x_0) if the reachability map δ_{x_0} is onto. A system $\mathscr{S}$ is called *observable*, if the map $x \mapsto f_x$ is one-to-one. These two concepts, reachability (also called controllability) and observability are due to Kalman [5]. Thinking of a system as of a model of some physical process, one may say that obtaining f_{x_0} means a *simulation* of that process. The realization problem is, in a sense, inverse to the problem described so far. More specifically, we are only given a behavior $f : U^* \to Y$, and we look for a system $\mathscr{S}$, whose behavior (from some initial state x_0) is exactly f.

As opposed to simulation, this inverse problem is referred to as *modeling*. As is to be expected, the "model" of a given behavior need not be unique. The uniqueness is achieved only by imposing some "optimality of structure" for the model. Then, the (essentially) unique "optimal model" is called *minimal realization* (of the given behavior). In precise mathematical terms, "optimal" means reachable and observable.

Suppose that we are given the input space U, the output space Y, and a function $f : U^* \to Y$, which describes the behavior of some process. The classical minimal realization is given by the following result.

Theorem 9.1 *There exists a reachable and observable system*

$$\mathscr{S}_f = (X_f, U, Y, \delta_f, \beta_f, x_0)$$

whose behavior from x_0 is exactly f. Moreover, any two such systems are isomorphic.

The proof can be found in Arbib and Zeiger [1] and is omitted here. We mention, however, that the state space X_f is obtained via the *Nerode equivalence* in U^*:

$$\theta_1 N \theta_2 \iff f(\theta_1 \theta) = f(\theta_2 \theta), \forall \theta \in U^*$$

and $X_f = U^*/N$ (the quotient set). This result can be put into a categorical framework, by defining the category of behaviors and the category of systems. Goguen [3] showed that there is a pair of adjoint functors between these categories.

In the following sections, we will embed the minimal realization into a more abstract framework. Our three "abstract models" for system identification are related to the work of Gaines [2] about complexity and admissibility. The main difference is that the relationship

$$\text{behavior} \longleftrightarrow \text{optimal model}$$

is replaced by the relationship

$$\text{model} \longleftrightarrow \text{optimal model}$$

Indeed, Theorem 9.1 may be thought of in another way as follows: start with a system $\mathscr{S}$, compute its behavior $f = f_{x_0}$, and then look for the minimal realization $\mathscr{S}_f$. The system $\mathscr{S}_f$ will be "optimal" since it will be reachable and observable.

3 Different Types of Identification Models

This section describes several models of system identification.

3.1 Relational Models

Our first model for system identification starts with a pair (X, R) where X is a set and R is an equivalence relation on X. The idea is to define some subsets of X which contain the "optimal systems". More precisely:

Definition 9.1 A subset $A \subset X$ is called admissible, if for any $x \in X$, there exists an $a \in A$, such that $a R x$.

We will denote by $A(X)$ the set of admissible subsets of X. The admissible subsets are, in a sense, "too large". We need some sort of "minimality" which is made precise in the following.

Definition 9.2 A subset $M \subset X$ is called minimal admissible, if for any $x \in X$, there exists a unique $m \in M$, such that $m R x$.

We denote by $M(X)$ the set of all minimal admissible subsets of X, and it is clear that $M(X) \subset A(X)$. Our main goal is to characterize the elements of $M(X)$. First, we make the following, simple remarks:

1. $X \in A(X)$
2. $A \in A(X), A \subset B \Longrightarrow B \in A(X).$

We prove now that each admissible set contains an admissible subset, which is the "best one" in some sense:

Proposition 9.1 *For each $A \in A(X)$, there exists $A_0 \in A(X)$, $A_0 \subset A$ such that is $a, b \in A_0$, then aRb is false.*

Proof Consider the quotient set A/R. Select one and only one element from each class of equivalence $[a] \in A/R$. The collection of these elements is denoted by $A_0 \subset A$. It is obvious that for any elements $a, b \in A_0, aRb$ is false. We have to show that $A_0 \in A(X)$. If $x \in X$, there exists $a \in A$, such that xRa. Thus $x \in [a]$, but in A_0 we have an element from each class $[a]$, say a_0. Thus $a_0 \in [a]$.

Corollary 9.1 *For each $A \in A(X)$, there exists $M \in M(X)$, $M \subset A$.*

The proof is obvious.

Remark 9.1 If $M \in M(X)$ is a minimal admissible subset, for any $x \in X$, we may call the unique $m \in M(X)$ with mRx, the *minimal realization* of x.

Before characterizing the minimal realization we provide two examples.

Example 9.1 In this example we show that the minimal realization of deterministic systems is included in our abstract model. Let us denote by Sys the class of all deterministic systems (as in Sect. 2). For $S \in Sys$, denote by f_S its behavior (from the initial state of S). Consider the pair (Sys, R), where R is the equivalence relation defined by:

$$S, S' \in Sys, S \, R \, S' \iff f_S = f_{S'}$$

Denote now by $Sys(r, o)$ the subset of reachable and observable system, $Sysy(r, o) \subset Sys$. The classical minimal realization theory proves that $Sys(r, o)$ is an *admissible* subset of Sys. In other words:

$$(\forall) S \in Sys \longrightarrow (\exists) S_m \in Sys(r, o), \text{ such that } f_{S_m} = f_S.$$

We mention that any other subset of Sys, $C \supset Sys(r, o)$, is also admissible. Note that $Sys(r, o)$ is not minimal admissible, but minimal admissibility may be achieved by identifying isomorphic systems.

Example 9.2 This is a simple, algebraic example, showing that other situations may be included in our relational model. Consider the set of integers $\mathbb{Z}$, and let R be the "congruence modulo n" ($n > 2$ fixed):

$$p, q \in \mathbb{Z}, \, pRq \iff p - q \text{ is divisible by } n$$

An admissible subset is $\mathbb{N} = \{0, 1, 2, 3, \ldots\}$, the set of nonnegative integers. The set $\mathscr{N}^* = \{1, 2, 3, \ldots\}$ is not admissible. By applying Proposition 9.1, we may obtain the minimal admissible subsets.

Let us turn now our attention to the general minimal admissible sets. In order to characterize them, we need the concept of a *section of a map*.

Definition 9.3 A map $A \xleftarrow{g} B$ is a section of the map $A \xrightarrow{f} B$ if $f \circ g = I_B$. Note that section means right-inverse.

For a pair (X, R), $X \xrightarrow{\phi} X/R$ denotes the quotient map $\phi(x) = [x]$.

Theorem 9.2 *There is a bijection between $M(X)$ and the set of all sections of the map $X \xrightarrow{\phi} X/R$.*

Proof Letting $Sec = \{s : X/R \to X/(\phi \circ s) = I_{X/R}\}$, we must define two functions, $M(X) \xrightarrow{\Phi} Sec$ and $Sec \xleftarrow{\Psi} M(X)$ such that $\Phi \circ \Psi = id$ and $\Psi \circ \Phi = id$. For each $[x] \in X/R$ and $M \in M(X)$, $[x] \cap M$ contains a unique element m_x. Define $\Phi(m) = s_M$, where $s_M([x]) = m_x$. Let now $s \in Sec$ and set $\Psi(s) = \text{Im } s = s(X/R)$. It is easy to prove that $\Psi(s)$ is indeed admissible, and then that $\Phi \circ \Psi = \text{id}$, $\Psi \circ \Phi = \text{id}$.

We will see in the next sections that the other 'abstract models'for systems identification are also related to such right or left inverses. In Sect. 3.2, for example, the admissible sets will be defined by an adjunction.

3.2 Order Models

The second type of models is based on order relations. The motivation for introducing ordering to replace equivalence, is that we look sometimes not for systems that are equivalent to a given one, but which 'approximate', or are 'less complex'than a given one. To quote Gaines [2], the equivalence relation is 'too powerful a notion for a theory that needs instead, some concept of *degree of approximation*, and order relation that allows us to say that one model accounts for the behavior *better* than the other does'.

We start with a pair $(X, \leq)$, where X is a set and $\leq$ is an ordering in X. Our terminology will be reminiscent of Sect. 3.1 although in a different context.

Definition 9.4 A subset $A \subset X$ is called *admissible*, if:

$$\forall x \in X, \exists a_x \in A \text{ such that } a_x \leq x \tag{1}$$

$$\forall x \in X \text{ and a} \in A, \text{ with } a \leq x, \text{ we have } a \leq a_x. \tag{2}$$

Remarks:

1. The ordering $\leq$ may be thought of as "complexity": if $x \leq y$, we say that "x is less complex than y". It is clear now that an admissible subset contains all "optimal models", in the sense that, for a given $x \in X$, a_x is the most complex model among those which are less complex than x.
2. To make the previous point even more precise, note that for any $x \in X$, $a_x = \sup\{a \in A \mid a \leq x\}$. Indeed from (2) it is clear that $a_x \geq a$ for any $a \leq x$. Thus, $a_x \geq \sup\{a \in A \mid a \leq x\}$.
3. It is clear now that for any $x \in X$, a_x is *uniquely* defined by (1) and (2). We may call a_x the *minimal realization* of x.

Our next goal is to characterize admissible subsets of X. To this end, we need the concept of a *retract*, which is dual to that of *section*, used above.

Definition 9.5 A map $M \xleftarrow{g} N$ is a *retract* of the map $M \xrightarrow{f} N$ if $g \circ f = I_M$. That is, the retract means the left-inverse.

Denote now by $A(X)$ the set of all admissible subsets of X. For $A \in A(X)$, $A \xrightarrow{i} X$ denotes the inclusion map.

Theorem 9.3 *The following statements are equivalent:*

(i) $A \in A(X)$

(ii) There exists an isotone retract, j of i, $i : A \longrightarrow X$, $j : X \longrightarrow A$, with $j \circ i = 1_A$, and $i \circ j \leq 1_X$.

Proof

(i) $\implies$ **(ii).** If $A \in A(X)$, define $X \xrightarrow{j} A$ by $j(x) = a_x$. It is easy to show that j is isotone, $j \circ i = I_A$, and $i \circ j \leq 1_X$.

(ii) $\implies$ **(i).** For any $x \in X$, we define $a_x = j(x) \in A$ and from properties of j it is easy to check that A is an admissible subset.

Curiously enough, there is some kind of duality between order models and relational models described in Sect. 3.1. While admissible sets are characterized in the order-theoretical setting by retracts of the one-to-one map $A \xrightarrow{i} X$, they were characterized by using section of the onto map $X \xrightarrow{\phi} X/R$. In the following, we give three examples of how admissible subsets look like, in specific situations.

Example 9.3 Consider $X = Sys$, the class of all systems. The ordering is the *inclusion* of systems. More explicitly, if $S = (X, U, Y, \delta, \beta, x_0)$, $S' = (X', U', Y', \delta', \beta', x_0')$, we say that S' is a subsystem of S, and denote by $S' \subset S$, if: $X' \subset X$, $U' \subset U$, $Y' \subset Y$, $\delta|_{X' \times U'} = \delta'$, $\beta|_{X'} = \beta'$, $x_0' = x_0$. We claim that the reachable systems $Sys(r)$ form an admissible subset of Sys. Indeed, if $S = (X, U, Y, \delta, \beta, x_0) \in Sys$, consider the reachability map $\delta_{x_0} : U^* \longrightarrow X$. Denote by $X_0 = \text{Im}, \delta_{X_0} = \delta_{x_0}(U^*)$ and consider the system $S_0 = (X_0, U, Y, \delta_0, \beta_0, x_0)$, where $\delta_0 = \delta|_{X_0 \times U}$, $\beta_0 = \beta|_{X_0}$. It is easy to show that $S_0 \in Sys(r)$, $S_0 \subseteq S$, and if $S' \in Sys(r)$ with $S' \subseteq S$, then $S' \subseteq S_0$. Thus, S_0 is the "best approximation" of S by a reachable system.

Example 9.4 Consider $X = \mathbb{R}$, the real numbers, and let $\leq$ be the usual ordering. Then $\mathbb{Z}$, the integers, form an admissible subset of $\mathbb{R}$. Comparing with Theorem 9.3, we find that the retract $j : \mathbb{R} \longrightarrow \mathbb{Z}$ is $j(x) = \lfloor x \rfloor$ (the greatest integer not exceeding x). Generally, if X is sup-complete, then each subset A which is itself sup-complete, is admissible. This is clear since, for $x \in X$ we have $a_X = \sup\{a \in A \mid a \leq x\} \in A$.

Example 9.5 This example is related to the problem of approximating fuzzy sets. The reader is referred to Negoita and Ralescu for the relevant definitions and results [8]. Let X be $X = F(M) = \{f \mid f : M \longrightarrow [0, 1]\}$, the set of all fuzzy subsets of M. The ordering $\leq$ is the *inclusion* of fuzzy sets, and is defined by:

$$f, g \in F(M), f \leq g \iff f(m) \leq g(m), \text{ for any } m \in M.$$

Consider $P(M) = \{f \mid f : M \longrightarrow \{0, 1\}\}$, the set of all subsets of M (identified by their characteristic functions). We claim that $P(M)$ is an admissible subset of $F(M)$. This is clear from the remark at the end of Example 9.4. Indeed, $F(M)$ is a complete lattice, and $P(M)$ is also complete. More precisely still, if we denote for each $f \in F(M)$, by $A_f = \{m \in M \mid f(m) = 1\}$, then the "best approximation" of f is $f_{A_f} \in P(M)$ (the characteristic function of A_f). This is easily proved: if $f \in F(M)$, then $f_{A_f} = \sup\{g \mid g \in P(M), g \leq f\}$.

One conclusion which may be drawn from these examples is that the ordering $\leq$ should be thought of as "approximation", and an admissible subset contains all the "best approximations".

3.3 Categorical Models

The last model for systems identification considered here is based on some concepts of category theory. Although much of what is needed to develop this model is presented in this section, for a more extensive/complete treatment the reader is referred to Mac Lane [7].

Definition 9.6 A category C is a concept consisting of a class $|C|$ of objects and, for any two objects $X, Y \in |C|$, a set of morphisms, $C(X, Y)$. The statement $f \in$

$C(X, Y)$ is denoted by $X \xrightarrow{f} Y$. Together with object and morphism, there is given a law of composition for morphisms, and the following three axioms:

(C1) If $f \in C(X, Y), g \in C(Y, Z)$, then there is a unique $g \circ f \in C(X, Z)$;

(C2) If $f \in C(X, Y), g \in C(Y, Z), h \in C(Z, T)$, then $h \circ (g \circ f) = (h \circ g) \circ f$ (associativity) ;

(C3) There is a unique morphism, $1_X \in C(X, X)$, such that $f \circ 1_X = f$, $1_X \circ g = g$, for any $f \in C(X, Y), g \in C(Z, X)$ (identity).

Among the examples of categories are: the category of sets, SET (its objects are sets, morphisms are functions), or the category of systems, Sys, described in Sect. 2. If C a and C' are two categories, a (covariant) *functor* from C to C' is denoted by $C \xrightarrow{U} C'$. This is an assignment of objects $|C| \longrightarrow |C'|$ and, for each $X, Y \in |C|$, an assignment of morphisms $C(X, Y) \longrightarrow C'(UX, UY)$ such that:

(F1) $U(1_X) = 1_{UX}$

(F2) $U(g \circ f) = U(g) \circ U(f)$, whenever the composition makes sense.

If $U, V : C \longrightarrow C'$ are two functors, a *natural transformation* from U into V is denoted by $U \xrightarrow{\alpha} V$, and is a collection of morphism $\alpha = (\alpha_X)_{X \in |C|}$, $UX \xrightarrow{\alpha_X} VX$, such that for each $X, Y \in |C|$ and $f \in C(X, Y)$, the following diagram commutes:

$$
\begin{array}{ccc}
UX & \xrightarrow{\alpha_X} & VX \\
{\scriptstyle Uf}\downarrow & & \downarrow{\scriptstyle Vf} \\
UY & \xrightarrow{\alpha_Y} & VY
\end{array}
$$

A natural transformation α is an *isomorphism* of functors, denoted by $U \cong V$, whenever α_X are isomorphisms for each $X \in |C|$. If C is a category, its *opposite* category, denoted by C^{op} has the same objects, while morphisms are "reversed", i.e., $f \in C^{op}(X, Y)$ if and only if $f \in C(Y, X)$. If C and C' are categories, the *product category*, denoted by $C \times C'$, is defined as follows: its objects are pairs (X, X'), with $X \in |C|, X' \in |C'|$; a morphism $(X, X') \longrightarrow (Y, Y')$ in $C \times C'$, is a pair (f, f') with $f \in C(X, Y)$, and $f' \in C'(X', Y')$. The last concept needed is that of an adjoint pair of functors. This concept will play an important role in the development of the third, categorical model for system identification. If C and C' are two categories, $U : C \longrightarrow C', V : C' \longrightarrow C$ a pair of functors, two new functors can be built as follows:

$$
C(\cdot, V\cdot) : C^{op} \times C' \longrightarrow SET \quad \text{and} \quad C'(U\cdot, \cdot) : C^{op} \times C' \longrightarrow SET
$$

For example, $C(\cdot, V\cdot)(X, Y) = C(X, VY)$ (this is a set!). If

$$
(X, Y) \xrightarrow{(f,g)} (X_1, Y_1)
$$

is a morphism in $C^{op} \times C'$, then

$$C(\cdot, V\cdot)(f, g) : C(X, VY) \longrightarrow C(X_1, VY_1)$$

is defined by $C(\cdot, V\cdot)(f, g)(\phi) = V_g \circ \phi \circ f$. The functor $C'(U\cdot, \cdot)$ is defined in an analogous manner. Note that "arrows" (morphisms) in C^{op} are reversed.

Definition 9.7 V is a *right adjoint* of U (or U is a *left adjoint* of V), if the functors $C(\cdot, V\cdot)$ and $C'(U\cdot, \cdot)$ are isomorphic.

Although the concept of an adjoint functor is quite complex, many examples arise and, in fact, it can be argued that any "natural structure" provides an adjoint pair of functors. Such a pair will be described later in an example. The idea of defining a categorical model for system identification is the following: obtain an abstract setting for the systems with "optimal structure". An admissible subcategory will be characterized by the property that it contains the "optimal objects". Let C be a fixed category.

Definition 9.8 A *realization subcategory* of C, is $A \subset C$, such that the inclusion functor $I : A \longrightarrow C$ has a left adjoint $J : C \longrightarrow A$.

In the pair (I, J), J is called the realization functor; for each $X \in |C|$, $JX \in |A|$ is called *minimal realization* of X. The following example shows that the classical minimal realization can be recaptured in this way.

Example 9.6 We restrict our attention to reachable systems. In Example 9.3 it was seen that a system, even when not reachable, contains a reachable subsystem (its reachable part). Let us consider $Sys(r)$, the category of reachable systems. A morphism, $S \longrightarrow S'$, is a triplet (f, g, h) with the second component g (which operates on the input spaces) being *onto*. The definition and relevance of such morphisms to system theory first appeared in Goguen [3]. Let $Sys(r, o)$ be the subcategory of $Sys(r)$ whose objects are reachable and observable systems. We show that $Sys(r, o)$ is a *realization subcategory* of $Sys(r)$. To do this, one must build a functor $Sys(r) \overset{J}{\longrightarrow} Sys(r, o)$ which is left adjoint of the inclusion functor $Sys(r, o) \overset{I}{\longrightarrow} Sys(r)$, proceeding as follows: let $S \in |Sys(r)|$, $S = (X, U, Y, \delta, \beta, x_0)$. Define the equivalence relation:

$$x, x' \in X, x \cong x' \iff f_x = f_{x'}.$$

It is easy to prove that the system

$$S_m = (X/_{\cong_1}, U, Y, \delta_m, \beta_m, x_0)$$

where $X/_{\cong_1}$ is the quotient set, $\delta_m([x], u) = [\delta(x, u)]$, $\beta_m([x]) = \beta(x)$, is reachable and observable. Therefore, S_m is the minimal realization of S. The functor J is defined by $J(S) = S_m$. Note that there is a morphism $S \overset{\Phi}{\longrightarrow} S_m$, $\Phi = (\phi, 1_U, 1_Y)$, where

$X \xrightarrow{\phi} X/_{\cong}$ is the quotient map. We can also prove that the following *universality property* holds: for each $S' \in |Sys(r, o)|$, and for each morphism $S \xrightarrow{a} S'$, there exists a unique morphism $S_m \xrightarrow{b} S'$ which makes the following diagram commutative:

$$
\begin{array}{ccc}
S & \xrightarrow{\Phi} & S_m \\
& {\scriptstyle a} \searrow & \downarrow {\scriptstyle b} \\
& S' &
\end{array}
$$

Indeed, let $a = (f, g, h)$. Set $b = (\overline{f}, \overline{g}, \overline{h})$, where $\overline{g} = g$, $\overline{h} = h$, and $\overline{f}$ results from the diagram

$$
\begin{array}{ccc}
X & \xrightarrow{\phi} & X/_{\cong} \\
& {\scriptstyle f} \searrow & \swarrow {\scriptstyle \overline{f}} \\
& X' &
\end{array}
$$

that is, $\overline{f}([x]) = f(x)$, for all $[x] \in X/_{\cong}$. The main point is to show that $\overline{f}$ is *well defined*. This means to show that $x \cong x'$ implies $f(x) = f(x')$ or, equivalently, that $f_x = f_{x'}$ implies $f'_{f(x)} = f'_{f(x')}$, where f_x, $f_{x'}$ are behaviors of S, while $f'_{f(x)}, f'_{f(x')}$ are behaviors of S'. The last statement is proved by considering the commutative diagram

$$
\begin{array}{ccc}
U^* & \xrightarrow{f_x} & Y \\
{\scriptstyle g^*} \downarrow & & \downarrow {\scriptstyle h} \\
U'^* & \xrightarrow{f'_{f(x)}} & Y'
\end{array}
$$

where g^* denotes the extension of $U \xrightarrow{g} U'$ to the free monoids $U^* \xrightarrow{g^*} U'^*$, $g^*(u_1 u_2 \ldots u_p) = g(u_1)g(u_2) \cdots g(u_p)$. Thus $h \circ f_x = f'_{f(X)} \circ g^*$ and, in the same way, we get $h \circ f_{X'} = f'_{f(X')} \circ g^*$. It follows that $f_X = f_{X'}$ implies $f'_{f(X)} \circ g^* = f'_{f(X')} \circ g^*$. Since g is *onto*, g^* is also onto and, therefore, we get $f'_{f(X)} = f'_{f(X')}$. To end the proof, recall that the system S' is observable, thus $f'_{f(X)} = f'_{f(X')}$ implies $f(x) = f(x')$, as desired. From the above universality property, by using a standard technique of category theory, it is easy to prove that the functors $Sys(r, o) \xrightarrow{I} Sys(r)$, $Sys(r) \xrightarrow{J} Sys(r, o)$ form an adjoint pair. Details of this general technique may be found in Mac Lane [7].

Remark 9.2 It can be seen from this example that the minimal realization is *left adjoint* to inclusion, restricting our attention to reachable systems and morphisms whose input component is onto. The inclusion functor "forgets" the observability property.

4 Concluding Remarks

Three "abstract" models for systems identification were presented in this paper. The categorical model applies to system theory in the following sense: proving that an inclusion functor has a left adjoint, proves the existence of minimal realization. All these models are deeply related to systems theory, either in which the "best realization" of a behavior is concerned, or in the problem of approximating a system by one with a better structure. The order model may also be embedded into a categorical framework. Indeed, we may think of a pair $(X, \leq)$ as being a category, whose objects are $x \in X$ and a unique morphism exists $x \longrightarrow y$ if and only if $x \leq y$. Then admissible subsets (Sect. 3.3) are exactly subcategories A of X for which the inclusion functor $A \overset{i}{\longrightarrow} X$ has a right adjoint. Even if these models are powerful enough to include classical minimal realization, some of the ideas may not appear as realistic, especially when we think of large-systems behavior. More precisely, starting with a system S, we may look for a system S' whose behavior is "very close" to that of S. In this case, minimal realization is no more precisely defined: it becomes a *fuzzy concept*. Another case, still: we look for "optimality of structure" in the sense of reachability and observability. But these requirements might be too strong in some applications. A possibility of future investigation is to define some "weak" or "approximate" concepts of reachability and/or observability, and to restate the problem of finding the "best system" for a given behavior.

References

1. Arbib, M.A., Zeiger, H.P.: On the relevance of abstract algebra to control theory. Automatica **5**(5), 589–606 (1969)
2. Gaines, B.R.: System identification, approximation and complexity. Int. J. Gen. Syst. **3**(3), 145–174 (1977)
3. Goguen, J.A.: Minimal realization of machines in closed categories. Bull Am. Math. Soc. **78**(5), 777–783 (1972)
4. Goguen, J.A.: Realization is universal. Math. Syst. Theory **6**(4), 359–374 (1972)
5. Kalman, R.E.: Lectures on controllability and observability. Stanford University CA Department of Operations Research, Technical report (1970)
6. Kalman, R.E., Falb, P.L., Arbib, M.A.: Topics in mathematical system theory, vol. 1. McGraw-Hill, New York (1969)
7. Mac Lane, S.: Categories for the working mathematician, vol. 5. Springer Science & Business Media (2013)
8. Negoita, C., Ralescu, D.: Applications of fuzzy sets to systems analysis. Stuttgart, Basel (1975)
9. Negoita, C.V., Ralescu, D.A., Ratiu, T.: Relations on monoids and realization theory. In: Proceedings of the Third International Congress of Cybernetics and Systems Modern Trends in Cybernetics and Systems (in Three Volumes), 1975, p. 337. Springer, Bucharest, Romania (1977)
10. Ralescu, D.: Fuzzy subobjects in a category and the theory of c-sets. Fuzzy Sets Syst. **1**(3), 193–202 (1978)
11. Zadeh, L.A.: From circuit theory to system theory. Proc. IRE **50**(5), 856–865 (1962)

Fuzzy Systems Interpretability: What, Why and How

Luis Magdalena

Abstract Interpretability has been always present in Machine Learning and Artificial Intelligence. However, it is difficult to measure it (even to define it), and quite commonly it collides with other properties as accuracy, with a clear meaning and well defined metrics. This situation has reduced its influence in the area. But due to different external reasons, interpretability is now gaining importance in Artificial Intelligence, and particularly in Machine Learning. This new situation has two effects on the field of fuzzy systems. First, considering the capability of the fuzzy formalism to describe complex phenomena in terms that are quite close to human language, fuzzy systems have gained significant presence as an interpretable modeling tool. Second, the attention paid to interpretability of fuzzy systems, that grew during the first decade of this century and then experienced a certain decay, is growing again. The present paper will consider four questions regarding interpretability: what is, why is it important, how to measure it, and how to achieve it. These questions will be first introduced in the general framework of Artificial Intelligence, to be then focused from the point of view of fuzzy systems.

1 Introduction

The idea of interpretability has been always present in the fields of Artificial Intelligence (AI) and Machine Learning (ML). It is clear in concepts as Michalski's "Comprehensibility Postulate" [26]:

> The results of computer induction should be symbolic descriptions of given entities, semantically and structurally similar to those a human expert might produce observing the same entities. Components of these descriptions should be comprehensible as single "chunks" of information, directly interpretable in natural language, and should relate quantitative and qualitative concepts in an integrated fashion.

L. Magdalena (✉)
Escuela Técnica Superior de Ingenieros Informáticos, Universidad Politécnica de Madrid,
Campus de Montegancedo, 28660 Boadilla del Monte, Madrid, Spain
e-mail: luis.magdalena@upm.es

© The Editor(s) (if applicable) and The Author(s), under exclusive license
to Springer Nature Switzerland AG 2021
M.-J. Lesot and C. Marsala (eds.), *Fuzzy Approaches for Soft Computing and Approximate Reasoning: Theories and Applications*, Studies in Fuzziness and Soft Computing 394,
https://doi.org/10.1007/978-3-030-54341-9_10

111

It is also present in definitions as that of knowledge discovery in databases [16]

> the non-trivial process of identifying valid, novel, potentially useful, and ultimately understandable patterns in data.

As a result, the comprehensibility/understandability, and the predictive accuracy of the model were somehow considered as two orthogonal performance indicators. But opposite to accuracy, that was easily defined and measured, interpretability (or any other of the similar terms considered) was not a clearly defined concept, and lacked from widely accepted measures to evaluate it. Consequently, interpretability analysis did not receive significant attention, since it was assumed to be an intrinsic property of some knowledge representation formalism (rule sets or decision trees), widely studied due to this characteristic. So, the generation of interpretable models was directly linked to the use of *interpretable* knowledge representation formalisms.

In that sense, fuzzy systems (systems using fuzzy rules) were widely acknowledged as a powerful modeling tool [6, 8, 29, 35], having one of its main strengths on the capability to represent knowledge in a form close to natural language. This characteristic helps to improve human machine interaction, a question that is particularly important in those situations where there will be a human in the loop, being either an expert providing knowledge, a designer building the system or an end user exploiting it. Prof. Zadeh generalized this idea under the term of *Humanistic Systems*, described as those where human judgment, perception, and emotions play an important role [39]. This ability to represent knowledge in a form close to natural language has received strong attention since the very early stage of fuzzy systems. In fact, concepts as those of linguistic variables [38] and linguistic models are central in fuzzy systems.

But the main role of a model is the representation of a system, a phenomenon, a situation, in summary, a part of the world of particular interest to solve a certain problem. Being a representation of the reality, the matching between representation and reality is extremely important. Consequently, any model will be primarily evaluated in terms of how accurately represents the fragment of reality if was designed for. The main effect of this situation was that in the early nineties, different methods for knowledge induction and accuracy improvement were developed in the field of fuzzy systems. Methods that produced some side effects on the *intrinsic* interpretability previously mentioned, so that there was a need for explicitly analyze interpretability of fuzzy systems. As a consequence, the question of interpretability started to be a matter of study in the field of fuzzy systems since the beginning of the century. On the other hand, the growing presence of new models being extremely precise as well as completely unintelligible (e.g. deep learning models), rises now the need for an explicit consideration of interpretability at a general level in ML and AI.

The rest of the paper will consider interpretability from a dual perspective considering both, the general point of view of AI and ML, and the particular situation of fuzzy systems. To do so, four questions regarding interpretability will be considered: what is, why is it important, how to measure it, and how to achieve it. Some conclusions will complete the paper.

Prior to focus on interpretability, some basic concepts of fuzzy systems that could be of interest for the rest of the paper will be briefly introduced.

2 Some Basic Concepts on Fuzzy Systems

It has been said before that systems using fuzzy rules were widely acknowledged as a powerful modeling tool well suited for producing interpretable models. This kind of systems are usually known as Fuzzy (Rule-Based) Systems (FRBSs) [22]. They constitute an extension of classical Rule-Based Systems, considering "*IF-THEN*" rules with the following structure:

$$IF \ X_1 \ is \ LT_1 \ and \ \ldots \ and \ X_n \ is \ LT_n \ THEN \ Y \ is \ LT_o, \qquad (1)$$

where X_i and Y are the input and output linguistic variables.

A linguistic variable [38] is a variable that takes values from a set of linguistic terms (in the previous representation LT_i are linguistic terms associated to variable X_i). The linguistic terms are symbols that usually describe natural language terms (in most cases adjectives or equivalent concepts). As an example, a linguistic variable could be *Temperature*, taking values from the term set {Very high, High, Medium, Low}.

The linguistic terms are then represented by means of fuzzy sets, where a fuzzy set is identified by a membership function being a mapping from the *Universe* of the variable to the interval [0, 1]. The fuzzy sets representing the different terms of the term set, constitute a fuzzy partition of the Universe of the linguistic variable.

The output of the fuzzy system is computed by processing the information related to the current value of the input variables, jointly with the knowledge represented by the fuzzy rules and the linguistic variables (including their term sets and fuzzy partitions) of the system. This process, usually known as fuzzy inference, involves the application of the compositional rule of inference as well as the a use of fuzzification, defuzzification and aggregation operators [37].

From the point of view of interpretability, is the use of (linguistic) rules and linguistic variables what makes fuzzy systems highly interpretable. But not every linguistic variable, rule or fuzzy system is equally interpretable. As an example, Mamdani rules that use linguistic/fuzzy variables in premise and consequent are more interpretable than TSK rules that use functions in the consequent. In addition, linguistic rules (those based on fuzzy partitions) are more interpretable than fuzzy rules (those where each fuzzy set in a rule is independently defined and modified).

This kind of question will be considered again in Sect. 5 when defining methods to compare fuzzy systems from the point of view of interpretability as well as measures of interpretability.

3   What Is Interpretability?

There is not a single answer to this question, and even, there is not a single name for the concepts underlying the term *interpretability*. Terms as comprehensibility or understandability have been considered in previous paragraphs. Even if we concentrate in the field of fuzzy systems, other expressions as readability, or even transparency [7, 32, 33] are also present in literature to describe quite similar concepts.

It is possible to consider the formalization of the concept of interpretability in other areas as the starting point to define interpretability in AI. As an example, in mathematical logic, interpretability is the relation between two formal theories (T and S), where T is said to be interpretable in S iff the language of T can be translated into the language of S in such a way that S proves the translation of every theorem of T [36].

We probably can adapt this idea by saying that an AI model is interpretable in human terms if the language of the model can be translated into the language of the human interpreter. This is a point of view that has already been considered in fuzzy logic. Mencar et al. [23] state that interpretability can be assessed by measuring the complexity of making the translation from the model description based on Fuzzy Logic to the model explanation based on Natural Language.

Within AI, comprehensibility of symbolic knowledge is viewed as one of the defining factors which distinguishes logic-based representations from statistical or neural ones, i.e., those techniques where the translation to human language is an easy task, from those others where it is much more complex.

The underlying idea in all cases relates to the possibility of a human to somehow capture the meaning of the system under consideration, and to the complexity of that process. And in that sense, it is clear that not every symbolic model is equally interpretable. Obviously, not even every fuzzy system is equally interpretable. So, let's concentrate now on the meaning of interpretability in the field of fuzzy systems

In addition to the previous definition by Mencar et al. [23], Bodenhofer and Bauer [9] state that *Interpretability means the possibility to estimate the system's behavior by reading and understanding the rule base only*. In general, these definitions involve different aspects that are usually grouped under two main dimensions: semantic and structural interpretability [20, 24].

On the one hand, semantic interpretability is mostly affected by the proximity between the objects building up the fuzzy model (linguistic variables, aggregation operators, etc) and the concepts (mental representations) considered by the human interpreter. This is somehow connected to the idea of cointension, firstly defined by Zadeh [40] as a qualitative measure of the proximity of the meanings of two objects, and then considered by Mencar and his coauthors [24] to state that a knowledge base is interpretable if its semantics is cointensive with the knowledge acquired by the user when reading it.

On the other hand, the limited human capability in what concerns the amount of information to be stored in short-term memory [27] and processed to capture the *meaning* of the system under analysis, imposes additional restrictions to interpretability. Interpretability is not only a matter of the semantics of information, but it is also a question of volume. A large number of rules, rules jointly considering many variables, or linguistic variables with large term sets, will negatively affect interpretability. Consequently, structural complexity of the rule based system should also be taken into account.

Summarizing all previous ideas we can say that:

> A system is said to be interpretable if its reduced complexity and clear semantics, make possible for us, by reading it, to understand and explain its behavior.

Given this sort of definition, it is important to notice that the idea of *explaining the behavior* could be considered either in a broad sense (understanding and explaining the overall behavior), or in a much more specific sense by focusing on explaining why a certain output happens in a specific situation. This later approach is quite connected to questions as the *right to an explanation* that is being introduced at present in different regulations and laws. It will be further considered in next section.

Finally, it is important to notice that, probably due to the characteristics of fuzzy systems, interpretability in the fuzzy field has been mostly understood as *linguistic interpretability*, i.e., interpretation based on a linguistic description, but this is not the only option. As an example, it is also possible to achieve a visual interpretation of the information [41], to produce an explanation by describing the contribution of each attribute to the produced output [31], or even to use inverse methods that identify the minimum changes required to modify the output associated to an specific situation [21]. This latest concept will transform the process of explaining why a client is not eligible for a certain benefit, to the process of describing what should she/he change to become eligible.

In addition, interpretability has been usually considered as achieved by design. The idea is that the human user will be able to understand the set of rules, the decision tree or the knowledge (whatever its representation was) involved in the process. But it is also possible to work out interpretability techniques that were not based on the model. As an example, to explain a certain output, an interpretable (linear) model can be build around that point, to locally approximate how the global (non-interpretable) model behaves in the region [30].

Having a preliminary idea of its meaning, and the different facets it has, it is time to explore the importance of interpretability.

4 Why Do We Need Interpretable Systems?

In the early years of expert systems, the proximity between its symbolic representation of knowledge and the natural language (being a key aspect of interpretability) was mainly considered as a significant design advantage. In fact, that proximity was the support for an easier knowledge extraction, the process to transfer human expert knowledge to the expert system. At present, in the age of data abundance, there has been a significant paradigm shift in what concerns knowledge extraction. Expert knowledge has been massively replaced by knowledge extracted from data as the primary source. This situation seems to significantly reduce the importance of interpretability, once it has almost lost its role in knowledge extraction. So, Why do we still need interpretable systems?

According to [3], the reasons justifying a choice focusing on interpretability include, but are not limited to: integration, validation, interaction and trust.

- **Integration** concerns knowledge extraction, considering that even being in the age of data, domain knowledge of an expert could still play a role in merging knowledge from different sources or even to integrate expert and induced knowledge into a single knowledge base [5]. As far as a human was involved in this knowledge integration process, interpretability will be required.
- **Validation** is an important step in the design process. Once a preliminary knowledge base has been produced, or even after merging knowledge from different sources, it is important to check knowledge validity against domain and common-sense knowledge, as well as validating its semantics (searching for redundancies, inconsistencies, etc). This validation process is even more important when considering knowledge extracted from data, and will obviously be easier in interpretable systems.
- **Interaction** between the user and the system will still be an important aspect on the exploitation phase. And interpretability remains as a key factor to ease this interaction.
- **Trust** is at present one of the most influential aspects to convince users to adopt (and accept) a system. Interpretable systems have the capability to *explain its reasoning process* in terms that are understandable to the user, so that the user could be confident on the way the system produces its output.

This final idea of *trusting the system* plays a central role in the growing concept of *explainable artificial intelligence* [2]. It is important to notice that in a time where big data and data analysis are central players in many business areas, the need for companies to trust the data driven systems they use for their daily operations is crucial.

And going one step forward, trusting could be interpreted not only from the point of view of the company using the system, but also from the point of view of the human or the community being affected by the decisions of that system. This question is directly related to the idea of the *right to an explanation*. This primarily refers to the right of individuals to be given an explanation for decisions significantly affecting them, e.g. a denied loan or insurance.

Moreover, this question is moving from a matter of trusting to what at present is becoming a legal matter as shown by the *General Data Protection Regulation* of the European Union [1] that in Recital 71 states:

> The data subject should have the right not to be subject to a decision, which may include a measure, evaluating personal aspects relating to him or her which is based solely on automated processing and which produces legal effects concerning him or her or similarly significantly affects him or her, such as automatic refusal of an online credit application or e-recruiting practices without any human intervention.

...

In any case, such processing should be subject to suitable safeguards, which should include specific information to the data subject and the right to obtain human intervention, to express his or her point of view, to obtain an explanation of the decision reached after such assessment and to challenge the decision.

In summary, interpretability empowers the human with the capacity to understand the system, to interact with it, to have an opinion about its components and results and to validate or challenge them.

Interpretability so represents the bridge between the human and the machine that takes decisions affecting him or her. In a time when machines are gaining more and more presence, interpretability is a tool to maintain their power *under human supervision*.

Any idea presented in this section can be applied either to AI in general or to fuzzy systems in particular. We need interpretable fuzzy systems because we need interpretable systems, and fuzzy systems are extremely well suited for that.

To this point we have a view of the meaning and importance of interpretability. But this property will only be useful if we are able to choose a more interpretable system over a less interpretable one. And that requires to be able to compare and state that system A is more interpretable than system B. Consequently, the next question is: Is it possible to measure interpretability? Or at least, is it possible to compare two fuzzy systems regarding their interpretability?

Evaluating or measuring interpretability will be the next step in this analysis.

5 How to Evaluate Interpretability?

As said before, Machine Learning was initially defined in terms of two orthogonal axes of performance: predictive accuracy and comprehensibility of generated hypotheses. But comprehensibility was not easily measurable, and subsequent definitions tended to adopt a one-dimensional approach to ML, considering only accuracy. Consequently, the presence and importance of interpretability is clearly related to the capability (or not) to measure it.

The parallel analysis relating the general case (AI and ML) and the particular situation (fuzzy) that has been considered in previous sections will not be continued. Considering both in parallel makes sense when working at a conceptual level (as in the two previous sections) where the problems, needs and ideas are quite similar. Once we enter the operational/functional level, questions are much more linked to the specific technique or approach that is under consideration. In this case, the measures and the design approaches are clearly adapted to fuzzy modeling. Consequently, from this point the paper will concentrate on fuzzy systems interpretability.

Most authors agree on the fact that interpretability is a subjective concept [4] difficult to evaluate or maximize, however, many interpretability measures are described in fuzzy literature. These measures consider quite different aspects and properties of the fuzzy systems to *compute* a value describing its interpretability.

As a first approach, it has been previously said that interpretability is a question of semantics, but, at the same time, is strongly connected to complexity. Consequently, interpretability analysis can be focused on from two different points of view: semantic interpretability and structural interpretability [18]. But this is not the only option. Fuzzy systems are knowledge based systems with a knowledge base made up of, at least, two different kinds of knowledge: linguistic variables and fuzzy rules. This division produces interpretability measures and criteria involving properties related to either variables or rules, being so considered as two different levels [4, 42].

Semantic interpretability is mostly concerned with the quality of the pieces of information applied by the model. In that sense, semantic interpretability at the lower level considers questions as normality (for any fuzzy set there exists at least one element with full membership), distinguishability (different terms associated to a linguistic variable should be represented by clearly distinguishable fuzzy sets) or coverage (each possible input value for a variable should be well covered by at least one fuzzy set related to that variable).

Properties as completeness (the set or rules covers any possible situation) or consistency (there are no contradictory rules), correspond to semantic interpretability at higher level.

On the other hand, the effect of complexity on interpretability mostly relates to the amount of information to be managed in the process of interpretation, i.e., is not a question of quality but of quantity of information. Questions as the number of features, or the cardinality of the fuzzy partitions (number of terms), are considered at the lower level; and compactness (the total number of rules must be small) or average rule length (the average number of conditions per rule should be as small as possible) are evaluated at the higher level.

The two axes (semantics/complexity and variables/rules) produce four boxes or categories to group the many proposed measures/indexes. In [20] a significant number of them is listed, described and categorized, showing a clearly larger number of indexes under the categories related to complexity measures. This situation probably relates to the fact that those measures are easier to formalize under the form of an objective value, while some semantic aspects drive to subjective considerations that are hardly formalized. However, there are several recent attempts to define semantics based measures considering fuzzy partitions [10, 19] as well as the fuzzy rule base [25, 28].

In summary, there is a panoply of interpretability indexes [20] considering questions that range from the number of variables to the number of rules, from the distinguishability of the fuzzy sets to the inconsistency of fuzzy rules, from the coverage of the fuzzy partitions to the cofiring of rules. And those indexes consider these (and other) properties either as single measures or in an aggregated way merging different aspects.

Assuming now that interpretability deserves attention, that some applications *require* the use of interpretable fuzzy systems, and that several fuzzy systems considering a similar problem can be ranked according to its interpretability, the final question is how to define design processes promoting interpretability.

6 How to Build Interpretable Fuzzy Systems?

As a first step, and considering the existence of different types of fuzzy systems, not every option produces similar levels of interpretability. In general, Mamdani rules (using linguistic/fuzzy variables in premise and consequent) are more interpretable than TSK rules (using functions in the consequent), and linguistic rules (those based on fuzzy partitions) are more interpretable than fuzzy rules (those where each fuzzy set in a rule is independently defined and modified). In addition, the design process should favor the fulfillment of some semantic constraints jointly with a reduced complexity. But obviously, it has to be done without missing the fact that accuracy is also important. In fact, most design approaches focused on producing interpretable fuzzy systems are at the end defined to achieve a good balance between accuracy and interpretability [11, 12].

In the quest for this trade-off, the most common method was to take one of the components as the main one (either interpretability or accuracy) and start with an approach focused on it. Then, some additional steps are added to *improve* the behavior related to the other component, trying not to lose too much of the initial qualities. When considering first interpretability and then accuracy, the effect will be the generation of linguistic models with improved accuracy. One of the many options to do so is the generation of an initial model based on a set of preliminary fuzzy partitions, followed by a refinement process where the fuzzy sets [14] or the fuzzy partitions as a whole [15] are *tuned* to improve accuracy. If accuracy is considered first, the result will be a fuzzy model with improved interpretability. In this case, most of the approaches consider a first step to produce an accurate model, plus a second one to reduce its complexity (improving interpretability from the structural point of view). To do so, some fuzzy sets or some fuzzy rules could be merged, or even directly removed from the system [34].

The recent advances in multiobjective optimization techniques has lead to a growing number of fuzzy systems that consider both questions (interpretability and accuracy) as the two objectives of a multiobjective design problem [17]. This option offers the way to simultaneously achieve good levels of interpretability and accuracy through an automatic design process.

So, at the end, having so many tools for automatic design of fuzzy systems [13], the question of achieving interpretability can be reduced to the selection of a suitable structure for the fuzzy system, plus the choice (to guide the design process) of the interpretability measure that better covers those aspects being more interesting for our specific application.

7 Conclusions

At present, our live is being continuously affected by automatic decisions or actions made by machines. In the age of data, the *algorithms* producing those decisions and actions are in many cases automatically generated and applied. Understanding and trusting those algorithms and automatic decisions is crucial for humans, and interpretability is a quite important tool for that.

Interpretability should be considered as a significant characteristic of fuzzy systems, being one of its main advantages when comparing with other modeling and decision techniques. Even considering that the fuzzy formalism is well adapted for interpretability, a careless design process can negatively affect it. The use of design approaches ensuring a good level of interpretability (according to a suitable interpretability measure) will maintain fuzzy systems as a powerful modeling tool.

References

1. EU General Data Protection Regulation. https://eugdpr.org/. Accessed 30 Sept 2018
2. Explainable Artificial Intelligence (XAI). https://www.darpa.mil/program/explainable-artificial-intelligence. Accessed 30 Sept 2018
3. Alonso, J., Castiello, C., Mencar, C.: Interpretability of fuzzy systems: current research trends and prospects. In: Kacprzyk, J., Pedrycz, W. (eds.) Springer Handbook of Computational Intelligence, pp. 219–237. Springer (2015)
4. Alonso, J.M., Magdalena, L., González-Rodríguez, G.: Looking for a good fuzzy system interpretability index: an experimental approach. Int. J. Approx. Reason. **51**(1), 115–134 (2009)
5. Alonso, J.M., Magdalena, L., Guillaume, S.: HILK: a new methodology for designing highly interpretable linguistic knowledge bases using the fuzzy logic formalism. Int. J. Intell. Syst. **23**(7), 761–794 (2008)
6. Babuska, R.: Fuzzy Modeling and Control. Kluwer, Norwell, MA (1998)
7. Babuska, R.: Data-driven fuzzy modeling: transparency and complexity issues. In: Proceedings of the European Symposium on Intelligent Techniques ESIT'99. ERUDIT, Crete, Greece (1999)
8. Bardossy, A., Duckstein, L.: Fuzzy Rule-Based Modeling with Application to Geophysical, Biological and Engineering Systems. CRC Press (1995)
9. Bodenhofer, U., Bauer, P.: A formal model of interpretability of linguistic variables. In: Casillas, J., Cordón, O., Herrera, F., Magdalena, L. (eds.) Interpretability Issues in Fuzzy Modeling. Studies in Fuzziness and Soft Computing, vol. 128, pp. 524–545. Springer, Berlin, Heidelberg (2003)
10. Botta, A., Lazzerini, B., Marcelloni, F., Stefanescu, D.C.: Context adaptation of fuzzy systems through a multi-objective evolutionary approach based on a novel interpretability index. Soft Comput. **13**(5), 437–449 (2009)
11. Casillas, J., Cordón, O., Herrera, F., Magdalena, L.: Accuracy improvements to find the balance interpretability-accuracy in linguistic fuzzy modeling: an overview. In: Casillas, J., Cordon, O., Herrera, F., Magdalena, L. (eds.) Accuracy Improvements in Linguistic Fuzzy Modeling, pp. 3–24. Springer, Berlin, Heidelberg (2003)
12. Casillas, J., Cordón, O., Herrera, F., Magdalena, L.: Interpretability improvements to find the balance interpretability-accuracy in fuzzy modeling: an overview. In: Casillas, J., Cordón, O., Herrera, F., Magdalena, L. (eds.) Interpretability Issues in Fuzzy Modeling, pp. 3–22. Springer, Berlin, Heidelberg (2003)
13. Cordon, O.: A historical review of evolutionary learning methods for Mamdani-type fuzzy rule-based systems: designing interpretable genetic fuzzy systems. Int. J. Approx. Reason. **52**, 894–913 (2011)
14. Cordón, O., Herrera, F.: A three-stage evolutionary process for learning descriptive and approximate fuzzy logic controller knowledge bases from examples. Int. J. Approx. Reason. **17**(4), 369–407 (1997)

15. Cordon, O., Herrera, F., Magdalena, L., Villar, P.: A genetic learning process for the scaling factors, granularity and contexts of the fuzzy rule-based system data base. Inf. Sci. **136**(1–4), 85–107 (2001)
16. Fayyad, U., Piatetsky-Shapiro, G., Smyth, P.: From data mining to knowledge discovery in databases. AI Mag. **17**(3), 37–53 (1996)
17. Fazzolari, M., Alcala, R., Nojima, Y., Ishibuchi, H., Herrera, F.: A review of the application of multi-objective evolutionary fuzzy systems: current status and further directions. IEEE Trans. Fuzzy Syst. **21**(1), 45–65 (2013)
18. Gacto, M.J., Alcala, R., Herrera, F.: A multi-objective evolutionary algorithm for tuning fuzzy rule-based systems with measures for preserving interpretability. In: Proceedings of IFSA-EUSFLAT 2009, pp. 1146–1151. Lisbon, Portugal (2009)
19. Gacto, M.J., Alcala, R., Herrera, F.: Integration of an index to preserve the semantic interpretability in the multiobjective evolutionary rule selection and tuning of linguistic fuzzy systems. IEEE Trans. Fuzzy Syst. **18**(3), 515–531 (2010)
20. Gacto, M.J., Alcala, R., Herrera, F.: Interpretability of linguistic fuzzy rule-based systems: an overview of interpretability measures. Inf. Sci. **181**(20), 4340–4360 (2011)
21. Laugel, T., Lesot, M., Marsala, C., Renard, X., Detyniecki, M.: Comparison-based inverse classification for interpretability in machine learning. Commun. Comput. Inf. Sci. **853** (2018)
22. Magdalena, L.: Fuzzy rule based systems. In: Kacprzyk, J., Pedrycz W. (eds.) Springer Handbook of Computational Intelligence, pp. 203–218. Springer (2015)
23. Mencar, C., Castellano, G., Fanelli, A.M.: Some fundamental interpretability issues in fuzzy modeling. In: Proceedings—4th Conference of the European Society for Fuzzy Logic and Technology and 11th French Days on Fuzzy Logic and Applications, EUSFLAT-LFA 2005 Joint Conference, pp. 100–105 (2005)
24. Mencar, C., Castiello, C., Cannone, R., Fanelli, A.: Design of fuzzy rule-based classifiers with semantic cointension. Inf. Sci. **181**(20), 4361–4377 (2011)
25. Mencar, C., Castiello, C., Cannone, R., Fanelli, A.: Interpretability assessment of fuzzy knowledge bases: a cointension based approach. Int. J. Approx. Reason. **52**(4), 501–518 (2011)
26. Michalski, R.S.: A theory and methodology of inductive learning. Artif. Intell. **20**(2), 111–161 (1983)
27. Miller, G.A.: The magical number seven, plus or minus two: some limits on our capacity for processing information. Psychol. Rev. **63**, 81–97 (1956)
28. Pancho, D., Alonso, J.M., Cordon, O., Quirin, A., Magdalena, L.: Fingrams: visual representations of fuzzy rule-based inference for expert analysis of comprehensibility. IEEE Trans. Fuzzy Syst. **21**(6), 1133–1149 (2013)
29. Pedrycz, W.: Fuzzy Modelling: Paradigms and Practice. Kluwer Academic Press (1996)
30. Ribeiro, M.T., Singh, S., Guestrin, C.: "Why should I trust you?": explaining the predictions of any classifier. In: Proceedings of the 22nd ACM SIGKDD International Conference on Knowledge Discovery and Data Mining, KDD '16, pp. 1135–1144. ACM, New York, NY, USA (2016)
31. Robnik-Šikonja, M., Likas, A., Constantinopoulos, C., Kononenko, I., Štrumbelj, E.: Efficiently explaining decisions of probabilistic RBF classification networks. Lecture Notes in Computer Science (including subseries Lecture Notes in Artificial Intelligence and Lecture Notes in Bioinformatics), vol. 6593 LNCS, PART 1 (2011)
32. Roubos, H., Setnes, M.: Compact and transparent fuzzy models and classifiers through iterative complexity reduction. IEEE Trans. Fuzzy Syst. **9**(4), 516–524 (2001)
33. Setnes, M., Babuska, R., Verbruggen, H.: Rule-based modeling: precision and transparency. IEEE Trans. Syst. Man Cybern. Part C (Appl. Rev.) **28**(1), 165–169 (1998)
34. Setnes, M., Babuška, R., Verbruggen, H.B.: Complexity reduction in fuzzy modeling. Math. Comput. Simul. **46**(5–6), 509–518 (1998)
35. Takagi, T., Sugeno, M.: Fuzzy identification of systems and its applications to modeling and control. IEEE Trans. Syst. Man Cybern. **15**(1), 116–132 (1985)
36. Tarski, A., Mostowsk, A., Robinson, R.: Undecidable Theories. North-Holland (1953)

37. Zadeh, L.A.: Outline of a new approach to the analysis of complex systems and decision processes. IEEE Trans. Syst. Man Cybern. **SMC-3**(1), 28–44 (1973)
38. Zadeh, L.A.: The concept of a linguistic variable and its application to approximate reasoning— I. Inf. Sci. **8**, 199–249 (1975)
39. Zadeh, L.A.: Fuzzy systems theory: a framework for the analysis of humanistic systems. In: Systems Methodology in Social Science Research. Frontiers in Systems Research (Implications for the social sciences), vol. 2. Springer (1982)
40. Zadeh, L.A.: Is there a need for fuzzy logic? Inf. Sci. **178**(13), 2751–2779 (2008)
41. Zhang, Q.S., Zhu, S.C.: Visual interpretability for deep learning: a survey. Front. Inf. Technol. Electron. Eng. **19**(1), 27–39 (2018)
42. Zhou, S., Gan, J.: Low-level interpretability and high-level interpretability: a unified view of data-driven interpretable fuzzy system modelling. Fuzzy Sets Syst. **159**(23), 3091–3131 (2008)

Fuzzy Clustering Models and Their Related Concepts

Mika Sato-Ilic

Abstract This chapter describes fuzzy clustering models. Fuzzy clustering models are typical examples of model-based clustering. The purpose of the model-based clustering is to obtain the optimal partition of objects by fitting the model to the observed similarity (or dissimilarity) of objects. The merit of model-based clustering is that we can obtain a mathematically clearer solution as the clustering result, because we know the mathematical features of the model. However, when we observe a large amount of complex data, it is difficult to fit the simple model to the data to obtain a useful result. In order to solve this problem, we have extended the model in the framework of fuzzy clustering models to adjust to the complexity caused from the recent variety of vast amounts of data. This chapter describes how we extend the fuzzy clustering models, along with the mathematical features of the several fuzzy clustering models. In particular, we describe the novel generalized aggregation operator defined on the product space of linear spaces and the generalized aggregation operator based nonlinear fuzzy clustering model.

1 Introduction

Model-based clustering is a typical technique in the area of the cluster analysis. There are many methods in model-based clustering. For example, model-based clustering based on within-class dispersion and between-class dispersion [12], mixture model-based clustering [1], model-based clustering based on maximum likelihood approach [3, 28], a latent class model [9, 10], as well as a latent structure model [14, 15], and a latent class DEDICOM [29] for asymmetric similarity data. The framework of the model-based clustering is proposed in [6]. These models are closely related with conventional multivariate analysis, such as homogeneity analysis, principal component

---

M. Sato-Ilic (✉)
Faculty of Engineering, Information and Systems, University of Tsukuba, Tennodai 1-1-1, Tsukuba, Ibaraki 305-8573, Japan
e-mail: mika@risk.tsukuba.ac.jp

© The Editor(s) (if applicable) and The Author(s), under exclusive license to Springer Nature Switzerland AG 2021
M.-J. Lesot and C. Marsala (eds.), *Fuzzy Approaches for Soft Computing and Approximate Reasoning: Theories and Applications*, Studies in Fuzziness and Soft Computing 394, https://doi.org/10.1007/978-3-030-54341-9_11

123

analysis, correspondence analysis, quantification method III, dual scaling, biplot, and multidimensional scaling. In addition, these are closely related with several models for multi-way data such as INDSCAL [7, 11, 24] and INDCLUS [8].

These models have a clearer obtained solution as the data partition within the scope of mathematical provision when compared with the obtained solution of exploratory methods. However, if the data is a complex situation, obtaining the efficient solutions from the data by using a simple model is difficult. For example, in the case of classification, it is often difficult to classify the objects by using a simple model which assumes that one object belongs to a single cluster. In addition, the assumption results in a discrete optimization problem and becomes a factor on the difficulty of deriving solutions. Also, the stochastic idea assuming the distribution for each cluster is often too strict for the state of data in the cluster.

In order to solve these problems, we have proposed fuzzy clustering models in which we assume that one object can belong to multiple number of clusters with the degree of the belongingness of objects to clusters. In addition, to obtain feasible clustering results considering the variety of observed data, we have extended the basic fuzzy clustering model to one with higher versatility. These versatile models have higher flexibility for adapting to complex data in the real world.

In this chapter, several fuzzy clustering models are enumerated along with a description of how generalization of the models has been realized. This chapter consists as follows: In Sect. 2, we describe the Additive Clustering Model [27] which is a foundation of the clustering models described in this chapter. In Sect. 3, the Additive Fuzzy Clustering Model [19, 20], which includes the ordinary Additive Clustering Model shown in Sect. 2 as a special case, is explained by including the idea of fuzzy clustering. In Sect. 4, the Kernel Fuzzy Clustering Model [21, 22] is discussed for the inclusion of interaction of different clusters in a higher dimensional space in a framework of nonlinear clustering model. We show that this model includes the Additive Fuzzy Clustering Model shown in Sect. 3 as a special case. In Sect. 5, the Generalized Fuzzy Clustering Model [22] is shown, in which the Kernel Fuzzy Clustering Model shown in Sect. 4 is included as a special case. This model considers a feasibility of observed similarity between a pair of objects over the clusters by using aggregation operator under the assumption of the additivity over the clusters in the framework of the nonlinear clustering model. In Sect. 6, the Generalized Aggregation Operator based Nonlinear Fuzzy Clustering Model [23] is described to avoid the restriction of additivity over the clusters in the Generalized Fuzzy Clustering Model shown in Sect. 5 in order to adjust the variety of similarity of a pair of objects. The Kernel Fuzzy Clustering Model is included as a special case to this model. In Sect. 7, several conclusions are stated.

2 Additive Clustering Model

The Additive Clustering Model [27] is a typical model-based clustering and is defined as follows:

$$s_{ij} = \sum_{k=1}^{K} w_k p_{ik} p_{jk} + \varepsilon_{ij}, \quad i, j = 1, \ldots, n, \quad i \neq j. \tag{1}$$

In Eq. (1), s_{ij} is observed similarity between objects i and j. Here we consider the case when the following condition is assumed:

$$s_{ij} \in [0, 1]. \tag{2}$$

p_{ik} satisfies the following condition which shows the status of clustering, that is, if an object i belongs to a cluster k, then p_{ik} is 1, otherwise it is 0.

$$p_{ik} \in \{0, 1\}. \tag{3}$$

w_k shows nonnegative weight representing salience of a cluster k and from the condition (2), we assume $\sum_{k=1}^{K} w_k \leq 1$. n is the number of objects and K is the number of clusters. The purpose of this model shown in Eq. (1) is to estimate the p_{ik} which minimizes the sum of squared errors ε_{ij}. In model (1), from Eq. (3), only the case when both p_{ik} and p_{jk} are 1, w_k is remained, so the model is essentially represented as follows:

$$s_{ij} = w_{k_1} + w_{k_2} + \cdots + w_{k_l}, \tag{4}$$

where p_{ik_t} and p_{jk_t} are 1, $k_t \in \{1, \ldots, K\}$, $t = 1, \ldots, l$. From Eq. (4), when we consider the degree of freedom for the possibly taking values of s_{ij}, which is shown in Eq. (2), and the right side of the model shown in Eq. (4), the number of clusters tends to increase to fit the similarity data to the model.

3 Additive Fuzzy Clustering Model

If the number of clusters increases for obtaining satisfactory fitness between similarity data and the model, then the model tends to be unstable due to increase in the number of parameters. So, in order to solve the problem of the Additive Clustering Model which is what the number of clusters tends to increase to fit the similarity data to the model, the Additive Fuzzy Clustering Model [19, 20] is defined with an inclusion of an idea of fuzzy clustering [5, 18, 32] to the Additive Clustering Model.

$$s_{ij} = \rho(\mathbf{u}_i, \mathbf{u}_j) + \varepsilon_{ij}, \quad i, j = 1, \ldots, n, \quad i \neq j. \tag{5}$$

In Eq. (5), $\mathbf{u}_i$ is the following vector

$$\mathbf{u}_i = (u_{i1}, \ldots, u_{iK}),$$

where u_{ik} shows a degree of belongingness of an object i to a cluster k which satisfy the following conditions:

$$u_{ik} \in [0, 1], \quad \sum_{k=1}^{K} u_{ik} = 1. \tag{6}$$

In the model shown in Eq. (5), ρ is an aggregation operator defined as follows:

Definition 1 An aggregation operator is a binary operator ρ on the unit interval $[0, 1]$, that is a function $\rho: [0, 1] \times [0, 1] \rightarrow [0, 1]$, such that $\forall a, b, c, d \in [0, 1]$, the following conditions are satisfied:

(1) $\rho(a, 0) = \rho(0, a) = 0, \quad \rho(a, 1) = \rho(1, a) = a$. (Boundary Condition)
(2) $\rho(a, c) \leq \rho(b, d)$, whenever $a \leq b, \ c \leq d$. (Monotonicity)
(3) $\rho(a, b) = \rho(b, a)$. (Symmetry)

Where $[0, 1] \times [0, 1]$ shows a product space.

T-norms [13, 16, 25] defined in the statistical metric space are typical examples which satisfy the conditions in Definition 1 and an additional condition, that is a condition of associativity. In T-norms, for any real number x, distribution function is defined as follows:

$$F_{pq}(x) \equiv \Pr\{d_{pq} < x\}, \quad \forall p, q \in S. \tag{7}$$

In Eq. (7), d_{pq} shows a distance between points p and q which are in a set S. Then $F_{pq}(x)$ satisfies the following conditions:

$$F_{pp}(x) = 1, \quad \forall x > 0. \tag{8}$$

$$F_{pq}(x) < 1, \quad p \neq q, \quad \exists x > 0. \tag{9}$$

$$F_{pq}(x) = F_{qp}(x). \tag{10}$$

Then T-norms are defined as an operator T which satisfy the following condition:

$$F_{pr}(x + y) \geq T(F_{pq}(x), F_{qr}(y)), \quad r \in S. \tag{11}$$

The above conditions (8), (9), (10), and (11) are similar conditions of the distance axiom as follows.

$$d_{pp} = 0.$$

$$d_{pq} > 0, \quad p \neq q.$$

$$d_{pq} = d_{qp}.$$

$$d_{pr} \leq d_{pq} + d_{qr}.$$

A typical example of the T-norm is algebraic product. Therefore, a special case of the Additive Fuzzy Clustering Model shown in (5) is defined as follows:

$$
\begin{aligned}
s_{ij} &= \langle \mathbf{u}_i, \mathbf{u}_j \rangle + \varepsilon_{ij} \\
&= \sum_{k=1}^{K} u_{ik} u_{jk} + \varepsilon_{ij}, \quad i, j = 1, \ldots, n, \quad i \neq j.
\end{aligned}
\tag{12}
$$

From Eqs. (1) and (12), and the conditions shown in Eqs. (3) and (6), it can be seen that the Additive Clustering Model shown in Eq. (1) is a special case of the Additive Fuzzy Clustering Model shown in Eq. (12). In addition, from the conditions shown in Eqs. (3) and (6), the Additive Fuzzy Clustering Model can fit the observed similarity data shown in Eq. (2) by using fewer clusters when compared with the Additive Clustering Model.

4 Kernel Fuzzy Clustering Model

In the Additive Fuzzy Clustering Model shown in Eq. (12), $u_{ik} u_{jk}$ shows a common property of objects i and j to a cluster k. However, in this case, we cannot consider the interaction of different clusters. That is, we cannot consider the common property of objects i and j to different clusters k and l ($k \neq l$), which is represented as $u_{ik} u_{jl}$. In order to include the interaction of different clusters to the Additive Fuzzy Clustering Model, the Kernel Fuzzy Clustering Model [21, 22] has been proposed as follows:

$$s_{ij} = \kappa(\mathbf{u}_i, \mathbf{u}_j) + \varepsilon_{ij}, \quad i, j = 1, \ldots, n, \quad i \neq j. \tag{13}$$

In the model shown in Eq. (13), κ is the kernel function from $R^K \times R^K$ to R which satisfy the following conditions [26]:

$$\kappa(\mathbf{u}_i, \mathbf{u}_j) = \kappa(\mathbf{u}_j, \mathbf{u}_i). \tag{14}$$

$$\sum_{i=1}^{n} \sum_{j=1}^{n} \kappa(\mathbf{u}_i, \mathbf{u}_j) u_i u_j \geq 0, \quad \mathbf{u}_1, \mathbf{u}_2, \ldots, \mathbf{u}_n \in R^K. \tag{15}$$

Since the function κ satisfy the conditions (14) and (15), there exists a function Φ which satisfies the following:

$$\kappa(\mathbf{u}_i, \mathbf{u}_j) = \langle \Phi(\mathbf{u}_i), \Phi(\mathbf{u}_j) \rangle = \sum_{m=1}^{M} \phi_m(\mathbf{u}_i) \phi_m(\mathbf{u}_j), \quad \forall \mathbf{u}_i, \mathbf{u}_j \in R^K, \tag{16}$$

where

$$\Phi(\mathbf{u}_i) = (\phi_1(\mathbf{u}_i), \ldots, \phi_M(\mathbf{u}_i)),$$

and

$$\Phi : R^K \to R^M, \quad K < M, \tag{17}$$

where $\langle \cdot, \cdot \rangle$ shows inner product on Hilbert space. An example of κ which satisfies Eqs. (14) and (15) is as follows:

$$\begin{aligned} s_{ij} &= \kappa(\mathbf{u}_i, \mathbf{u}_j) + \varepsilon_{ij} \\ &= \langle \mathbf{u}_i, \mathbf{u}_j \rangle^\alpha + \varepsilon_{ij} \\ &= \left(\sum_{k=1}^{K} u_{ik} u_{jk} \right)^\alpha + \varepsilon_{ij}, \end{aligned} \tag{18}$$

$$\alpha \geq 1, \quad i, j = 1, \ldots, n, \quad i \neq j.$$

When $\alpha = 1$, Eq. (18) is equivalent to Eq. (12). That is, the Additive Fuzzy Clustering Model shown in Eq. (12) is a special case of the Kernel Fuzzy Clustering Model shown in Eq. (18). In addition, from Eq. (17), the Kernel Fuzzy Clustering Model can estimate the solution which is the clustering result in a higher dimensional space than the K dimensional space.

5 Generalized Fuzzy Clustering Model

With the inclusion of a function ϕ and the aggregation operator shown in Definition 1, the Generalized Fuzzy Clustering Model [22] has been proposed as follows:

$$s_{ij} = \phi \circ g_\rho(\mathbf{u}_i, \mathbf{u}_j) + \varepsilon_{ij}, \quad i, j = 1, \ldots, n, \quad i \neq j. \tag{19}$$

Then, when

$$g_\rho(\mathbf{u}_i, \mathbf{u}_j) \equiv \sum_{k=1}^{K} \rho(u_{ik}, u_{jk}), \quad \mathbf{u}_i = (u_{i1}, \ldots, u_{iK}), \quad \mathbf{u}_j = (u_{j1}, \ldots, u_{jK}),$$

Eq. (19) can be rewritten as

$$s_{ij} = \phi \left(\sum_{k=1}^{K} \rho(u_{ik}, u_{jk}) \right) + \varepsilon_{ij}, \quad i, j = 1, \ldots, n, \quad i \neq j. \tag{20}$$

When $\alpha = 1$ in Eq. (18), Eq. (20) is equivalent to Eq. (18) when $\rho(u_{ik}, u_{jk}) = u_{ik}u_{jk}$ and ϕ is an identity mapping. That is, the Kernel Fuzzy Clustering Model shown in Eq. (18) is a special case of the Generalized Fuzzy Clustering Model shown in Eq. (20).

6 Generalized Aggregation Operator Based Nonlinear Fuzzy Clustering Model

We have extended the Additive Clustering Model shown in Eq. (1) to the Generalized Fuzzy Clustering Model shown in Eq. (20) to obtain an adaptable clustering result which is adjusted to the complex similarity data, considering feasibility of the variety of the observed similarity data. However, in the model shown in Eq. (20), we still need to assume a restriction, that is the additivity over the clusters. This means that we assume the independence of each cluster and the similarity of a pair of objects is represented only by the base sum of common properties of each cluster. In order to represent the variety of similarity of a pair of objects, we have defined the generalized aggregation operator [23]. This aggregation operator $\tilde{\rho}$ is defined as the function on a product space of linear spaces as follows:

$$\tilde{\rho} : X \times X \rightarrow [0, 1], \quad \mathbf{u}_i, \mathbf{u}_j \in X, \tag{21}$$

where X is a linear space. Therefore, the value of $\tilde{\rho}(\mathbf{u}_i, \mathbf{u}_j)$ is common degree of objects i and j to "K clusters". On the other hand, the ordinary aggregation operator shown in Definition 1 is a binary operator on $[0, 1]$ which is shown as

$$\rho : [0, 1] \times [0, 1] \rightarrow [0, 1], \quad u_{ik}, u_{jk} \in [0, 1].$$

Therefore, the ordinary aggregation operator shows the variety of common degree of a pair of objects to each cluster. That is, the value of $\rho(u_{ik}, u_{jk})$ shows common degree of objects i and j to "a cluster k".

Some research exists that is related with the generalized aggregation operator. For example, topology for probabilistic metric spaces including features of T-norm in a semi-probabilistic metric space have been discussed [30]. In addition, several features of the product in several metric spaces has been discussed [31]. From the aspect of engineering, the multidimensional case of T-norm have been discussed [2, 4, 17].

The generalized aggregation operator shown in Eq. (21) is also a function on a product space of linear spaces which is similar with the previously researched aggregation functions. However, there is a difference from previously proposed functions and the generalized aggregation operator shown in Eq. (21). That is, the generalized aggregation operator exploits the features of fuzzy clustering results which is represented as the conditions shown in Eq. (6). Moreover, the generalized aggregation operator uses the conventional vector operation which satisfies similar conditions of the aggregation operator shown in Definition 1.

Let X be a real linear space. Then an inner product space (pre-Hilbert spaces) is defined as follows.

Definition 2 A linear space X is called an inner product space if for each pair of elements $\mathbf{x}, \mathbf{y} \in X$ there is defined a real scalar denoted $\langle \mathbf{x}, \mathbf{y} \rangle$ having the following properties. $\exists \mathbf{x}, \mathbf{y}, \mathbf{z} \in X, \ \alpha \in R,$

(1) $\langle \mathbf{x}, \mathbf{x} \rangle \geq 0,$
(2) $\langle \mathbf{x}, \mathbf{x} \rangle = 0$ if and only if $\mathbf{x} = \mathbf{0},$
(3) $\langle \mathbf{x}, \mathbf{y} \rangle = \langle \mathbf{y}, \mathbf{x} \rangle,$
(4) $\langle \alpha \mathbf{x}, \mathbf{y} \rangle = \alpha \langle \mathbf{x}, \mathbf{y} \rangle,$
(5) $\langle \mathbf{x} + \mathbf{y}, \mathbf{z} \rangle = \langle \mathbf{x}, \mathbf{z} \rangle + \langle \mathbf{y}, \mathbf{z} \rangle.$

Then a generalized aggregation operator is defined as follows:

Definition 3 A generalized aggregation operator is a function $\tilde{\rho} \colon X \times X \to [0, 1]$, such that $\forall \mathbf{a}, \mathbf{b}, \mathbf{c}, \mathbf{d}, \mathbf{0}, \mathbf{1} \in X$, where $\mathbf{a} = (a_1, \ldots, a_K)$, $\mathbf{b} = (b_1, \ldots, b_K)$, $\mathbf{c} = (c_1, \ldots, c_K)$, $\mathbf{d} = (d_1, \ldots, d_K)$, $\mathbf{0} = (0, \ldots, 0)$, $\mathbf{1} = (1, \ldots, 1)$, $a_k, b_k, c_k, d_k \in [0, 1]$, $k = 1, \ldots, K$, the following conditions are satisfied:

(1) $\tilde{\rho}(\mathbf{a}, \mathbf{0}) = \tilde{\rho}(\mathbf{0}, \mathbf{a}) = 0, \ \ \tilde{\rho}(\mathbf{a}, \mathbf{1}) = \tilde{\rho}(\mathbf{1}, \mathbf{a}) = \alpha, \ \alpha \in [0, 1].$
(2) $\tilde{\rho}(\mathbf{a}, \mathbf{c}) \leq \tilde{\rho}(\mathbf{b}, \mathbf{d}),$ whenever $\mathbf{a} \leq \mathbf{b}, \ \mathbf{c} \leq \mathbf{d}.$
Where the equivalence relation is assumed: $\mathbf{a} \leq \mathbf{b} \ \leftrightarrow \ a_k \leq b_k, \ k = 1, \ldots K.$
(3) $\tilde{\rho}(\mathbf{a}, \mathbf{b}) = \tilde{\rho}(\mathbf{b}, \mathbf{a}).$

Then the following theorem is proved.

Theorem 1 *The operators shown in Eqs. (23)–(25) are the generalized aggregation operator if the following condition shown in Eq. (22) is satisfied.*

$$\sum_{k=1}^{K} a_k = \sum_{k=1}^{K} b_k = \sum_{k=1}^{K} c_k = \sum_{k=1}^{K} d_k = 1, \quad 2 \leq K \leq n. \tag{22}$$

- *Generalized Algebraic Product:*

$$\tilde{\rho}(\mathbf{a}, \mathbf{b}) = \mathbf{ab}^t. \tag{23}$$

- *Generalized Hamacher Product*

$$\tilde{\rho}(\mathbf{a}, \mathbf{b}) = \frac{\mathbf{ab}^t}{\mathbf{a1}^t + \mathbf{b1}^t - \mathbf{ab}^t}. \tag{24}$$

- *Generalized Einstein Product*

$$\tilde{\rho}(\mathbf{a}, \mathbf{b}) = \frac{\mathbf{a}\mathbf{b}^t}{2\mathbf{1}^t - (\mathbf{a}\mathbf{1}^t + \mathbf{b}\mathbf{1}^t - \mathbf{a}\mathbf{b}^t)} = \frac{\mathbf{a}\mathbf{b}^t}{2K - (\mathbf{a}\mathbf{1}^t + \mathbf{b}\mathbf{1}^t - \mathbf{a}\mathbf{b}^t)}. \tag{25}$$

The brief proof of Theorem 1 is shown in our previous paper [23]. Equation (22) is the condition of fuzzy clustering result shown in Eq. (6), so this theorem means that under the conditions of fuzzy clustering, the operators shown in Eqs. (23), (24), and (25) satisfy the conditions of generalized aggregation operator shown in Definition 3.

By including the generalized aggregation operator $\tilde{\rho}$ defined in Definition 3 to a model, the Generalized Aggregation Operator based Nonlinear Fuzzy Clustering Model [23] is defined as follows:

$$s_{ij} = \phi \circ \tilde{\rho}(\mathbf{u}_i, \mathbf{u}_j) + \varepsilon_{ij}, \quad i, j = 1, \ldots, n, \quad i \neq j. \tag{26}$$

The following models shown in Eqs. (27), (28), and (29) are examples of the Generalized Aggregation Operator based Nonlinear Fuzzy Clustering Model shown in Eq. (26) using the generalized aggregation operators shown in Eqs. (23), (24), and (25).

$$s_{ij} = \langle \mathbf{u}_i, \mathbf{u}_j \rangle^\alpha + \varepsilon_{ij}. \tag{27}$$

$$s_{ij} = \frac{\langle \mathbf{u}_i, \mathbf{u}_j \rangle^\alpha}{\langle \mathbf{u}_i, \mathbf{1} \rangle^\alpha + \langle \mathbf{u}_j, \mathbf{1} \rangle^\alpha - \langle \mathbf{u}_i, \mathbf{u}_j \rangle^\alpha} + \varepsilon_{ij}. \tag{28}$$

$$s_{ij} = \frac{\langle \mathbf{u}_i, \mathbf{u}_j \rangle^\alpha}{2K - (\langle \mathbf{u}_i, \mathbf{1} \rangle^\alpha + \langle \mathbf{u}_j, \mathbf{1} \rangle^\alpha - \langle \mathbf{u}_i, \mathbf{u}_j \rangle^\alpha)} + \varepsilon_{ij}. \tag{29}$$

Equation (27) is a Generalized Aggregation Operator based Nonlinear Fuzzy Clustering Model that used the generalized algebraic product shown in Eq. (23). This Eq. (27) is the same as the Kernel Fuzzy Clustering Model shown in Eq. (18). That is, the Kernel Fuzzy Clustering Model is a special case of the Generalized Aggregation Operator based Nonlinear Fuzzy Clustering Model.

Figure 1 shows the relationship of the clustering models which are the Additive Clustering Model, the Additive Fuzzy Clustering Model, the Kernel Fuzzy Clustering Model, the Generalized Fuzzy Clustering Model, and the Generalized Aggregation Operator based Nonlinear Fuzzy Clustering Model. In this figure, the Additive Clustering Model shows the base of these clustering models and the Additive Fuzzy Clustering Model includes the Additive Clustering Model by using an idea of continuous values of degree of belongingness of objects to clusters. These models have a linear structure of the common properties of objects with respect to clusters. The Kernel Fuzzy Clustering Model is defined to include the interaction of different clusters and obtain the clustering result in a higher dimensional space. This model includes the Additive Fuzzy Clustering Model as a special case. The Generalized Fuzzy Clustering Model and the Generalized Aggregation Operator based Nonlinear

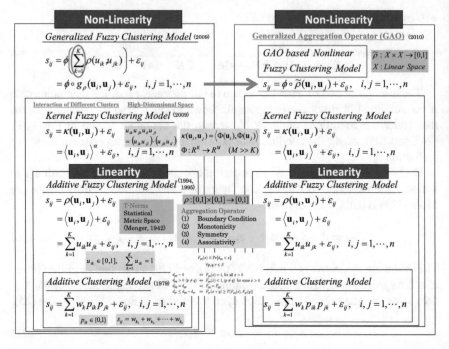

Fig. 1 Relationship of clustering models

Fuzzy Clustering Model include the Kernel Fuzzy Clustering Model as the special case. These models are different with each other on the presence or absence of the restriction of additivity over the clusters. In addition, these three clustering models which are the Kernel Fuzzy Clustering Model, the Generalized Fuzzy Clustering Model, and the Generalized Aggregation Operator based Nonlinear Fuzzy Clustering Model, have nonlinear structure of the common properties of objects with respect to clusters. These extensions of models have the merit of adjusting to a variety of recent complex similarity data.

7 Conclusion

A fuzzy clustering model has been proposed to realize a more natural classification in a clustering model and make it applicable to complicated data. Furthermore, in the fuzzy clustering model, to apply it to various data, more generalized fuzzy clustering models have been proposed. In this chapter, we describe several fuzzy clustering models and how mathematically the generalization has been done to utilize the fuzzy clustering models for the recent vast amount of complex data.

References

1. Anderson, T.W.: An Introduction to Multivariate Statistical Analysis. Wiley, New York (1958)
2. Ai-Ping, L., Yan, J., Quan-Yuan, W.: Harmonic triangular norm aggregation operators in multicriteria decision systems. J. Converg. Inf. Technol. **2**(1), 83–92 (2007)
3. Banfield, J.D., Raftery, A.E.: Model-based Gaussian and non-Gaussian clustering. Biometrics **49**, 803–821 (1993)
4. Beliakov, G.: How to build aggregation operators from data. Int. J. Intell. Syst. **18**, 903–923 (2003)
5. Bezdek, J.C.: Pattern Recognition with Fuzzy Objective Function Algorithms. Plenum Press, New York (1981)
6. Bock, H.H.: Probabilistic models in cluster analysis. Comput. Stat. Data Anal. **23**, 5–28 (1996)
7. Carroll, J.D., Chang, J.J.: Analysis of individual differences in multidimensional scaling via an N-generalization of the Eckart-Young decomposition. Psychometrika **35**, 283–319 (1970)
8. Carroll, J.D., Arabie, P.: INDCLUS: an individual differences generalization of ADCLUS model and the MAPCLUS algorithm. Psychometrika **48**, 157–169 (1983)
9. Clogg, C.C.: Some latent structure models for the analysis of Likert-type data. Soc. Sci. Res. **9**, 287–301 (1979)
10. Clogg, C.C.: Latent structure models of mobility. Am. J. Sociol. **86**(4), 836–868 (1981)
11. Cox, T.F., Cox, M.A.A.: Multidimensional Scaling. Chapman & Hall (2001)
12. Friedman, H.P., Rubin, J.: On some invariant criteria for grouping data. J. Am. Stat. Assoc. **62**, 1159–1178 (1967)
13. Klement, E.P., Mesiar, R., Pap, E.: Triangular Norms. Kluwer Academic Publications (2000)
14. Lazarsfeld, P.F.: The Interpretation and Mathematical Foundation of Latent Structure Analysis, Measurement and Prediction, pp. 413–472. Princeton University Press (1950)
15. Lazarsfeld, P.F., Henry, N.W.: Latent Structure Analysis. Houghton-Mifflin, Boston (1968)
16. Menger, K.: Statistical metrics. Proc. Natl. Acad. Sci. U.S.A. **28**, 535–537 (1942)
17. Merigo, J.M., Casanovas, M.: Fuzzy generalized hybrid aggregation operators and its application in fuzzy decision making. Int. J. Fuzzy Syst. **12**(1), 15–24 (2010)
18. Ruspini, E.H.: A new approach to clustering. Inf. Control **15**, 22–32 (1969)
19. Sato, M., Sato, Y.: An additive fuzzy clustering model. Jpn. J. Fuzzy Theory Syst. **6**, 185–204 (1994)
20. Sato, M., Sato, Y.: On a general fuzzy additive clustering model. Int. J. Intell. Autom. Soft Comput. **1**(4), 439–448 (1995)
21. Sato-Ilic, M., Ito, S.: Kernel fuzzy clustering model. In: 24th Fuzzy System Symposium, pp. 153–154 (2008) (in Japanese)
22. Sato-Ilic, M., Ito, S., Takahashi, S.: Nonlinear kernel-based fuzzy clustering model. In: Viattchenin, D.A. (ed.) Developments in Fuzzy Clustering, pp. 56–73. VEVER, Minsk (Belarus) (2009)
23. Sato-Ilic, M.: Generalized aggregation operator based nonlinear fuzzy clustering model. In: Intelligent Engineering Systems Through Artificial Neural Networks, New York, USA, vol. 20, pp. 493–500 (2010)
24. Schonemann, P.H.: An algebraic solution for class of subjective metrics models. Psychometrika **37**, 441–451 (1972)
25. Schweizer, B., Sklar, A.: Probabilistic Metric Spaces. Dover Publications (2005)
26. Shawe-Taylor, J., Cristianini, N.: Kernel Methods for Pattern Analysis. Cambridge University Press (2004)
27. Shepard, R.N., Arabie, P.: Additive clustering: representation of similarities as combinations of discrete overlapping properties. Psychol. Rev. **86**(2), 87–123 (1979)
28. Scott, A.J., Symons, M.J.: Clustering methods based on likelihood ratio criteria. Biometrics **27**, 387–397 (1971)
29. Takane, Y., Kiers, H.A.L.: Latent class DEDICOM. J. Classif. 225–247 (1997)
30. Tardiff, R.M.: Topologies for probabilistic metric spaces. Pac. J. Math. **65**(1), 233–251 (1976)

31. Tardiff, R.M.: On a functional inequality arising in the construction of the product of several metric spaces. Aequationes Math. **20**, 51–58 (1980)
32. Zadeh, L.A.: Fuzzy Sets. Inf. Control **8**, 338–353 (1965)

Fast Cluster Tendency Assessment for Big, High-Dimensional Data

Punit Rathore, James C. Bezdek, and Marimuthu Palaniswami

Abstract Assessment of clustering tendency is an important first step in crisp or fuzzy cluster analysis. One tool for assessing cluster tendency is the Visual Assessment of Tendency (VAT) algorithm. The VAT and improved VAT (iVAT) algorithms have been successful in determining potential cluster structure in the form of visual images for various datasets, but they can be computationally expensive for datasets with a very large number of samples and/or dimensions. Scalable versions of VAT/iVAT, such as sVAT/siVAT, have been proposed for iVAT approximation, but they also take a lot of time when the data is large both in the number of records and dimensions. In this chapter, we introduce two new algorithms to obtain approximate iVAT images that can be used to visually estimate the potential number of clusters in big data. We compare the two proposed methods with the original version of siVAT on five large, high-dimensional datasets, and demonstrate that both new methods provide visual evidence about potential cluster structure in these datasets in significantly less time than siVAT with no apparent loss of accuracy or visual acuity.

1 Introduction

Data clustering [4] is an essential method of exploratory data analysis in which data are partitioned into several subsets of similar objects. A natural question that comes before applying any clustering method on a dataset is- "Does the data contain any

P. Rathore (✉) · M. Palaniswami
Department of Electrical and Electronic Engineering, The University of Melbourne, Melbourne, Australia
e-mail: prathore@student.unimelb.edu.au

M. Palaniswami
e-mail: palani@unimelb.edu.au

J. C. Bezdek
School of Computing and Information Systems, The University of Melbourne, Melbourne, Australia
e-mail: jcbezdek@gmail.com

135

M.-J. Lesot and C. Marsala (eds.), *Fuzzy Approaches for Soft Computing and Approximate Reasoning: Theories and Applications*, Studies in Fuzziness and Soft Computing 394, https://doi.org/10.1007/978-3-030-54341-9_12

inherent grouping structure or clusters?" A major problem in unsupervised machine learning, specifically in cluster analysis, is that all clustering methods will partition the data into groups even if no "actual" clusters exist in the data. Therefore, prior to clustering or as a pre-clustering step, it is important to decide whether the data contains meaningful clusters (i.e., non-random structure) or not. If yes, then how many clusters, k? This problem is called *assessment of clustering tendency or clusterability*.

Many popular clustering algorithms including hard k-means (HKM) and *fuzzy k-means* (FKM) require the number of clusters (k) as an input. A common technique for determining k is to use a *cluster validity index* (CVI). This post-clustering procedure is used to evaluate the quality of partitions obtained at different values of k and choose the partition that best fits the data in the sense of the CVI.

Unlike CVI, most clustering tendency assessment techniques attempt to determine a reasonable value of k *before* applying a clustering algorithm to the input data. There are mainly two types of approaches to evaluate clustering tendency: (i) statistical methods, that measure the probability that a given dataset is generated by a uniform data distribution to test the spatial randomness of the data. A popular statistical approach in this genre is Hopkins statistic [16]; and (ii) visual methods, which provide a visualization of the input data to inspect for apparent clustering tendency of the dataset. These methods reorder the dissimilarity matrix of the input data, and visually estimate the number of clusters that appear as the dark blocks along the diagonal of a *reordered dissimilarity image* (RDI), also known as a cluster heat map. Some popular approaches in this category are the *visual assessment of cluster tendency* (VAT) [5], *improved VAT* (iVAT) [11] algorithms and their variants.

In many applications such as biomedical imaging, sequencing, and time series matching, the dataset consists of millions of instances in hundreds to thousands of dimensions [7]. The two most important ways a dataset can be big are: (1) it has a very large number (N) of instances, and (2) each instance has many features (p) i.e. it is high-dimensional data. Scalable versions of VAT/iVAT (sVAT/siVAT) are available in [10, 12] for visual assessment of cluster tendency for large datasets. However, sVAT/siVAT take a lot of time when the data is large jointly in the number of instances (N) and the number of dimensions (p).

To deal with large amounts of high-dimensional data, this chapter introduces two fast, approximate scalable algorithms, called siVAT1 and siVAT2, that combine the visual assessment of cluster tendency (iVAT) algorithm [11] with a smart sampling strategy, called *Maximin* (MM) [10, 19], and *random projection* (RP) of the input data to a lower dimension. Maximin sampling [19] picks distinguished objects (or MM objects) from the dataset, hence it requires relatively very few samples (compared to random sampling) to yield a diverse subset of the big data, that represents the cluster structure in the original (big) dataset. The siVAT1 and siVAT2 algorithms produce cluster heat map images that approximate the uncomputable iVAT image of the big data in significantly less time than siVAT.

2 Preliminaries

In this section, we briefly discuss the random projection and VAT algorithms.

2.1 Random Projection

Random projections are based on the *Johnson-Lindenstrauss* (JL) lemma [13], which states that a set of N points $X = \{x_1, \ldots, x_N\} \subset \mathbb{R}^p$ (denoted as the 'upspace') can be linearly projected (with approximate preservation of distances in probability) into a set of points $Y = \{y_1, \ldots, y_N\} \subset \mathbb{R}^q, q \ll p$, (denoted as the 'downspace') using a random projection matrix $T \subset \mathbb{R}^{p \times q}$. In this paper, we use a variant of the JL lemma proposed by Achlioptas in [1]. The theorem is as follows:

Theorem 1 *Given a set of N points $X \subset \mathbb{R}^p$, and $\varepsilon, \beta > 0$, for any integer q*

$$q \geq q_0 = (4 + 2\beta)(\varepsilon^2/2 - \varepsilon^3/3)^{-1}log(N), \tag{1}$$

let $Y = \frac{1}{\sqrt{q}}XT$ be the projection (matrix) of the N points in $\mathbb{R}^q$, where T is a $p \times q$ random projection matrix, whose entries t_{ij} are independently and identically distributed (i.i.d) with $t_{ij} = +1$ or -1, with equal probability. Let $f : \mathbb{R}^p \to \mathbb{R}^q$ map the ith row of X to the ith row of Y. Then for any $u, v \in X$ with probability at least $1 - N^{-\beta}$, we have

$$(1 - \varepsilon)||u - v||_2 \leq ||f(u) - f(v)||_2 \leq (1 + \varepsilon)||u - v||_2 \qquad \square$$

The parameter ε controls the accuracy of distance preservation, while β controls the probability that distance preservation is within $1 \pm \varepsilon$. According to Theorem 1, if the reduced (downspace) dimension q obeys inequality (1), then pairwise Euclidean distances are preserved (in probability) within a multiplicative factor of $1 \pm \varepsilon$, and we say that Y has a *JL certificate*. There have been studies (cf. [6, 17]) that assert that the JL result often holds for $q \ll q_0$. Projections to dimensions less than q_0 are called "*rogue random projections*" (RRP). We will study the use of rogue random projections in our experiments.

2.2 Visual Assessment of Tendency (VAT) and Its Relative Methods

Consider a set of N objects $O = \{o_1, o_2, \ldots, o_N\}$, partitioned into $k \in \{2, \ldots, N - 1\}$ subsets of similar objects, where each object o_i is defined by a p-dimensional feature vector, $\mathbf{x}_i \in \mathbb{R}^p$. The data can also be presented in the form of dissimilarity matrix $D_N = [d_{ij}]$, where d_{ij} represents dissimilarity between o_i and o_j.

The VAT [5] algorithm is based on (but not identical to) Prim's algorithm for finding the *minimum spanning tree* (MST) of a weighted undirected graph. It is essentially a single-linkage (SL) based approach which proceeds by connecting the next nearest vertex to the current edge until the complete MST is formed. The VAT algorithm reorders the dissimilarity matrix D_N to D_N^* using edge insertion ordering of the vertices added to the MST and specifies either end of the longest edge as the initial vertex for MST formation. The intensity of each pixel in the VAT (or RDI) image $I(D_N^*)$, reflects the dissimilarity between the corresponding row and column objects. In a grayscale image of RDI, white pixels represent high dissimilarity, while black represents no dissimilarity. A "useful" RDI, $I(D_N^*)$, highlights potential clusters as a set of "dark blocks" along the diagonal of the image.

An improved VAT (iVAT) algorithm [11] provides a much sharper reordered diagonal matrix image by replacing the input distance matrix D by the distance matrix $D_N'^* = [d_{ij}']$, where

$$d_{ij}' = \min_{r \in P_{ij}} \max_{1 < h < |r|} D_{r[h]r[h+1]}, \tag{2}$$

where $r \in P_{ij}$ is an acyclic path in the set of all acyclic paths between object i and j in O. Both VAT and iVAT suffer from resolution and memory constraints that limit their usefulness to input matrices on the order of size 10^5 or so. To overcome these limitations, *scalable iVAT* sVAT/siVAT [10, 12] were proposed based on a hybrid strategy that combines *Maximin (MM) and Random Sampling*, called MMRS sampling. These algorithms first find $n \ll N$ samples using MMRS, and then construct an image of this sample using distance matrix $D_n'^*$. This image $I(D_n'^*)$ usually provides a useful estimate of k without the need to calculate the very large distance matrix, D_N of the big dataset, and circumvents the problem that $I(D_N^*)$ is not computable. The siVAT scheme requires the user to supply only two parameters: n the desired sample size, and k', an overestimate of k, the assumed number of clusters, to obtain k' distinguished objects in the sample.[1]

Figure 1 illustrates VAT, iVAT, and siVAT for a 2D synthetic dataset. View (a) is the scatterplot of 5000 data points randomly drawn from five *Gaussian mixture* (GM) components with equal prior probabilities. Its VAT and iVAT images are shown in Views (b) and (c). While both VAT and iVAT images show five dark blocks along the diagonal corresponding to the five clusters in the dataset, dark blocks in the iVAT image are much sharper than the VAT image. View (d) shows the siVAT image of $n = 500$ samples (10% of the total dataset) which was made in $1/1000$ fraction of the time taken to compute the full iVAT image.

Figure 1e–h illustrate VAT, iVAT, and siVAT for a big dataset ($N = 1,000,000$) extracted from same five GM components with equal probabilities. In this case, the VAT and iVAT images cannot be generated due to their high computational complexity and memory constraints, indicated by question marks (?) in Views (f)

---

[1]$n = a\ few\ hundred\ samples$ is a good choice for most datasets. In [18], k' and n are randomly chosen between $2k$ and $4k$, and $10k$ and $30k$ respectively, where $k', n \in \mathbb{Z}$, and k is the number of labeled subsets in the ground truth data. The k' is an overestimate of k i.e., $k' > k$.

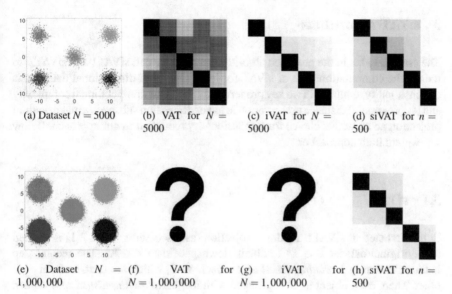

(a) Dataset $N = 5000$ (b) VAT for $N = 5000$ (c) iVAT for $N = 5000$ (d) siVAT for $n = 500$

(e) Dataset $N = 1,000,000$ (f) VAT for $N = 1,000,000$ (g) iVAT for $N = 1,000,000$ (h) siVAT for $n = 500$

Fig. 1 Data scatterplot, VAT, iVAT, and siVAT images for a small (top) ($N = 5000$) and a big dataset (bottom) ($N = 1,000,000$)

and (g). However, siVAT extracts a small size ($n = 500$) MMRS sample from this big data and produces an approximate iVAT image which suggests that five clusters are present in the big dataset.

The most essential step in sVAT/siVAT is MMRS sampling. Maximin (MM) sampling starts at a random point, and then chooses as the first MM sample (distinguished point) the point which is furthest from the initial point with respect to a chosen measure of distance on the set being sampled. The second object selected maximizes the distance from the first point. The third object selected maximizes the distance from both of the first two points. This process continues until the desired number (k') of *distinguished MM objects* are chosen. Then, each object in O is grouped with its nearest *distinguished MM object*. This stage divides the entire dataset O into k' groups. Then, a sample S of size n (just a small fraction of N), is built by selecting a specified number of random data points from each of the k' groups using *random sampling* (RS). Hence the term MMRS is used for the overall process.

The computational complexities in the first and second stages of Maximin sampling in siVAT are $O(pk'N)$, and the last stage requires $O(pn^2)$ operations to build sample S. Despite its computational efficiency in low dimensions, MMRS sampling is computationally expensive in high dimensions. Therefore, siVAT is adequate for large sample size datasets, but it becomes computationally expensive when the dataset is large both in N and p.

3 siVAT Algorithms

The essential idea in the proposed algorithms, which we call siVAT1 and siVAT2, to reduce the computation time of siVAT for large size, high-dimensional datasets, is dimensionality reduction. Two key properties, namely low computational complexity and (approximate) distance preservation in lower dimension subspaces, make random projection an attractive choice for dimensionality reduction in our approach. Below we explain both approaches.

3.1 siVAT1

In the first step of siVAT1, random projection (as discussed in Sect. 2.1) is applied to an original dataset $X \in \mathbb{R}^p$ to obtain downspace data $Y \in \mathbb{R}^q$. The second step is the selection of k' *distinguished MM objects* in Y, which are furthest from each other. Then, each object in O is grouped with its nearest *distinguished MM object*. This stage divides the entire dataset O into k' groups, $\{Z_i\}_{i=1}^{k'}$ by associating $|Z_i|$ objects to the i-th *distinguished MM object*, which provides a representation of each of the k' clusters. This grouping task requires the computation of a $k' \times N$ matrix, now done in downspace ($\mathbb{R}^q$), which reduces the computational time that would be needed for the calculations of a $k' \times N$ distance matrix of $p-$dimensional feature vectors. Finally, the downspace (subscript d) sample S_d of size n is built by selecting random data points from each of the k' clusters $\{Z_i\}_{i=1}^{k'}$. The number of points, n_i extracted from cluster Z_i is proportional to the number of datapoints in Z_i, namely, $n_i = \lceil n \times |Z_i|/N \rceil$, where $\lceil \cdot \rceil$ denotes the ceiling function.

The approximate distance preservation (within $1 \pm \varepsilon$) property of randomly projected pairs from X asserted by Theorem 1 supports a belief that if the distinguished MM objects in Y are generated by applying MM sampling to it, beginning with the same initial point, that the MM samples in Y should be the same or nearly the same (due to approximation distance error) as the k' MM points in X (upspace) that would be produced by MM sampling in the upspace. Two results from [10] about this procedure provide some justification for believing this.

Proposition 1 *Let O be a finite set of distinct objects that can be partitioned into k compact-separated (CS) [9] clusters and let $k' \geq k$, then*

A. Step (a) of Algorithm 1 (the first step of the MMRS sampling algorithm) selects at least one distinguished object from each cluster.

B. In addition, if $n_i = n \times |Z_i|/N$ (Step (c) in Algorithm 1) is an integer for $i = 1, 2, \ldots, k'$ then the proportion of the objects in the MMRS sample from cluster $O^{(j)}$ equals the proportion of objects from the same cluster $O^{(j)}$ in the original data, for $j = 1, 2, \ldots, k$.

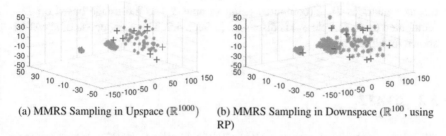

(a) MMRS Sampling in Upspace ($\mathbb{R}^{1000}$)   (b) MMRS Sampling in Downspace ($\mathbb{R}^{100}$, using RP)

Fig. 2 Scatterplots of upspace and downspace MMRS samples in $\mathbb{R}^3$ (using PCA), where green circles represent samples, and red plus represent *distinguished MM objects*. The parameters in MMRS sampling are as: $k' = 10$, and $n = 200$ samples, from $N = 1,000,000$ data points

Proof See [10] for proof.

In siVAT1, MMRS sampling is performed in the randomly projected lower dimensional space Y (downspace). Therefore, if dataset X has k CS clusters and $k' \geq k$, and if downspace data Y has k CS clusters, and carries a JL certificate ($q \geq q_0$) as in Theorem 1, then Proposition 1A guarantees that MMRS sampling in Y will select at least one distinguished object from each of the k clusters, and Proposition 1B assures us that the proportion of the objects in each cluster in the MMRS sample will be similar to the proportion of objects in each subset in the original data.

For empirical validation of our hypothesis, we performed MMRS sampling on a synthetic dataset, GM (see Sect. 4.1) with original dimension ($p = 1000$), and on a randomly projected downspace dataset in the reduced dimension ($q = 100$). The scatter plots of the samples of $X \in \mathbb{R}^{1000}$ and $Y \in \mathbb{R}^{100}$ are shown in $\mathbb{R}^3$ (using the *principal components analysis* (PCA)) in Fig. 2, which shows that cluster distributions including $k' = 10$ *distinguished MM objects* (red plus) in both samples are similar.

The previous step provides n samples in the downspace, $S_d \subset \mathbb{R}^q$, which can be used to build an $n \times n$ distance matrix $D_{n,d}$, and subsequently, to obtain an approximate iVAT image by applying VAT/iVAT to $D_{n,d}$. We need a reliable iVAT image in order to estimate the number of clusters in the big data. The VAT/iVAT image provides a subjective visual assessment of potential cluster substructure based on how distinctive the dark blocks (clusters) appear in the image. However, the quality of the image of the reordered distance matrix $D_{n,d}^{'*}$, obtained by applying VAT/iVAT to $D_{n,d}$, often turns out to be very poor (see Sect. 4.2) due to the unpredictable nature of random projection.

To overcome this problem, the $n \times n$ distance matrix is computed in the upspace. First, the MMRS samples S_d are back-projected to the upspace (subscript u) using the sample indices in S_d to identify the corresponding samples S_u in X, and then they are used to compute a distance matrix $D_{n,u}$. The advantage of using the upspace MMSR samples S_u to compute $D_{n,u}$ is that all the original features are utilized in the distance matrix computation, i.e., no information is lost. Moreover, the computation of $D_{n,u}$ in the upspace is not very time consuming as the number of MMSR samples n is typically in order of 10^2. The comparison of iVAT images obtained using the distance matrices $D_{n,d}$ (computed using sample S_d) to the image based on $D_{n,u}$ (computed using sample S_u) is discussed in Sect. 4.2. The pseudocode of siVAT1 is shown in Algorithm 1.

3.2 siVAT2

We tested another architecture, called siVAT2, similar to siVAT1, which is also a reasonable alternative for siVAT. Unlike siVAT1 in which RP is applied only once to obtain MMRS samples, multiple RPs are used in siVAT2. In this approach, Q MMRS samples, each of size $\lceil n/Q \rceil \in \mathbb{Z}$, are generated using multiple RPs. First, Q random projections of $X \subset \mathbb{R}^{N \times p}$ produce multiple downspace datasets $\{Y_i\}_{i=1}^{Q} \subset \mathbb{R}^{N \times q}$. Second, MMRS sampling is applied to each Y_i with sub-sample size $n_{s_i} = \lceil n/Q \rceil$. Then, a sample of size n is formed as the union of $\{n_{s_i}\}_{i=1}^{Q}$ samples, and subsequently used to compute the upspace distance matrix $D_{n,u}$ as discussed in siVAT1. The rationale for siVAT2 is that different RPs and random initializations of the MMRS algorithm in each run will yield a diverse subset, which essentially overcomes the problem caused by a single bad random projection of X.

Cluster Assessment and Single Linkage Clustering
The VAT/iVAT algorithm is applied to distance matrix $D_{n,u}$, which returns a reordered matrix, $D'^{*}_{n,u}$ and the cut magnitudes of the k longest edges in the MST links (represented by c). The visualization of $D'^{*}_{n,u}$ using $I(D'^{*}_{n,u})$ suggests the number of clusters k present in the dataset. Since *single-linkage* (SL) clusters are always diagonally aligned in iVAT ordered images, we merely have to cut the largest $(k-1)$ edges in the MST built by iVAT to form the corresponding k aligned partition [14, 15] of sample S.

4 Experiments

We performed two sets of experiments on five datasets, that are big in sample size (N) as well as in dimension (p). In the first experiment, we explore the capability of siVAT1 and siVAT2 to visually suggest the number of clusters in big datasets. In the second experiment, we compare the performance of siVAT1 and siVAT2 with siVAT based on the *partition accuracy* (PA). For quantitative comparison, we partition the sample S into k subsets with SL and compare their PAs with respect to the ground truth labels of sample S.

Algorithm 1 siVAT1

Input: Dataset $X = \{x_1, .., x_N\} \subset \mathbb{R}^p$, downspace dimension q, Overestimate of true number of clusters k', number of approximated samples n

Output: $D'^*_{n,u}$ - iVAT reordered dissimilarity matrix of $D_{n,u}$.

Dataset generation in downspace.

Generate downspace dataset $Y \subset \mathbb{R}^{N \times q}$ using $Y = \frac{1}{\sqrt{q}}XT$, where $T \in \mathbb{R}^{p \times q}$ is the random matrix as discussed in Section 2.1.

MMRS Sampling on Y.

a. *Select the indices m of k' distinguished objects.*

Randomly select the first distinguished object x_{m_0}.

Distance of x_{m_0} from N points; $R = \{dist(x_{m_0}, x_1), ..., dist(x_{m_0}, x_N)\} = (r_{m_0 1}, ...r_{m_0 N}) = (R_1, ...R_N)$.

for $i = 1$ **to** k' **do**
    $R \leftarrow (min\{R_1, r_{m_{i-1}1}\}, ..., min\{R_N, r_{m_{i-1}N}\})$
    $m_i = \arg\max_{1 \le j \le N} R_j$
end for

b. *Group each object in O with its nearest distinguished object.*

$Z_1 = Z_2.... = Z_{k'} = \emptyset$.

for $t = 1$ **to** N **do**
    $l = \arg\min_{1 \le i \le k'}\{r_{m_i t}\}$
    $Z_l = Z_l \cup \{t\}$
end for

c. *Randomly select data near each distinguished point to obtain the n number of samples.*

$n_i = \lceil n \times |Z_i|/N \rceil \quad i = 1, 2, ..., k'$.

Draw n_i unique random indices from Z_i to build sample Z'_i.

$$S_d = \bigcup_{i=1}^{k'} Z'_i$$

d. Back project $S_d \subset \mathbb{R}^q$ to $S_u \subset \mathbb{R}^p$ $(S_d \rightarrow S_u)$ using sample indices.

e. Compute $D_{n,u}$ using S_u, and apply VAT/iVAT on $D_{n,u}$ returning $D'^*_{n,u}$, P, c.

f. Choose the number of clusters k using image of $D'^*_{n,d}$.

4.1 Datasets

We performed our experiments on the following datasets.

4.1.1 Synthetic Datasets

A synthetic dataset GM, having $n = 1,000,000$ data points in $p = 1000$ dimensions, was constructed by drawing an equal number of labeled samples from each component of a mixture of $k = 3$ Gaussian distributions. GM has overlapping clusters. The mean of components were $(-2, \ldots, -2)_{1000}$, $(0, \ldots, 0)_{1000}$, and $(2, \ldots, 2)_{1000}$ and the standard deviations were $(1, \ldots, 1)_{1000}$, $(2, \ldots, 2)_{1000}$, and $(3, \ldots, 3)_{1000}$.

4.1.2 Real Datasets

Four publicly available[2] real, high-dimensional (large volumes) labeled datasets were chosen to demonstrate the applicability of our approach. The details of all real datasets are given in Table 1. The KDD and FOREST datasets are labeled: the US Census and BigCross dataset s are not.

For the GM, KDD and FOREST datasets, the quality of the output crisp SL partition of sample S for siVAT, siVAT1 and siVAT2 is assessed using ground truth information, U_{gt}. The similarity of computed partitions with respect to ground truth labels is measured using the *partition accuracy* (PA). The PA of a clustering algorithm is the ratio of the number of objects with matching ground truth and algorithmic labels to the total number of objects in the sample S. The value of PA ranges from 0 to 1, and a higher value implies a better match to the ground truth partition. Before the PA can be calculated, it is necessary to ensure that the algorithmic labels of the SL clusters in S correspond to the same subsets in the ground truth.

To address the label correspondence issue, we transformed the cluster labels based on the class of majority of points within each cluster. Our implementation first sorts the cluster labels based on the number of points in each cluster. Then, for each cluster, starting from the first cluster in the sorted labels, we compute the class (ground truth label) of the majority of samples within the cluster, and assign it to that cluster. This method works correctly only when the number of clusters is equal to the number of classes. So, our clustering algorithms are performing a classification task via SL clustering with predefined (ground truth) class numbers.

Since, the ground truth information is not available for the US Census and BigCross datasets, we use an internal cluster validity index, *Dunn's Index* (DI) [9], to evaluate the quality of output partitions for these two datasets. DI is a metric of how well a set of clusters represent *compact separated* (CS) clusters. Dunn defined CS clusters in X with a distance criteria, and showed that X contains CS clusters if and only there is a partition U^* of X for which $DI > 1$. The DI of the ground truth partition for all KDD and FOREST datasets is shown in Table 1.

The experiments were performed using MATLAB software on a normal PC with the following configurations; OS: Windows 7 (64 bit); processor: Intel(R) Core(TM) i7-4770 @3.40GHz; RAM: 16GB. In all the experiments, siVAT, siVAT1 and siVAT2 parameters k' and n are 50 and 500, respectively. For siVAT1, we chose the downspace dimension q as 100 for synthetic dataset GM and $q = 5$ for the four other datasets. For the choices of $\varepsilon = \beta = 0.5$, and $n = 1,000,000$ (GM dataset), $q_0 = 828$ (using (1)) and the probability of distance preservation $= 0.99$, so the chosen q values are well below the JL bound, and the projections for this pair ($\varepsilon = \beta = 0.5$) are the rogue random projections. The existence of a JL certificate for Y clearly depends on the choices for ε and β. And even when a choice for q satisfies Theorem 1, the JL guarantee is still "in probability," so this aspect of the experiments is not very important. For

---

[2]These datasets can be found at the UCI machine learning data repository [2, 3]. The features were normalized to the interval [0, 1] by subtracting the minimum and then dividing by the subsequent maximum so that they all had the same scale.

Table 1 Properties of real datasets

Dataset	N	p	k	$DI(k, U_{gt})$
KDD	4898431	41	23	0 (Non-CS)
FOREST	581012	54	7	0.002 (Non-CS)
US Census	2458285	68	Unknown	Unknown
BigCross	11,600,320	57	Unknown	Unknown

siVAT2, we chose the number of random projections $Q = 10$. All the experiments were performed 20 times on each dataset and the average results are reported.

4.2 Distance Matrix in Downspace Versus Distance Matrix in Upspace

In this experiment, we choose $n = 500$ and compare the quality of iVAT images obtained by applying VAT/iVAT to distance matrix $D_{n,d}$, computed using S_d, to the iVAT image of the distance matrix $D_{n,u}$, computed using S_u for siVAT, siVAT1 and siVAT2. We also compare the PA values of SL partitions of samples. Figure 3a–c show three 500×500 iVAT images $I_{a,b,c}(D'^*_{n,d})$ for three different random projections of the GM dataset, which has three labeled subsets. It is clear from these three iVAT images and their corresponding PA values that the quality of the image varies due to the random nature of the downspace projection, and none of them strongly suggests that the number of clusters in GM may be $k = 3$. Figure 3d shows the iVAT image $I(D'^*_{n,u})$ for siVAT, which shows three clusters as dark blocks. Views 3 (e) and (f) show the images $I(D'^*_{n,u})$ for siVAT1 and siVAT2, which contain three dark blocks that are visible along the diagonal. Moreover, the PA values of siVAT, siVAT1 and siVAT2 are almost equal. The iVAT images in Fig. 3a and c appear very similar to the siVAT images in views (d), (e) and (f). However, the PA values for the three siVAT images are nearly equal (about 75%), and are quite a bit higher than the PA values for the three iVAT images. This shows that the siVAT schemes work better that iVAT does on a naive RP of GM.

4.3 Cluster Assessment Using siVAT, siVAT1 and siVAT2

The approximate iVAT images $(I(D'^*_N))$ for all four big datasets obtained using the siVAT, siVAT1 and siVAT2 algorithms are shown in Fig. 4. The ordering of the dark blocks in siVAT/siVAT1/siVAT2 images can be different for the same dataset based on the (random) initialization of the first distinguished MM object in Algorithm 1, but the size and the number of dark blocks should be (ideally) same.

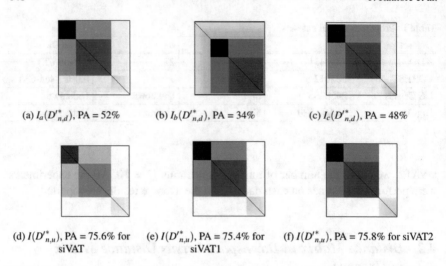

(a) $I_a(D'^*_{n,d})$, PA = 52% (b) $I_b(D'^*_{n,d})$, PA = 34% (c) $I_c(D'^*_{n,d})$, PA = 48%

(d) $I(D'^*_{n,u})$, PA = 75.6% for (e) $I(D'^*_{n,u})$, PA = 75.4% for (f) $I(D'^*_{n,u})$, PA = 75.8% for siVAT2
siVAT   siVAT1

Fig. 3 iVAT images $I(D'^*_{n,d})$ (**a–c**) obtained using distance matrices computed using S_d and iVAT images (**d–f**) $I(D'^*_{n,u})$ obtained using siVAT, siVAT1 and siVAT2, respectively

The KDD dataset has 23 labeled subsets (22 simulated attacks and a normal) that fall into four main categories: DOS, R2L, U2R, and probing. Figure 4a, e, i show the siVAT, siVAT1 and siVAT2 images for KDD. All three images suggest four major clusters, with a tiny subcluster in the uppermost block, in which, the middle dark block represents the "smurf" attack (60% of dataset) in the DOS category. Figure 4b, f, j show the corresponding images for the FOREST dataset. All three show 9–10 dark blocks (of moderate size) in medium resolution, and 15–16 tiny dark blocks at high-resolutions. This is a case where the physically labeled subsets do NOT form well-defined clusters, at least not in the sense of SL distance, the basis of the siVAT/siVAT1/siVAT2 images.

Figure 4c, g, k show the siVAT, siVAT1, and siVAT2 images for the US Census dataset, which all indicate two distinguished dark blocks along the diagonal in which the small dark block (lower) comprises two smaller (mini) dark blocks. This suggests that there are two major clusters and that the lower block represents two subclusters in this data. Previous research [8] also suggests $k = 2$ or 3 as the best estimate of the number of clusters for this dataset. The siVAT/siVAT1/siVAT2 images for BigCross are very similar to the siVAT/siVAT1/siVAT2 images of FOREST dataset, and show 18–20 (tiny) dark blocks. This is very reasonable because BigCross is the Cartesian product of the FOREST with another (TOWER) dataset [2]. Experiments in [2] also suggest 20–30 clusters for this dataset. The overall conclusions that can be made from Fig. 4 are: (i) siVAT, siVAT1 and siVAT2 contain essentially the same visual information about cluster structure in the samples processed; and (ii) siVAT1/siVAT2 produces this information 5–52 times faster than siVAT.

Table 2 compares siVAT with siVAT1 and siVAT2 based on CPU times and the partition accuracy for labeled data and DI on unlabeled data on SL partitions of sample.

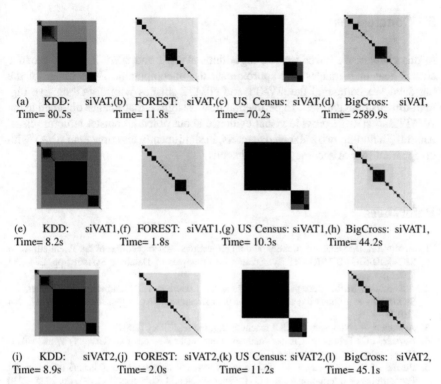

(a) KDD: siVAT,(b) FOREST: siVAT,(c) US Census: siVAT,(d) BigCross: siVAT,
Time= 80.5s   Time= 11.8s   Time= 70.2s   Time= 2589.9s

(e) KDD: siVAT1,(f) FOREST: siVAT1,(g) US Census: siVAT1,(h) BigCross: siVAT1,
Time= 8.2s   Time= 1.8s   Time= 10.3s   Time= 44.2s

(i) KDD: siVAT2,(j) FOREST: siVAT2,(k) US Census: siVAT2,(l) BigCross: siVAT2,
Time= 8.9s   Time= 2.0s   Time= 11.2s   Time= 45.1s

Fig. 4 siVAT (first row), siVAT1 (second row) and siVAT2 (last row) images of D'^*_n for each of the datasets

Table 2 Average PA (%) values (DI for US Census and BigCross) and run-time (in seconds) on all the datasets

Methods	Datasets							
	KDD		FOREST		US Census		BigCross	
	PA	Time	PA	Time	DI	Time	DI	Time
siVAT	**91.2**	81.2	48.5	10.9	**1.28**	65.6	**0.70**	2505.3
siVAT1	90.5	8.3	**48.7**	0.9	**1.28**	**8.2**	**0.70**	**44.8**
siVAT2	90.7	8.8	**48.7**	0.9	**1.28**	**8.2**	**0.70**	**44.8**

The listed average values indicate that siVAT, siVAT1 and siVAT2 achieve approximately equal values (either partition accuracy if the input data are labeled, or Dunn's index for unlabeled input data) for all the datasets, but significantly, siVAT1/siVAT2 achieves the same results about 5–52 times faster than the siVAT algorithm. This demonstrates that both siVAT1 and siVAT2 produce approximate cluster heat maps, that are quantitatively equivalent to siVAT based on SL partitions of the big datasets, in significantly less time than the siVAT.

5   Conclusions

In this chapter, we introduce two algorithms siVAT1 and siVAT2, which produce cluster heat map images that approximate the uncomputable iVAT image of the big data. We compared the siVAT1 and siVAT2 images with siVAT on five big, high-dimensional datasets. Experimental results suggest that images obtained using siVAT1 and siVAT2 provide visual evidence about potential cluster structure in all datasets, including two unlabeled datasets, in significantly less time than siVAT with no apparent loss of accuracy or visual acuity.

References

1. Achlioptas, D.: Database-friendly random projections. In: Proceedings of the Twentieth ACM SIGMOD-SIGACT-SIGART Symposium on Principles of Database Systems, pp. 274–281 (2001)
2. Ackermann, M.R., Märtens, M., Raupach, C., Swierkot, K., Lammersen, C., Sohler, C.: Streamkm++: a clustering algorithm for data streams. J. Exp. Algorithmics (JEA) **17**, 2–4 (2012)
3. Asuncion, A., Newman, D.: UCI machine learning repository (2007)
4. Bezdek, J.C.: Primer on Cluster Analysis: Four Basic Methods that (Usually) Work, vol. 1. First Edition Design Publishing (2017)
5. Bezdek, J.C., Hathaway, R.J.: VAT: a tool for visual assessment of (cluster) tendency. In: Proceedings of International Joint Conference on Neural Networks (IJCNN), pp. 2225–2230 (2002)
6. Bezdek, J.C., Ye, X., Popescu, M., Keller, J., Zare, A.: Random projection below the JL limit. In: Proceedings of International Joint Conference on Neural Network (IJCNN), pp. 2414–2423 (2016)
7. Bingham, E., Mannila, H.: Random projection in dimensionality reduction: applications to image and text data. In: Proceedings of the Seventh ACM SIGKDD International Conference on Knowledge Discovery and Data Mining, pp. 245–250. ACM (2001)
8. Chen, K., Liu, L.: Detecting the change of clustering structure in categorical data streams. In: Proceedings of the 2006 SIAM International Conference on Data Mining, pp. 504–508. SIAM (2006)
9. Dunn, J.C.: A fuzzy relative of the isodata process and its use in detecting compact well-separated clusters. J. Cybern. **3**(3), 32–57 (1973)
10. Hathaway, R.J., Bezdek, J.C., Huband, J.M.: Scalable visual assessment of cluster tendency for large data sets. Pattern Recognit. **39**(7), 1315–1324 (2006)
11. Havens, T.C., Bezdek, J.C.: An efficient formulation of the improved visual assessment of cluster tendency (iVAT) algorithm. IEEE Trans. Knowl. Data Eng. **24**(5), 813–822 (2012)
12. Havens, T.C., Bezdek, J.C., Palaniswami, M.: Scalable single linkage hierarchical clustering for big data. In: IEEE Eighth International Conference on Intelligent Sensors, Sensor Networks and Information Processing, pp. 396–401. IEEE (2013)
13. Johnson, W.B., Lindenstrauss, J.: Extensions of Lipschitz mappings into a Hilbert space. Contemp. Math. **26**(189–206), 1 (1984)
14. Kumar, D., Bezdek, J.C., Palaniswami, M., Rajasegarar, S., Leckie, C., Havens, T.C.: A hybrid approach to clustering in big data. IEEE Trans. Cybern. **46**(10), 2372–2385 (2016)
15. Kumar, D., Palaniswami, M., Rajasegarar, S., Leckie, C., Bezdek, J.C., Havens, T.C.: clusiVAT: a mixed visual/numerical clustering algorithm for big data. In: IEEE International Conference on Big Data, pp. 112–117. IEEE (2013)

16. Lawson, R.G., Jurs, P.C.: New index for clustering tendency and its application to chemical problems. J. Chem. Inf. Comput. Sci. **30**(1), 36–41 (1990)
17. Rathore, P., Bezdek, J.C., Erfani, S.M., Rajasegarar, S., Palaniswami, M.: Ensemble fuzzy clustering using cumulative aggregation on random projections. IEEE Trans. Fuzzy Syst. **26**(3), 1510–1524 (2018)
18. Rathore, P., Kumar, D., Bezdek, J.C., Rajasegarar, S., Palaniswami, M.S.: A rapid hybrid clustering algorithm for large volumes of high dimensional data. IEEE Trans. Knowl. Data Eng. (2018)
19. Thorndike, R.L.: Who belongs in the family? Psychometrika **18**(4), 267–276 (1953)

An Introduction to Linguistic Summaries

Ronald R. Yager

Abstract First we provide the fundamental ideas involved in linguistic summaries, next we look at the quantification of the information contained in a linguistic summary and finally we look at the formulation of summaries involving rich concepts.

1 Introduction

In [1, 2] Yager introduced the idea of linguistic summaries as a user-friendly method of summarizing information in a database. Kacprzyk, Bouchon-Meunier and other researchers [3–16] have made considerable use of the idea of linguistic summary. Here we look at a number of ideas related to linguistic summaries. We first present the fundamental ideas involved in linguistic summaries, we next look at the quantification of the information contained in a summary and finally we look at the formulation of summaries involving rich concepts.

2 Basic Ideas of Linguistic Summaries

We briefly review some basic ideas associated with linguistic summaries. Assume we have a database $Y = \{y_1, \ldots, y_n\}$ where the y_i are the objects in the database. Assume V is some attribute, having as its domain X, associated with the objects in the database. For example, if each y_i is a person then V could be their age. Here then for each y_i we have a value $V(y_i) = a_i$ where $a_i \in X$. Associated with attribute V is a data set $D = [a_1, \ldots, a_n]$, a bag, containing the values of V assumed by the objects in the database Y. We emphasize that a bag allows multiple elements with the same value.

R. R. Yager (✉)
Iona College, New Rochelle, NY, USA
e-mail: yager@panix.com

M.-J. Lesot and C. Marsala (eds.), *Fuzzy Approaches for Soft Computing and Approximate Reasoning: Theories and Applications*, Studies in Fuzziness and Soft Computing 394,
https://doi.org/10.1007/978-3-030-54341-9_13

A linguistic summary associated with V is a global statement based on the values in D. If V is the attribute age some examples of simple linguistic summaries are: *Most people in the database are about 30 year old, Few people in the database are old, Nearly a quarter of the people in the database are middle aged.*

Formally a simple linguistic summary is a statement of the form.

Q objects in the database have V is **S**.

In the above **S** is called the summarizer and **Q** is called the quantity in agreement. Also associated with a linguistic summary is a measure of validity of the summary, τ. The value τ is used to indicate the truth of statement that **Q** objects have the property that V is **S** in the light of the data set D.

A fundamental characteristic of this formulation is that the summarizer and quantity in agreement are expressed in linguistic terms. One advantage of the use of linguistic summaries is that they provide statements about the dataset in terms that it is very easy for people to comprehend.

In [17] Yager showed that linguistic summaries are closely related to what Zadeh called Z-numbers [18].

Using fuzzy subsets we are able to provide a formal semantics for the terms used in a linguistic summary. In a procedure to be subsequently described, we shall use this ability to formalize the summarizer and quantity in agreement as fuzzy sets to enable us to evaluate the validity of a linguistic summary. This validation process will be based upon a determination of the compatibility of the linguistic summary with the data set D. It should be pointed out that for a given attribute we can conjecture numerous different summaries, then with the aid of the data set D we can obtain the validity, τ, of a conjectured linguistic summary.

In developing our approach to validating a linguistic summary considerable use will be made of the ability to represent a linguistic summarizer by a fuzzy subset over the domain of the attribute. If V is some attribute taking its value from the domain X and if **S** is some concept associated with this attribute we can represent **S** by a fuzzy subset S on X such that for each $x \in X$, $S(x) \in [0, 1]$ is the degree of compatibility of the value x with the concept **S**. If V is age and **S** is the concept middle age then S(40) indicates the degree to which 40 years old is compatible with the idea of middle age. Even in environments in which the underlying domain is non-numeric using this approach allows us to obtain numeric values for the membership grade in the fuzzy subset S corresponding to the concept **S**. For example if V is the attribute city of residence that takes as its domain the cities in the U.S. we can express the concept **S**, "lives near New York", as a fuzzy subset. The second component in our linguistic summary is the quantity in agreement **Q**. These objects belong to a class of concepts called linguistic quantifiers [19]. Examples of linguistic quantifiers are terms such as most, few, about half, all. Essentially linguistic quantifiers are fuzzy proportions, an alternative view of these subjects are generalized logical quantifiers. In [19] Zadeh suggested we could represent these linguistic quantifiers as fuzzy subsets of the unit interval. Using this representation the membership grade of any proportion r $\in [0, 1]$ in the fuzzy set Q corresponding to the linguistic quantifier **Q**, Q(r), is a measure of the compatibility of the proportion r with the linguistic quantifier we are representing by the fuzzy subset Q. For example if Q is the fuzzy set corresponding to the quantifier *Most* then $Q(0.9)$ represents the degree to which the proportion 0.9 satisfies the concept *Most*.

In [20] Yager identified three classes of linguistic quantifiers that cover most of those used in natural language. (1) Q is said to be monotonically non-decreasing if $r_1 > r_2 \Rightarrow Q(r_1) \geq Q(r_2)$, examples of this type of quantifier are *at least 30%, most, all*. (2) A quantifier Q is said to monotonically non-increasing if $r_1 > r_2 \Rightarrow Q(r_1) \leq Q(r_2)$, examples of this type of quantifiers are *at most 30%, few, none*. (3) A quantifier Q is said to be unimodal if there exist two values $a \leq b$ both contained in the unit interval such that for $r < a$, Q is monotonically non-decreasing, for $r > b$, Q is monotonically non-increasing and for $r \in [a, b]$, $Q(r) = 1$, an example of this type of quantifier is *about 0.3*.

An important idea that can be associated with a linguistic quantifier is the concept of an antonym. If Q is a linguistic quantifier its antonym is also a linguistic quantifier, denoted $\widehat{Q}$, such that $\widehat{Q}(r) = Q(1 - r)$. The operation of taking an antonym is involutionary, that is $\widehat{\widehat{Q}} = Q$. From this we see that antonyms come in pairs. Prototypical examples of antonym pairs are **all-none** and **few-many**. Consider the quantifier *at most 0.3* defined as $Q(r) = 1$ if $r \leq 0.3$ and $Q(r) = 0$ if $r > 0.3$. Its antonym has $\widehat{Q}(r) = 1$ if $r \geq 0.3$ and $\widehat{Q}(1 - r) = 0$ if $r \geq 0.3$. This can be seen to be equivalent to $\widehat{Q}(r) = 1$ if $r \geq 0.7$ and $\widehat{Q}(r) = 0$ if $r < 0.7$. Thus the antonym of *at most 0.3* is *at least 0.7*.

Care must be taken to distinguish between the antonym of a quantifier and its negation. We recall the negation of Q denoted $\overline{Q}$ is defined such that $\overline{Q}(r) = 1 - Q(r)$. We see that the negation of *at most 0.3* is $\overline{Q}(r) = 0$ if $r \leq 0.3$ and $\overline{Q}(r) = 1$ if $r > 0.3$, this corresponds *to at least 0.3"*.

Having discussed the concepts of summarizer and quantity in agreement we are now in a position to describe the methodology used to calculate the validity τ of a linguistic summary. Assume $D = [a_1, a_2, ..., a_n]$ is the collection of values that appear in the database for the attribute V. Consider the linguistic summary:

Q items in the database have value for V that are S.

The basic procedure to obtain the validity τ of this summary in the face of the data D is: (1) For each a_i in D, calculate $S(a_i)$, the degree to which a_i satisfies the summarizer S. (2) Let $r = \frac{1}{n} \sum_{i=1}^{n} S(a_i)$, the proportion of D that satisfy S.

(3) $\tau = Q(r)$, the grade of membership of r in the quantity in agreement.

A number of interesting properties can be associated with these summaries. Consider "*Q items have V is S*" and assume the data set D has cardinality n. The associated validity is $\tau = Q\left(\frac{\sum_i S(x_i)}{n}\right)$. Now consider the summary $\widehat{Q}$ *items have V is (notS)* where $\widehat{Q}$ is the antonym of Q. In this case the measure of validity $\tau_1 = \widehat{Q}\left(\frac{\sum_i \overline{S}(x_i)}{n}\right) = \tau$. These two linguistic summaries have the same measure of validity. The prototypical manifestation of this is that for a given piece of data D the summary *Many people are young* will have the same validity as *Few people are not young*.

The summary "Not Q objects are S." has validity $\tau_2 = \overline{Q}\left(\frac{\Sigma_i S(x_i)}{n}\right) = 1-\tau$. This statement has validity complementary to our original proposition. From this we see that. $\tau(S, Q) + \tau(S, \overline{Q}) = 1$.

Thus far we have considered linguistic summaries involving only one attribute. The approach described above can be extended to the case of multiple attributes from a database. We shall first consider linguistic summaries of this form *Most people in the database are tall and young*. Assume U and V are two attributes appearing in the database. Let R and S be concepts associated with each of these attributes respectively, then the generic form of the above linguistic summary is *Q objects in the database have U is R and V is S*.

In this case our data consists of a collection of pairs, $D = <(a_1, b_1), (a_2, b_2), ..., (a_n, b_n)>$ where $U(x_i) = a_i$ is a value of attribute U and $V(x_i) = b_i$ is a value for attribute V. Our procedure for obtaining the validity of the linguistic summary in this case is: 1. For each i calculate $R(a_i)$ and $S(b_i)$, 2. Let $r = \frac{1}{n}\sum_{i=1}^{n}(R(a_i)S(b_i))$ and 3. $\tau = Q(r)$.

We can consider linguistic summaries of the forms. *Q objects in the database have U is R or V is S*. The procedure is the same except in step two the product is replaced by a union operation, t-conorm, such as $Max(R(a_i), S(b_i))$ or $R(a_i) + S(b_i) - R(a_i)S(b_i)$.

Consider the class of summaries of the form *Most tall people in the database are young*. This form of linguistic summary is related to the type of association rule discovery that is of great interest in many applications of data mining. The linguistic summary expressed here can be equivalently expressed as.

In **most** cases of our data; if a person is **tall** then they are **young**

Here then we have an embedded association rule *if height is tall then age is young*. Furthermore we are qualifying our statement of this association rule with the quantifier *most*. In this case we have as our generic form.

Q of the U is **R** objects in the database have V is **S**

In the above we call R the **qualifier** of the summary. The procedure for calculating the validity of this type of summary is a similar three-step process:

1. For each data pair (a_i, b_i) calculate $R(a_i)$ and $S(b_i)$.

2. Calculate $r = \dfrac{\sum_{i=1}^{n} R(a_i)S(b_i)}{\sum_{i=1}^{n} R(a_i)}$ and

3. $\tau = Q(r)$

A fundamental distinction between this and the previous case is in step two, here instead of dividing by n, the number of objects in the database, we divide by the number of objects having R.

We note that we can naturally extend this procedure to handle summaries of the form.

Most **young** people in the database are **tall** and live **near New York**

Few **well-paid** and **young** people in the database live in the **suburbs**.

Consider the linguistic summary corresponding to the quantified association rule *QR are S* and the related summary $\widehat{Q}$ R are $\overline{S}$ where $\widehat{Q}$ is the antonym of Q and $\overline{S}$ is the negation of S. We can show the preceding two statements have the same validity for a given data set D.

3 Information Content of Linguistic Summaries

One purpose in providing linguistic summaries is to provide useful global information about the database. In determining the usefulness of a summary a reasonable measure is some indication of the amount of information conveyed by the summary. At first impulse it is natural to think that only the degree of validity of a summary indicates the information about usefulness of a summary. As the following situation illustrates this is not the case. Assume we have a database of employees and consider the summary *Most employees are over 10 years old*. The fact that this is valid really doesn't convey much information. As a matter of fact this is a manifestation of the following observation about the measure of validity. Consider two summaries *Q objects are S_1* and *Q objects are S_2* where Q is a monotonically increasing quantifier. It can be easily shown that if $S_1 \subseteq S_2$, $S_1(x) \leq S_2(x)$ for all x, then $\tau_2 \geq \tau_1$. Thus for simple monotonic increasing quantifiers we can always increase the validity of a summary by broadening the summarizer used in the summary. Similarly if Q_1 and Q_2 are two monotonically increasing quantifiers such that $Q_1 \subseteq Q_2$ then for a fixed S, $\tau(Q_1, S) \leq \tau(Q_2, S)$. For example the summary "at least 50% of the people are tall" will have a smaller degree of validity then the summary "at least 25% of the people are tall." We see that if we make the quantifier or summarizer broader we can increase the validity. However by broadening it too much we can reach a point where the content of the summary is vacuous. Thus the transmission of useful information by linguistic summaries requires some trade-off between the size of the quantifiers and summarizer used and the resulting validity.

In order to more formally discuss the idea of **informativeness** associated with a simple linguistic summary we introduce some concepts from fuzzy set theory. Assume V is some attribute with domain X. Let F_1, F_2, ..., F_q be a collection of fuzzy subsets corresponding to linguistic concepts associated with the attribute. In [21] we discussed the idea of the specificity of a fuzzy subset. Essentially the specificity measures the degree to which the fuzzy subset points to one element as the manifestation of that fuzzy subset. For example the concept 30 years old is more specific then "about thirty" which in turn is more specific then "at least 20." In [22] measures of specificity associated with a normal fuzzy subset over the space X were introduced. Here we shall use one of these.

Assume the domain of V is the interval X = [a, b] and let F be a normal fuzzy subset defined over X. Then the specificity is $Sp(F) = 1 - \frac{1}{b-a} \int_a^b F(x)dx$ it is the negation of the average membership grade.

It can be shown that this measure of specificity has the following properties

(1) $Sp(F) = 1$ if $F(x) = 1$ for exactly one element and $F(x) = 0$ for all other elements
(2) $Sp(F) = 0$ if $F = X$ and

(3) $Sp(F_1) \geq Sp(F_2)$ if $F_1 \subset F_2$.

For linguistic quantifiers where $X = [0, 1]$ we get $Sp(Q) = 1 - \int_0^1 Q(x)\,dx$, the negation of the area of the quantifier. We that since $\overline{Q}$ is the negation of Q then $Sp(\overline{Q}) = 1 - \int_0^1 \overline{Q}(x)\,dx = 1 - Sp(Q)$. On the other hand if is the antonym of Q, $\widehat{Q}(x) = Q(1 - x)$, we get that $Sp(\widehat{Q}) = Sp(Q)$.

We look at the specificity of a number of prototypical quantifiers. We start with non-decreasing quantifiers. Consider the quantifier "at least α", $Q(x) = 0$ if $x < \alpha$ and $Q(x) = 1$ if $x > \alpha$. In this case $Sp(Q) = \alpha$. Consider the quantifier $Q(x) = xx^\beta$, where $\beta > 0$, here $Sp(Q) = 1 - \int_0^1 x^\beta dx = \frac{\beta}{1+\beta}$. The specificity increases as β increases.

Defining "most" as $Q(r) = 0$ if $r < 0.5$ and $Q(r) = (2r - 1)^{1/2}$ if $r \geq 0.5$. For this quantifier we get $Sp(Q) = 2/3$. More generally we can consider the class $Q(r) = 0$ if $r < \alpha$ and $Q(r) = (\frac{r-\alpha}{1-\alpha})^\beta$ if $r \geq \alpha$, we assume $\alpha \leq 1$ and $\beta \geq 0$. In this case $Sp(Q) = \frac{\beta+\alpha}{\beta+1}$. To some extend $Sp(Q)$ provides a crisp approximation of the quantity in agreement. We can refer to this as the focus of the quantifier.

Let us now consider the decreasing quantifiers, those for which $Q(r_1) \geq Q(r_2)$ if $r_1 < r_2$. We recall that if Q is a decreasing quantifier then its antonym $\widehat{Q}$, where $\widehat{Q}(r) = Q(1 - r)$ is a non-decreasing quantifier. Thus we see that the non-increasing and non-decreasing quantifiers always correspond to antonym pairs. Furthermore since $Sp(Q) = Sp(\widehat{Q})$ these pairs have the same specificity. In particular we note that "at most α" has as its antonym "at least $(1 - \alpha)$" which has specificity $1 - \alpha$.

Finally we consider uni-modal quantifiers. One important class of uni-modal quantifiers are those centered about some value a and having spread $2b$, $Q(r) = 0$ if $0 \leq r \leq a - b$, $Q(r) = 1$ if $a - b \leq r \leq a + b$ and $Q(r) = 0$ if $r \geq a + b$, where we assume $a - b \geq 0$ and $a + b \leq 1$. In this case $Sp(Q) = 1 - \int_{a-b}^{a+b} dx = 1 - 2b.$. We note specificity is indifferent to value of a, the focus of the quantifier. Here we introduce the general idea of the focus of a uni-modal quantifier. If Q is uni-modal we can define the focus as $FOC(Q) = \frac{\int_0^1 Q(x)\,x\,dx}{\int_0^1 Q(x)\,dx}$.

Having introduced the idea of specificity we are now in a position to discuss the measure of information associated with a linguistic summary. Consider a typical simple linguistic summary involving a non-decreasing quantifier, Q_1objects have V is S_1. Assume that with respect to our database we can establish this summary has validity equal to τ_1. This type of statement provides more useful information if the validity is large. In addition informativeness is increased if S_1 is a narrow fuzzy subset, the specificity of S_1 is large. With respect to Q_1 we prefer a higher the focus, a large specificity which corresponds to a narrow fuzzy subset. Using this observation we can provide as a measure of useful information of this type of linguistic summary $I(Q_1, S_1) = \tau_1\, Sp(Q_1)\, Sp(S_1)$.

Consider the case in which we have a simple non-increasing quantifier, Q_2objects have V is S_2 which has validity τ_2. We see that this is equivalent to the following summary $\widehat{Q}_2$ objects have V is $\overline{S}_2$ where $\widehat{Q}_2$ is the antonym of Q_2. Since with Q_2 a non-increasing quantifier we have $\widehat{Q}_2$ is non-decreasing and hence we can express our measure of usefulness of summary using the preceding form $I(Q_2, S_2) = \tau_2$

$Sp(\hat{Q}_2) \, Sp(\overline{S}_2)$. Furthermore we have shown that for antonym $Sp(\hat{Q}_2) = Sp(Q_2)$ thus we can express the information as $I(Q_2, S_2) = \tau_2 \, Sp(Q_2) \, Sp(\overline{S}_2)$.

We now turn to unimodal statements such as Q_3 objects have V is S_3 with validity τ_3. The situation here is somewhat more complex then in the preceding cases. Again we note that informativeness is increased by making Q_3 and S_3 specific and τ_3 large, thus one form is $I(Q_3, S_3) = \tau_3 \, Sp(Q_3) \, Sp(S_3)$. However there is one other consideration, location. Consider the two summaries **About 75%** of the objects are <u>tall</u> and **About 25%** of the objects are <u>tall</u>. We can see that the first statement has provided more useful information. In order to introduce this aspect we must use the idea of the focus of the quantifier. Using this we suggest as a measure of usefulness of a uni-modal quantifier $I(Q_3, S_3) = Foc(Q_3) \, Sp(Q_3) \, \tau \, Sp(S_3)$.

Let us now turn to the informativeness of summaries involving association rules, those of the type *Q tall people are young*. Formally we can represent this as **QR are S**. First we shall consider the case in which Q is a non-decreasing quantifier. Again informativeness is improved if τ is large and Q and S are specific. We must consider the effect of R on informativeness. Consider the two propositions *All twenty year olds are tall* and *All people are tall*.

We easily see that the second summary has provided more information. More generally we can see that the broader the context of the antecedent, everything else being equal, the more informative the summary. Thus we suggest for the measure of informativeness of $Q_1 R_1$ are S_1 where Q_1 is non-decreasing.

$$I(Q_1, R_1, S_1) = \tau_1 \, Sp(Q_1) \, Sp(S_1) \, (1 - Sp(R_1))$$

Thus here we want wide antecedents, narrow consequents and large quantities in agreement as well as large validity.

It is interesting to note that if R_1 is the whole space then the above statement becomes QX are S_1 an unqualified summary. Here $R = X$, hence the specificity of R is $Sp(X) = 0$ thus we get $(1 - Sp(R)) = 1$ and hence our suggested measure reduces to our unqualified measure.

For qualified summaries involving non-increasing or unimodal quantifiers we extend the measure of informativeness by adding the term $(1 - Sp(R))$ as in the case of qualified summaries involving non-decreasing quantifiers.

4   Concepts and Hierarchies

In the preceding we considered the basic linguistic summary of the form "Q objects in the database have V is S". We see that the term V is S can be seen as some *property* associated with objects in the database. In this basic case the property is expressed via an attribute of V and an associated value S.

To determine the validity τ of this linguistic summary we needed to obtain the validity t_i of the property for each element y_i in the database. Here then if $V(y_i)$ is the value of the attribute V for object y_i then $t_i = S(V(y_i))$ is the degree to which object y_i has property S.

Thus t_i is the membership grade of $V(y_i)$ in the fuzzy subset S. We see this can also be viewed as the truth of the statement y_i has property V is S. Using this we then obtain $r = \frac{1}{n}\sum_{i=}^{n} t_i$, the proportion of objects in Y that have this property. Finally the validity of the statement "Q objects in the database have the property V is S" is obtained as $\tau = Q(r)$.

Our objective is to extend the range of the type of properties we can include in a linguistic summary from those properties based simply on a single attribute associated with objects in the database to a class of more sophisticated description of properties associated with objects in the database which we shall refer to as **rich concepts**. Thus we are interested in statements of the form,

Q objects in the database have the property "concept."

Some examples of this would be.

Most objects in the database are "upwardly mobile".

At least half the people in the database are "potential customers".

About half people in the database are "affluent".

A fundamental aspect of these concepts is that they are not necessarily a directly observable attribute of the objects but may involve the combination of attributes associated with the objects in the database. That is these concepts can involve a complex aggregation of attributes and other concepts associated with the objects in the database.

One feature required of these concepts is that if Con is a concept then for any object y_i we can obtain the degree to which object y_i satisfies the concept, $Con(y_i)$, such that $Con(y_i) \in [0, 1]$. Having this property allows us to determine the validity of linguistic summaries of the form "Q objects in the database have property Con." Here we can use the procedure developed earlier to obtain validity on this linguistic summary involving the property Con. We proceed as follows:

(1) For each object y_i in the database we obtain the degree to which it satisfies the concept Con, $t_i = Con(y_i)$.

(2) We then obtain the proportion of elements in the database satisfying the concept Con, $r = \frac{1}{n}\sum_{i=1}^{n} Con(y_i)$.

(3) Finally the validity of the linguistic summary is $\tau = Q(r)$.

Here then the central issue becomes the determination of $Con(y_i)$. We now look at this issue.

Let us formalize our ideas in the following. Assume our database is such that each object has attributes V_j for $j = 1$ to q and each attribute takes its value in its domain, X_j. Thus $V_j(y_i) \in X_j$ is the value of attribute V_j for object y_i.

We can associate with V_j a statement V_j *is* S_{jk} where S_{jk} is a fuzzy subset over the domain X_j representing a linguistic value associated with V_j. We shall refer to the statements of the form V_j *is* S_{jk} as an ***atomic concept*** because we can directly obtain its truth for object y_i as $t_i = S_{jk}(V_j(y_i)))$. We shall at times represent this atomic concept simply as a $Con = \langle S_{jk} \rangle$. Here the associated attribute V_j is implicit.

We can construct concepts from other concepts. Assume Agg is some aggregation operator [23] that can take values in the unit interval and return a value in the unit interval. We can use Agg to construct a concept from other concepts,

$$Con = Agg < Con_1, Con_2, \ldots Con_m > \quad (I)$$

We can obtain the validity of Con for a database object y_i as.

$$Con(y_i) = Agg(Con_1(y_1), Con_2(y_1), \ldots, Con_m(y_i))$$

Since each $Con_k(y_i) \in [0, 1]$ then $Con(y_i) \in [0, 1]$, thus Con is a valid concept.

For clarity in our following discussion we shall refer to Con as the composed concept and the Con_j for $j = 1$ to n as the constituent concepts.

In the special case where all the constituent concepts are atomic concepts, $Con = Agg(S_{j_1k_1}, S_{j_2k_2}, \ldots, S_{j_mk_m})$, then $Con(y_i) = Agg(S_{j_1k_1}(V_{j_1}(y_i)), \ldots, (S_{j_mk_m}(V_{j_m}(y_i)))$.

We note that while Agg must have some formal mathematical properties, the algorithmic process of combining the $Con_j(y_i)$, to obtain $Con(y_i)$ it is often based at a higher level on some cognitive/semantic structure that underlies the desired meaning of composed concept.

We can view formula (I) as a module of type shown in Fig. 1.

Using the preceding ideas we now provide a hierarchical framework that can be used for obtaining **rich concepts** from primal properties of the database objects. In the following we shall Fig. 2 useful in our discussion.

In the Fig. 2 a box or module is used to indicate a concept that is composed by the aggregation of its constituent concepts. A circle is used to indicate an atomic concept.

Here we construct the hierarchy starting with a concept of interest, concept 1, which is then defined in terms of the aggregation of other concepts, it's constituent concepts, in this case concepts 2 and concept 3. We continue in this fashion until we obtain a concept that is an atomic concept, which stops the processing. We shall refer

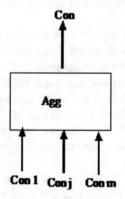

Fig. 1 Structure of composed concept

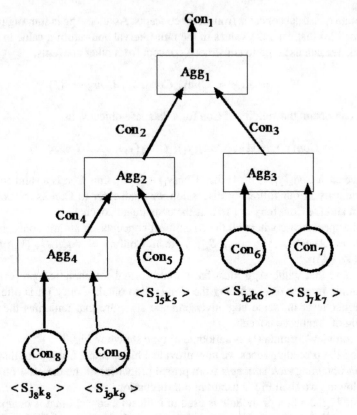

Fig. 2 Hierarchical framework for constructing *rich concepts*

to such a hierarchy as a "**rich concept**". Each branch in the hierarchy terminates in an atomic concept V_j is S_{jk} having the associated attribute V_j. We note that a_{ji} the value of the attribute V_j for object y_i, $V_j(y_i)$, can be obtained directly from the database. We further note that for object y_i the truth-value of this branch terminating atomic proposition, V_j is S_{jk}, can be obtained as $S_{jk}(a_{ji})$, the membership grade of a_{ji} in the fuzzy subset S_{jk}.

The determination of the degree of truth of a rich concept for a database object y_i is obtained in the opposite direction. We start at the bottom of the hierarchy and determine the truth of each terminal atomic concept for object y_i. As we just indicated these could be directly obtained using the attribute values associated with the object that are in the database. Once having these we can use the structure of the hierarchy to determine the degree to which y_i satisfies the rich concept defined by the hierarchy. Specifically the truth of a composed concept emanating from Agg module in the hierarchy is obtained using the prescribed aggregation of the truth-values of the input constituent concepts.

Thus using the structure of the hierarchical description of the rich concept along with the information contained in the database about with the objects, the y_i, we can obtain, t_i, the degree to which y_i satisfies the rich concept RC. Here then $RC(y_i) = t_i$. Using the collection of these t_i's from each of the elements in the database we are able to determine τ the validity of the linguistic summary "Q objects in the database have property **Rich Concept**."

Let us now consider qualified linguistic summaries that are exemplified by statements of the form **Most tall** people in the database are *young*. This statement has a generic form:

Q of the U is **R** objects in the database have V is **S**.

We see here that both U is R and V is S are examples of atomic concepts. This inspires us to consider linguistic summaries of the form.

Q of the objects in the database that have property **Concept-1** also have Property **Concept-2**.

Here then each of Concept 1 and Concept 2 can be defined using a concept hierarchy of the type described earlier which allows the determination of $Con_1(y_i)$ and $Con_2(y_1)$, the truths of **Concept-1** and **Concept-2** for each of the objects in Y. Using this we can the validity τ of the preceding linguistic summary as follows:

(1) For each y_i calculated $Con_1(y_1)$ and $Con_2(y_1)$

(2)

$$r = \frac{\sum_{i=1}^{n} Con_1(y_i)\,(Con_2(y_i)}{\sum_{i=1}^{n} Con_1(y_i)}.$$

(3) $\tau = Q(r)$

5 Conclusion

In the limited space available we provide the fundamental ideas involved in linguistic summaries and looked at the formulation of summaries involving rich concepts. What should be pointed out is that linguistic summaries can have a central role in the task of mining the immense amount of data that is now available. One potential step in this direction is the bringing together of the ideas inherent in the linguistic summary paradigm with the data intense technology of deep learning.

References

1. Yager, R.R.: On linguistic summaries of data. In: Proceedings of IJCAI Workshop on Knowledge Discovery in Databases, Detroit, pp. 378–389 (1989)

2. Yager, R.R.: On linguistic summaries of data. In: Piatetsky-Shapiro, G., Frawley, B. (eds.) Knowledge Discovery in Databases, pp. 347–363. MIT Press, Cambridge, MA (1991)
3. Kacprzyk, J., Yager, R.R., Zadrozny, S.: A fuzzy logic based approach to linguistic summaries in databases. Int. J. Appl. Mathematic. Comp. Sci. **10**, 813–834 (2000)
4. Kacprzyk, J., Yager, R.R.: Linguistic summaries of data using fuzzy logic. Int. J. Gen Syst **30**, 133–154 (2001)
5. Kacprzyk, J., Yager R.R., Zadrozny S.: Fuzzy linguistic summaries of databases for efficient business data analysis and decision support. In: Abramowicz, W., Zaruda, J. (eds.) Knowledge Discovery for Business Information Systems, pp. 129–152. Kluwer Academic Publishers, Hingham, MA (2001)
6. Kacprzyk, J., Zadrozny, S.: Computing with words is an implementable paradigm: fuzzy queries, linguistic data summaries, and natural-language generation. IEEE Trans. Fuzzy Syst. **18**, 461–472 (2010)
7. Bouchon-Meunier, B., Moyse, G.: Fuzzy linguistic summaries: Where are we, where can we go? In: Proceedings of the 2012 IEEE Conference on Computational Intelligence for Financial Engineering and Economics, pp. 1–8, CIFEr 2012, New York, NY, USA (2012)
8. Almeida, R., Lesot, M.-J., Bouchon-Meunier, B., Kaymak, U., Moyse, G.: Linguistic summaries of categorical time series for septic shock patient data. In: Fuzz-IEEE 2013 - IEEE International Conference on Fuzzy Systems, pp. 1–8, Hyderabad, India (2013)
9. Moyse, G., Lesot, M.-J., Bouchon-Meunier, B.: Linguistic summaries for periodicity detection based on mathematical morphology. In: IEEE Symposium Series on Computational Intelligence, pp. 106–113, Singapore (2013)
10. Moyse, G., Lesot, M.J., Bouchon-Meunier, B.: Mathematical morphology tools to evaluate periodic linguistic summaries," Flexible Query Answering Systems, vol. 8132, pp. 257–268. Lecture Notes in Computer Science, Granada, Spain (2013)
11. Moyse, G., Lesot, M.-J., Bouchon-Meunier, B.: Oppositions in fuzzy linguistic summaries. In: FUZZ-IEEE'15 - IEEE International Conference on Fuzzy Systems, pp. 1–8. Istanbul, Turkey (2015)
12. Lesot, M.J., Moyse, G., Bouchon-Meunier, B.: Interpretability of fuzzy linguistic summaries. Fuzzy Sets Syst. **292**, 307–317 (2016)
13. Moyse, G., Lesot, M.J.: Linguistic summaries of locally periodic time series. Fuzzy Sets Syst. **285**, 94–117 (2016)
14. Boran, F.E., Akay, D., Yager, R.R.: A probabilistic framework for interval type-2 fuzzy linguistic summarization. IEEE Trans. Fuzzy Syst. **22**, 1640–1653 (2014)
15. Boran, F.E., Akay, D., Yager, R.R.: An overview of methods for linguistic summarization with fuzzy sets. Expert Syst. Appl. **61**, 356–377 (2016)
16. Wilbik, A., Keller, J.M.: A fuzzy measure similarity between sets of linguistic summaries. IEEE Trans. Fuzzy Syst. **21**, 183–189 (2013)
17. Yager, R.R.: On Z-valuations using Zadeh's Z-numbers. Int. J. Intell. Syst. **27**, 259–278 (2012)
18. Zadeh, L.A.: A note on Z-numbers. Inf. Sci. **181**, 2923–2932 (2011)
19. Zadeh, L.A.: A computational approach to fuzzy quantifiers in natural languages. Comput Mathemat Applic **9**, 149–184 (1983)
20. Yager, R.R.: Reasoning with fuzzy quantified statements: part I. Kybernetes **14**, 233–240 (1985)
21. Yager, R.R.: On the specificity of a possibility distribution. Fuzzy Sets Syst. **50**, 279–292 (1992a)
22. Yager, R.R.: Default knowledge and measures of specificity. Inf. Sci. **61**, 1–44 (1992b)
23. Beliakov, G., Pradera, A., Calvo, T.: Aggregation functions: a guide for practitioners. Springer, Heidelberg (2007)

Graduality in Data Sciences: Gradual Patterns

Anne Laurent

Abstract Graduality is a key concept when dealing with imprecision and uncertainty. There would thus have been numerous manners to deal with graduality. This chapter however focuses on gradual patterns of the form of "the more/less A_1, ..., the more/less A_n". The study of such patterns has emerged very early with the study of rule-based systems but at this time they were provided by experts. The automatic extraction of gradual patterns from databases has then benefited from the progress of data mining (in the sense of pattern mining) and has then generated works from several teams.

1 Introduction

Gradual patterns [5, 11, 17] extract knowledge from numerical data bases as attribute co-variations allowing linguistic representations of the form "the more A increases, the more B increases", or equivalently, "the higher A, the higher B", where A and B are numerical attributes.

Whereas classic pattern mining applies to transactional data, described by binary attributes denoting the presence or absence of each item, gradual patterns (GP) are extracted from numerical data, described by real values associated to numerical features. Imposing constraints across the whole data set, GP must be distinguished from fuzzy gradual rules that impose constraints on the attribute values for each data point individually [6, 13, 15]. These rules are not considered further in this paper.

In the following, $\mathscr{D}$ denotes the considered data set, constituted of n objects described by m numerical attributes; A denoting an attribute, for any object $x \in \mathscr{D}$, $A(x)$ denotes the value A takes for x. A sample data, taken from [11] is given by Table 1.

A. Laurent (✉)
LIRMM, Université de Montpellier, CNRS, Montpellier, France
e-mail: anne.laurent@umontpellier.fr

163

M.-J. Lesot and C. Marsala (eds.), *Fuzzy Approaches for Soft Computing and Approximate Reasoning: Theories and Applications*, Studies in Fuzziness and Soft Computing 394,
https://doi.org/10.1007/978-3-030-54341-9_14

Table 1 Touristic sample table

	Hostel	Town	Pop. (10^3)	Dist. from centre	Price
X_{h_1}	h_1	Paris	2.1	0.3	82
X_{h_2}	h_2	New York	8.0	5	25
X_{h_3}	h_3	New York	8.0	0.2	135
X_{h_4}	h_4	Ocala	0.04	0.1	60

2 Definitions

As given by [5, 11], the formal definitions of gradual items and gradual patterns are as follows:

Definition 1 A *gradual item*, denoted by $A^\uparrow$ or $A^\downarrow$, is a pair made of an attribute A and a variation denoted by $\uparrow$ or $\downarrow$.

Note that other notations have been defined in the literature, such as A^+, A^- or $A^\geq$, $A^\leq$.

If A is an attribute corresponding to the user age for instance, $A^\uparrow$ and $A^\downarrow$ are the gradual items that can be linguistically expressed as *the older* and *the younger* respectively. If A corresponds to the last column of the sample data illustrated in Table 1, i.e. price, then $A^\uparrow$ and $A^\downarrow$ can respectively be linguistically expressed as *the more expensive* and *the less expensive*.

Definition 2 A *gradual pattern* M is a combination of several gradual items and denoted $M = \{A_i^{*_i}, i = 1, \ldots, k\}$, where $*_i \in \{\uparrow, \downarrow\}$ for all $i \in [1, k]$.

The number of attributes M involves, k, is called its length.

A gradual pattern is semantically interpreted as the conjunction of its gradual items: for instance $M = A^\uparrow B^\downarrow$ is interpreted as *the more A and the less B*.

A gradual pattern $M = \{A_i^{*_i}, i = 1, \ldots, k\}$ therefore imposes a variation constraint on several attributes simultaneously. It induces an order on objects, denoted $\preceq_M$, defined as $o \preceq_M o'$ iff $\forall i \in [1, k]$, $A_i(o) *_i A_i(o')$, where in this mathematical expression $\uparrow$ is replaced with $\geq$ and $\downarrow$ with $\leq$.

3 Quality Criterion

The quality of a pattern is measured as the extent to which it holds for a given data set, and is assessed as its *support*.

Two main approaches for the support definition can be distinguished in the case of gradual patterns: the first interpretation takes into account attribute values, e.g. performing regression analysis. The support of a candidate gradual pattern can then be measured as the quality of the regression, combined to the slope of the line in the case of a linear regression [16]. This approach requires to define numerical combinations of attribute values and in particular applies to fuzzy data, where the features correspond to membership degrees to various fuzzy modalities.

A second interpretation only considers the order induced by the attribute values, ignoring their values; it can, in turn, be decomposed into two main approaches. The compliant subset approach [10, 11] identifies data subsets $\mathscr{D}^*$ that can be ordered so that all couples from $\mathscr{D}^*$ satisfy the pattern induced order. Formally, the support is defined as

$$supp(M) = \frac{1}{|\mathscr{D}|} \max_{\mathscr{D}^* \in \mathscr{L}(M)} |\mathscr{D}^*| \tag{1}$$

where $\mathscr{L}(M)$ denotes the set of all maximal subsets $\mathscr{D}^* = \{x_1, \ldots, x_m\} \subseteq \mathscr{D}$ for which there exists a permutation π such that $\forall l \in [1, m-1], x_{\pi_l} \preceq_M x_{\pi_{l+1}}$.

The rank correlation approach [5, 17] considers a more local view, focused on data couples instead of data subsets: it counts the number of data couples that satisfy the order induced by the pattern. Formally, its support is defined as

$$supp(M) = \frac{|\{(x, x') \in \mathscr{D}^2 / x \preceq_M x'\}|}{|\mathscr{D}|(|\mathscr{D}| - 1)} \tag{2}$$

Despite their interpretation differences [7], all these support definitions satisfy the classic anti-monotonicity property, allowing for efficient algorithms to extract frequent gradual patterns.

Two specific features of gradual patterns as opposed to classic patterns must be underlined: both in terms of data and attributes, they focus on pairs and not individuals.

Indeed, as can be seen from the order they induce, gradual patterns apply to data pairs, which significantly increases the computational complexity of their processing: mining gradual patterns can be interpreted as mining classic patterns in a rewritten data base, transforming the data to a transactional form that contains a transaction for each data couple [5]. The approach explicitly building this transformed data set requires approximations to keep tractable extraction processes [5]. An alternative approach solves the crucial data representation issue exploiting a representation by means of concordance matrices, that indicate for each data couple whether it satisfies a considered gradual pattern. These matrices allow highly efficient process through bitmap operations [11, 17]. An example of such matrices is given by Fig. 1.

	h_1	h_2	h_3	h_4
h_1	0	1	1	0
h_2	0	0	1	0
h_3	0	1	0	0
h_4	1	1	1	0

Fig. 1 Binary matrix for $Pop^\uparrow$ from the database in Table 1

In addition to this data pair specificity, gradual patterns also focus on attribute pairs: elementary gradual patterns are actually of length 2. Indeed gradual patterns of length 1 do not impose constraints, as any object pair can be trivially ordered to satisfy them. As a consequence, in the explicit data transformation approach [5], items are built for all pairs of gradual items. This approach therefore altogether leads to a transformed transaction base with $n(n-1)/2$ rows, one for each data pair, and $m(m-1)$ columns, to represent all possible 2-gradual items $A^\uparrow B^\uparrow$ and $A^\uparrow B^\downarrow$ for each pair of attributes AB. Note that the gradual itemsets $A^\downarrow B^\downarrow$ and $A^\downarrow B^\uparrow$ can be considered as equivalent to $A^\uparrow B^\uparrow$ and $A^\uparrow B^\downarrow$ respectively, as they induce the reverse orders and are supported by the same data pairs.

4   Advanced Protoforms for Gradual Patterns

Many discussions can be opened, as for instance the semantic of the clauses, taken as covariations or taken as an acceleration [22]. "All the more clauses" have been considered in [7] for building patterns like "the closer the wall, the harder the brakes are applied, all the more the higher the speed".

Gradual patterns have also been extended for classification in [9].

Fuzzy attributes have been considered with fuzzy gradual patterns [4]. Hierarchies can be managed with multiple levels gradual patterns [1]. Stream data have been considered in [20].

The management of fuzzy orderings is considered in [14] because in many application domains, it is hardly possible to consider that data values are crisply ordered. When considering gene expression, it is not true from the biological point of view that Gene 1 is more expressed than Gene 2 if the levels of expression only differ from the tenth decimal. The work thus considers fuzzy orderings and fuzzy gamma rank correlation.

Closed gradual patterns have been studied in [3] with the aim of producing condensed representations of the sets of gradual patterns with good computation performances [12]. Such closed patterns are considered for dealing with ermerging gradual patterns [18], defined as gradual patterns that describe a data set *in opposition to* a reference data set, i.e. occur in a data set but not the other one. Such patterns aim at characterising the specificity of the considered data in terms of attribute co-variations and highlighting its differences with respect to reference data. In the case where the two data sets correspond to different dates, emerging gradual patterns allow to adapt to the data evolution over time and underline their changes.

Gradual patterns like "the higher the number of inhabitants, the higher the degree of humidity, the higher the number of cases of the disease" can be mixed with spatial maps so as to provide end-users with visualizations [2].

Castelltort et al. [8] proposes to exploit heterogeneous data, i.e. data described by both numerical and categorical features, so as to gain knowledge about the categorical attributes from the numerical ones. The work is based on gradual patterns and on mathematical morphology tools, as [21] also did some time before.

The advanced protoforms are still studied, the existing work having opened many research avenues.

5 Conclusion

Graduality is a key concept when dealing with data and knowledge and has attracted many works. Gradual patterns are one of the topics that have been addressed, first as rules being provided by experts and managed in expert systems, and then automatically extracted by data mining algorithms as presented in this chapter. Several fields must be crossed to deal with gradual pattern mining, from databases to cognitive science. Recent advances in big data and the need for interpretable models and methods in artificial intelligence may increase their use.

References

1. Aryadinata, Y.S., Castelltort, A., Laurent, A., Sala, M.: M2LFGP: mining gradual patterns over fuzzy multiple levels. In: Larsen, H.L., Martín-Bautista, M.J., Vila, M.A., Andreasen, T., Christiansen, H. (eds.) Flexible Query Answering Systems—Proceedings of the 10th International Conference, FQAS 2013, Granada, Spain, September 18–20, 2013. Lecture Notes in Computer Science, vol. 8132, pp. 437–446. Springer (2013). https://doi.org/10.1007/978-3-642-40769-7_38
2. Aryadinata, Y.S., Lin, Y., Barcellos, C., Laurent, A., Libourel, T.: Mining epidemiological dengue fever data from Brazil: a gradual pattern based geographical information system. In: Laurent, A. et al. [19], pp. 414–423. https://doi.org/10.1007/978-3-319-08855-6_42
3. Ayouni, S., Laurent, A., Yahia, S.B., Poncelet, P.: Mining closed gradual patterns. In: Rutkowski, L., Scherer, R., Tadeusiewicz, R., Zadeh, L.A., Zurada, J.M. (eds.) Artificial Intelligence and Soft Computing, 10th International Conference, ICAISC 2010, Zakopane, Poland, June 13–17, 2010. Part I, Lecture Notes in Computer Science, vol. 6113, pp. 267–274. Springer (2010). https://doi.org/10.1007/978-3-642-13208-7_34
4. Ayouni, S., Yahia, S.B., Laurent, A., Poncelet, P.: Fuzzy gradual patterns: what fuzzy modality for what result? In: Martin, T.P., Muda, A.K., Abraham, A., Prade, H., Laurent, A., Laurent, D., Sans, V. (eds.) Second International Conference of Soft Computing and Pattern Recognition, SoCPaR 2010, Cergy Pontoise/Paris, France, December 7–10, 2010, pp. 224–230. IEEE (2010). https://doi.org/10.1109/SOCPAR.2010.5686082
5. Berzal, F., Cubero, J.C., Sanchez, D., Vila, M.A., Serrano, J.M.: An alternative approach to discover gradual dependencies. Int. J. Uncertain. Fuzziness Knowl. Based Syst. (IJUFKS) 15(5), 559–570 (2007)
6. Bouchon-Meunier, B., Desprès, S.: Acquisition numérique/symbolique de connaissances graduelles. In: 3èmes Journées Nationales du PRC Intelligence Artificielle, pp. 127–138. Hermès (1990)
7. Bouchon-Meunier, B., Laurent, A., Lesot, M.J., Rifqi, M.: Strengthening fuzzy gradual rules through "all the more" clauses. In: FUZZ-IEEE 2010, Proceedings of the IEEE International Conference on Fuzzy Systems, Barcelona, Spain, 18–23 July, 2010, pp. 1–7. IEEE (2010). https://doi.org/10.1109/FUZZY.2010.5584858

8. Castelltort, A., Laurent, A., Lesot, M.J., Marsala, C., Rifqi, M.: Discovering ordinal attributes through gradual patterns, morphological filters and rank discrimination measures. In: Proceedings of the 12th International Conference on Scalable Uncertainty Management (SUM) (2018)
9. Choong, Y., Di Jorio, L., Laurent, A., Laurent, D., Teisseire, M.: CBGP: classification based on gradual patterns. In: Abraham, A., Muda, A.K., Herman, N.S., Shamsuddin, S.M., Choo, Y. (eds.) First International Conference of Soft Computing and Pattern Recognition, SoCPaR 2009, Malacca, Malaysia, December 4–7, 2009, pp. 7–12. IEEE Computer Society (2009). https://doi.org/10.1109/SoCPaR.2009.15
10. Di Jorio, L., Laurent, A., Teisseire, M.: Fast extraction of gradual association rules: a heuristic based method. In: Proceedings of the IEEE/ACM International Conferences on Soft Computing as a Transdisciplinary Science and Technology, CSTST'08 (2008)
11. Di Jorio, L., Laurent, A., Teisseire, M.: Mining frequent gradual itemsets from large databases. In: Proceedings of the International Conference on Intelligent Data Analysis, IDA'09 (2009)
12. Do, T.D.T., Termier, A., Laurent, A., Négrevergne, B., Omidvar-Tehrani, B., Amer-Yahia, S.: PGLCM: efficient parallel mining of closed frequent gradual itemsets. Knowl. Inf. Syst. **43**(3), 497–527 (2015). https://doi.org/10.1007/s10115-014-0749-8
13. Dubois, D., Prade, H.: Gradual inference rules in approximate reasoning. Inform. Sci. **61**(1–2), 103–122 (1992). https://doi.org/10.1016/0020-0255(92)90035-7
14. Flores, P.M.Q., Laurent, A., Poncelet, P.: Fuzzy orderings for fuzzy gradual patterns. In: Christiansen, H., Tré, G.D., Yazici, A., Zadrozny, S., Andreasen, T., Larsen, H.L. (eds.) Flexible Query Answering Systems—Proceedings of the 9th International Conference, FQAS 2011, Ghent, Belgium, October 26–28, 2011. Lecture Notes in Computer Science, vol. 7022, pp. 330–341. Springer (2011). https://doi.org/10.1007/978-3-642-24764-4_29
15. Hüllermeier, E.: Implication-based fuzzy association rules. In: Proceedings of PKDD'01, pp. 241–252 (2001)
16. Hüllermeier, E.: Association rules for expressing gradual dependencies. In: Proceedings of the 6th European Conference on Principles of Data Mining and Knowledge Processing, PKDD'02, pp. 200–211. Springer-Verlag (2002)
17. Laurent, A., Lesot, M.J., Rifqi, M.: GRAANK: exploiting rank correlations for extracting gradual itemsets. In: Proceedings of FQAS, pp. 382–393 (2009)
18. Laurent, A., Lesot, M.J., Rifqi, M.: Mining emerging gradual patterns. In: Alonso, J.M., Bustince, H., Reformat, M. (eds.) 2015 Conference of the International Fuzzy Systems Association and the European Society for Fuzzy Logic and Technology (IFSA-EUSFLAT-15), Gijón, Spain, June 30, 2015. Atlantis Press (2015)
19. Laurent, A., Strauss, O., Bouchon-Meunier, B., Yager, R.R. (eds.): Information processing and management of uncertainty in knowledge-based systems—Proceedings of the 15th International Conference, IPMU 2014, Montpellier, France, July 15–19, 2014. Part II, Communications in Computer and Information Science, vol. 443. Springer (2014). https://doi.org/10.1007/978-3-319-08855-6
20. Nin, J., Laurent, A., Poncelet, P.: Speed up gradual rule mining from stream data! A b-tree and OWA-based approach. J. Intell. Inf. Syst. **35**(3), 447–463 (2010). https://doi.org/10.1007/s10844-009-0112-9
21. Oudni, A., Lesot, M.J., Rifqi, M.: Characterisation of gradual itemsets based on mathematical morphology tools. In: Montero, J., Pasi, G., Ciucci, D. (eds.) Proceedings of the 8th Conference of the European Society for Fuzzy Logic and Technology, EUSFLAT-13, Milano, Italy, September 11–13, 2013. Atlantis Press (2013). https://doi.org/10.2991/eusflat.2013.122
22. Oudni, A., Lesot, M.J., Rifqi, M.: Accelerating effect of attribute variations: accelerated gradual itemsets extraction. In: Laurent, A. et al. [19], pp. 395–404. https://doi.org/10.1007/978-3-319-08855-6_40

Evolving Systems

Fernando Gomide, Andre Lemos, and Walmir Caminhas

Abstract Evolving systems emerge from the synergy between systems with adaptive structures, and the recursive methods of machine learning. Evolving algorithms construct models and derive decision patterns from stream data produced by dynamically changing environments. Different components can be chosen to assemble the system structure, rules, trees, and neural networks being amongst the most prominent. Evolving systems concern mainly with time-varying environments, and processing of nonstationary stream data using computationally efficient recursive algorithms. They are particularly suitable for on-line, real-time applications, and dynamically changing situations, and operating conditions. This chapter gives an overview of evolving systems focusing on the model components, learning algorithms, and illustrative applications. The aim of to introduce the main ideas and a state of the art view of the area.

1 Introduction

The behavior of evolving systems result from the interactions among local subsystems and components aiming at lifelong self-organization of the system structure and parameter estimation to adapt to unknown environments, and to detect temporal

F. Gomide (✉)
Department of Computer Engineering and Automation, University of Campinas, Campinas, Brazil
e-mail: gomide@dca.fee.unicamp.br

A. Lemos · W. Caminhas
Department of Electronic Engineering, Federal University of Minas Gerais, Belo Horizonte, Brazil
e-mail: andrepaim@gmail.com

W. Caminhas
e-mail: caminhas@cpdee.ufmg.br

M.-J. Lesot and C. Marsala (eds.), *Fuzzy Approaches for Soft Computing and Approximate Reasoning: Theories and Applications*, Studies in Fuzziness and Soft Computing 394, https://doi.org/10.1007/978-3-030-54341-9_15

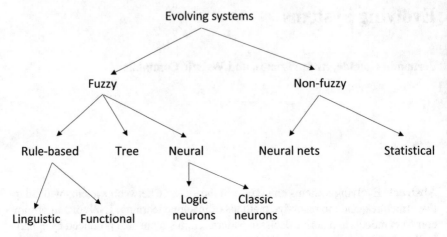

Fig. 1 Different types of evolving systems

shifts and drifts in input data. Applications prevails in modeling, control, prediction, classification, fault diagnosis, anomaly detection, and pattern recognition.

Mining nonstationary data streams bring unique issues and challenges for online data science and analytics. For instance, industrial machines suffer from stress, aging, and faults; economic performance indicators such as stock indices vary at different rates; communication systems transmission capacity and responsiveness also are subject to continuous changes.

Pioneering works in the area were focused in neural networks [6, 23], fuzzy rule-based systems [1, 15, 18], and neural-fuzzy hybrids [10, 16]. Recent advances, and generalizations of previous work include statistic-based models [7], granular computing models [11], and deep learning. Figure 1 shows the different categories of evolving systems that have been developed. Applications in intelligent sensing and actuators [2, 19], autonomous unmanned systems [3], process monitoring and control [5, 12, 18, 22], biomedic data processing have been reported [9].

The term *evolving* is used here to mean systems that simultaneously learn and adapt their structure and parameters endlessly, trading off short term adaptability and long term survivability.

One of the first evolving models is the evolving Takagi-Sugeno model [1], a functional fuzzy rule-based model whose structure and parameters are continuously adapted using recursive clustering, and recursive least squares, respectively. The idea is to assign to each cluster a local functional model. The collection of fuzzy rules and their parameters assemble the model. Many conceptually similar approaches are found in the literature [1, 10, 17, 18]. They differ mostly in way that recursive clustering is performed. A distinctive example is the participatory learning recursive clustering [17]. A multivariable evolving approach using the participatory learning approach is found in [15] to account for interactions amongst input variables, and to attenuate the curse of dimensionality during clustering. An evolving fuzzy regression tree model is given in [14]. The tree uses an affine function at each leaf whose parameters are estimated using the recursive least squares algorithm. The tree structure is updated using a statistical selection procedure based on a hypothesis test.

The remainder of the chapter is organized as follows. Next section gives examples of powerful rule-based evolving systems. Section 3 overviews evolving neural fuzzy systems architectures and learning approaches while Sect. 4 outlines an evolving fuzzy regression tree procedure. Section 5 concludes the chapter and elaborates on the role of evolving mechanisms in future intelligent data modeling and analysis systems.

2   Rule-Based Evolving Systems

The issue of adapting structure and parameters automatically dates early 90' especially in the field of neural networks [6]. Rule-based evolving systems have advantages over evolving black-box models because they are more understandable and can be linguistically explained.

Several works in literature develop evolving fuzzy models to address system identification, time series forecasting, pattern classification, and control. Many of these works [1, 10, 17] use functional fuzzy, Takagi-Sugeno rules as system components. Whenever data is input, evolving fuzzy rule-based models learn rule antecedents using unsupervised recursive clustering algorithms. Figure 2 shows the idea of structural adaptation in evolving systems. Functional fuzzy rule consequent parameters are estimated using recursive least squares or its variations.

For instance, functional affine fuzzy models are a set of rules of the form:

$$R_i \quad : \quad \text{If } x_1 \text{ is } A_{i1} \text{ and } \cdots \text{ and } x_m \text{ is } A_{im} \text{ then } y_i = \lambda_{i0}x_1 + \cdots + \lambda_{im}x_m$$

where R_i is the ith fuzzy rule, for $i = 1, \cdots, c^k$, c^k is the number of fuzzy rules at step k, x_j for $j = 1, \cdots, m$ are the input variables, A_{ij} are the antecedent fuzzy sets, y_i is the $i - th$ model output, and λ_{ij} are the parameters of the $i - th$ local model. The model output is the weighted average of the local models. The weight of each

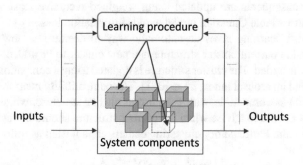

Fig. 2 Evolving systems

local model is found from the membership degrees of the input data in the fuzzy sets of the antecedent (If part) of the rule.

Unsupervised recursive clustering algorithms based on potential P can be used to determine the number of the fuzzy rules of a model. Each cluster dictates a fuzzy rule, the one cluster-one rule paradigm. Clustering partitions the data space and forms the fuzzy sets of rule antecedents, while local functions compose the rule consequents. Clustering is done in the input-output data space in which data points are $z = [x^T \ y]^T$. The potential $P(z^k)$ of a datum z^k is a measure of its distance to all other data samples at step k. It can be computed recursively [1]. The Algorithm 2 summarizes the functional system learning algorithm, where $\mathbf{V^k}$ is the $m \times c^k$ center matrix of the c^k clusters at k.

Algorithm 2 Evolving functional learning algorithm

   Input : z^k, $\mathbf{V^k}$
   Output: $\mathbf{V^{k+1}}$
1: Compute the new data sample potential $P(z^k)$
2: **for** $i = 1, \cdots, c^k$ **do**
3:    Compute the center v_i^k potential
4: **end for**
5: **if** $P(z^k) > P(v_i^k)$ $\forall i$ **then**
6:    **if** z^k is close enough to some cluster i **then**
7:       z^k replaces v_i^k as the center of cluster i
8:    **else**
9:       A new cluster is created centered in z^k
10:   **end if**
11: **else**
12:   Update the consequent parameters of the closest cluster
13: **end if**
14: **return** $\mathbf{V^{k+1}}$

A powerful evolving fuzzy model is the evolving multivariable Gaussian developed in [15]. The models uses a clustering algorithm derived from the participatory learning [24] paradigm. Differently of previous neural and rule-based models [1, 10, 18], the clustering procedure used here assumes interactive input variables, and trigger ellipsoidal clusters whose axes are not necessarily axis parallel. Coefficients of the rule consequents are updated using weighted recursive least squares. The evolving multivariable Gaussian model avoids information loss.

Participatory learning gives an automatic way to decide if a new observation lying far outside current cluster structure is a new cluster to be added, or if it is an outlier to be discarded. The cluster structure is updated using a compatibility measure $\rho_i^k \in [0, 1]$ and an arousal index, $a_i^k \in [0, 1]$. The compatibility measures how much an observation is compatible with the current cluster structure, while the arousal index acts as a critic to advise when the current structure should be revised in front of new input data. Participatory clustering centers are adjusted as follows:

$$v_i^{k+1} = v_i^k + \beta(\rho_i^k)^{1-a_i^k}(x^k - v_i^k) \tag{1}$$

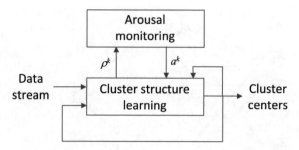

Fig. 3 Participatory clustering

where $\beta \in [0, 1]$ is the learning rate. Figure 3 illustrates the idea of participatory clustering.

The evolving multivariable Gaussian model is formed by functional fuzzy rules 2. The parameters of the consequent are updated using the weighted recursive least squares algorithm.

The main steps of the learning algorithm are summarized in Algorithm 3 where $\mathbf{V}^k$ is the $m \times c^k$ center matrix, $\mathbf{\Sigma}^k$ is the $m \times m \times c^k$ dispersion tensor, and $\mathbf{\Lambda}^k$ is the $m \times c^k$ consequent parameters matrix at step k.

Algorithm 3 Evolving Gaussian multivariable learning algorithm

---

Input : z^k, $\mathbf{V}^k$, $\mathbf{\Sigma}^k$, $\mathbf{\Lambda}^k$
Output: $\mathbf{V}^{k+1}$, $\mathbf{\Sigma}^{k+1}$, $\mathbf{\Lambda}^{k+1}$

1: Compute model output
2: Update the cluster structure
3: **if** Cluster was created **then**
4:    Create a new rule
5: **end if**
6: **if** Cluster was modified **then**
7:    Update antecedent parameters of the respective rule using cluster parameters
8:    Update consequent parameters of the respective rule using weighted least squares
9: **end if**
10: **if** Two cluster were merged **then**
11:    Merge corresponding rules
12: **end if**
13: **return** $\mathbf{V}^{k+1}$, $\mathbf{\Sigma}^{k+1}$, $\mathbf{\Lambda}^{k+1}$

---

3 Neural Evolving Fuzzy Systems

The dynamic evolving neural-fuzzy systems [10] is a neural like model functionally similar to evolving rule-based models. The neural structure is developed using a recursive evolving clustering algorithm based on the Euclidean distance. A new cluster is created whenever the distance between new input data and all existing

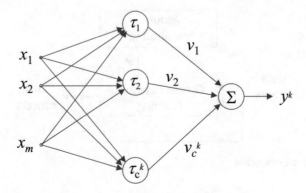

Fig. 4 Self-organizing fuzzy modified least-squares network

cluster centers is greater than a threshold. Clustering proceeds in the input space, and cluster centers is not a particular data sample.

The flexible fuzzy inference system [18] uses a recursive clustering algorithm, an incremental version of the vector quantization technique called evolving vector quantization). Similarly to dynamic evolving neural-fuzzy systems, it uses distances and thresholds to create new clusters. The threshold depends on the input space dimension to attenuate the effect of the data space dimension.

The self-organizing fuzzy modified least-squares network [21] employs an evolving nearest-neighborhood clustering algorithm. It structure is isomorphic to a collection of functional rules. Its topology is a two layer feedforward neural network. The first layer computes the activation degrees of the neurons, and the output layer produces the model output. Figure 4 shows the neural network structure. The self-organizing fuzzy modified least-squares model may remove inactive neurons using the idea of rule density.

The sequential adaptive fuzzy neural inference system also analogous to a functional fuzzy model. It uses the concept of *influence* of a fuzzy neuron to adapt the network structure creating, excluding, and adjusting neurons and respective weights. The influence of a neuron weights the contribution of a hidden layer neuron to the output.

Several other evolving neural fuzzy models can be found in literature with feedforward, and recurrent network structures [4, 16, 20].

4 Evolving Fuzzy Linear Regression Trees

A distinctive evolving fuzzy system is the one in which system components are nodes of a regression trees [14]. Fuzzy regression trees replaces binary splitting decisions at each tree node using pairs of membership functions to partition the input space into overlapping regions. Figure 5 shows an example of a regression tree.

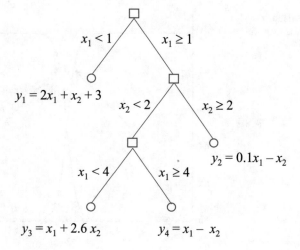

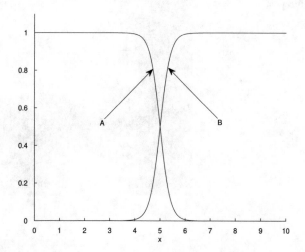

Fig. 5 Example regression tree

There are several algorithms to grow regression trees using incremental learning [8]. While most evolving fuzzy modeling methods use spatial information, the evolving fuzzy tree model granulates the data space selecting split points to improve the goodness of fit of the resulting models. Fuzzy regression trees replaces splitting decision tests by two membership functions to represent *less than* and *greater than*. All branches of the tree will fire to some degree. For instance, sigmoidal membership functions of Fig. 6 and the tree of Fig. 5 partition the input space as shown in Fig. 7.

The tree computes the output as the weighted average of the output of all local models assigned to the leaf nodes.

Fig. 6 Membership
functions for *less than* and
greater than 5

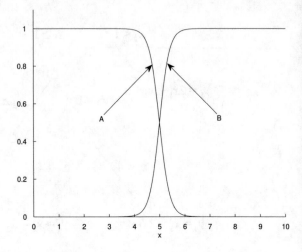

Fig. 7 Input space partition
produced by a fuzzy
regression tree

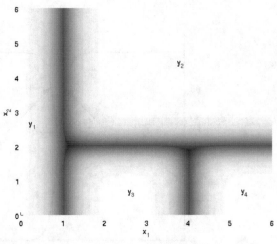

Learning of evolving fuzzy regression trees is incremental and starts with a single
leaf tree and its corresponding local model. The tree evolves as data are input replacing leaves by subtrees using a recursive statistical model selection test. Parameters of
the local regression models are updated using the weighted least squares algorithm.
Algorithm 4 summarizes the evolving tree learning algorithm where $\mathbf{V^k}$ is the $n \times c^k$
tree split points matrix, and $\mathbf{\Lambda}^k$ is the $m \times c^k$ consequent parameters matrix at step k.

Algorithm 4 Evolving regression tree learning algorithm

Input : z^k, $\mathbf{V^k}$, $\mathbf{\Lambda}^k$
Output: $\mathbf{v}^{k+1}$ $\mathbf{\Lambda}^{k+1}$

1: Compute the output and membership value of all leaves
2: Update the linear models
3: Select the leaf with the highest membership value
4: **for all** inputs (m) **do**
5:     **for all** candidate splits (k) **do**
6:         Estimate the output replacing the selected leaf with the candidate split
7:         Compute the *p-value* of the model selection test for the candidate split
8:     **end for**
9: **end for**
10: Select the candidate split with minimum *p-value* for confidence level α
11: **if** *p-value* $< \frac{\alpha}{k \times m}$ **then**
12:     Replace the selected leaf by the candidate split
13: **end if**
14: **return** $\mathbf{V}^{k+1}$ $\mathbf{\Lambda}^{k+1}$

A detailed description of all evolving models, algorithms, and their application in distinct domains is found in [13].

5 Conclusion

The chapter has introduced the main ideas and concepts of evolving systems. It has also overviewed the area focusing on its main system components, and learning algorithms representative of the current state of the art. Because of the steadily increasing growth rate, and the large amount of data in applications domains worldwide, it is safe to predict that soon evolving systems of different forms will become paramount to surpass computational processing power, and storage limits.

References

1. Angelov, P., Filev, D.: An approach to Online identification of Takagi-Sugeno fuzzy models. IEEE Trans. Syst. Man Cybern. B Cybern. **34**(1), 484–498 (2004)
2. Angelov, P., Kordon, A.: Adaptive inferential sensors based on evolving fuzzy models. IEEE Trans. Syst. Man Cybern. B Cybern. **40**(2), 529–539 (2010). https://doi.org/10.1109/TSMCB. 2009.2028315
3. Angelov, P., Ramezani, R., Zhou, X.: Autonomous novelty detection and object tracking in video streams using evolving clustering and takagi-sugeno type neuro-fuzzy system. In: IEEE International Joint Conference on Neural Networks, 2008. IJCNN 2008. (IEEE World Congress on Computational Intelligence), pp. 1456–1463 (2008)
4. Er, M., Wu, S.: A fast learning algorithm for parsimonious fuzzy neural systems. Fuzzy Sets Syst. **126**(3), 337–351 (2002)
5. Filev, D., Tseng, F.: Novelty detection based machine health prognostics. In: 2006 International Symposium on Evolving Fuzzy Systems, pp. 193–199 (2006)

6. Fritzke, B.: Growing cell structures: a self-organizing network for unsupervised and supervised learning. Neural Netw. **7**, 1441–1460 (1994). https://doi.org/10.1016/0893-6080(94)90091-4

7. Hisada, M., Ozawa, S., Zhang, K., Kasabov, N.: Incremental linear discriminant analysis for evolving feature spaces in multitask pattern recognition problems. Evol. Syst. **1**, 17–27 (2010)

8. Ikonomovska E., G.J.R., Gjorgjevik, D.: Regression trees from data streams with drift detection. In: Proceedings of the 12th International Conference on Discovery Science, DS '09, pp. 121–135. Springer-Verlag, Berlin, Heidelberg (2009). https://doi.org/10.1007/978-3-642-04747-3_12

9. Kasabov, N.: Evolving Connectionist Systems: The Knowledge Engineering Approach. Springer-Verlag, New York Inc, Secaucus, NJ, USA (2007)

10. Kasabov, N., Song, Q.: DENFIS: Dynamic evolving neural-fuzzy inference system and its application for time-series prediction. IEEE Trans. Fuzzy Syst. **10**(2), 144–154 (2002)

11. Leite, D., Costa, P., Gomide, F.: Granular approach for evolving system modeling. In: Hullermeier, E., Kruse, R., Hoffmann, F. (eds.) Computational Intelligence for Knowledge-Based Systems Design, Lecture Notes in Computer Science, vol. 6178, pp. 340–349. Springer Berlin, Heidelberg (2010). https://doi.org/10.1007/978-3-642-14049-5_35

12. Lemos, A., Caminhas, W., Gomide, F.: Fuzzy multivariable gaussian evolving approach for fault detection and diagnosis. In: Hullermeier, E., Kruse, R., Hoffmann, F. (eds.) Computational Intelligence for Knowledge-Based Systems Design, Lecture Notes in Computer Science, vol. 6178, pp. 360–369. Springer Berlin, Heidelberg (2010). https://doi.org/10.1007/978-3-642-14049-5_37

13. Lemos, A., Caminhas, W., Gomide, F.: Evolving intelligent systems: methods, algorithms and applications. In: Ramanna, S., Jain, L., Howlett, R. (eds.) Emerging Paradigms in Machine Learning. Springer (2013)

14. Lemos, A., Gomide, F., Caminhas, W.: Fuzzy evolving linear regression trees. Evol. Syst. **2**(1), 1–14 (2011). https://doi.org/10.1007/s12530-011-9028-z

15. Lemos, A., Gomide, F., Caminhas, W.: Multivariable gaussian evolving fuzzy modeling system. IEEE Trans. Fuzzy Syst. **19**(1), 91–104 (2011). https://doi.org/10.1109/TFUZZ.2010.2087381

16. Leng, G., McGinnity, T., Prasad, G.: An approach for on-line extraction of fuzzy rules using a self-organising fuzzy neural network. Fuzzy Sets Syst. **150**(2), 211–243 (2005)

17. Lima, E., Hell, M., Ballini, R., Gomide, F.: Evolving fuzzy modeling using participatory learning. In: Angelov, P., Filev, D., Kasabov, N. (eds.) Evolving Intelligent Systems: Methodology and Applications. Wiley-Interscience, IEEE Press (2010)

18. Lughofer, E.: FLEXFIS: A robust incremental learning approach for evolving Takagi-Sugeno fuzzy models. IEEE Trans. Fuzzy Syst. **16**(6), 1393–1410 (2008)

19. Macias-Hernandez J., A.P., Zhou, X.: Soft sensor for predicting crude oil distillation side streams using Takagi Sugeno evolving fuzzy models. In: IEEE International Conference on Systems, Man, and Cybernetics, pp. 3305–3310 (2007)

20. Rosa R., G.F.D.D., Skrjanc, I.: Evolving neural network with extreme learning for system modeling. In: IEEE Evolving and Adaptive Intelligent Systems, pp. 11–14 (2014)

21. Rubio, J.: Sofmls: Online self-organizing fuzzy modified least-squares network. IEEE Trans. Fuzzy Syst. **17**(6), 1296–1309 (2009). https://doi.org/10.1109/TFUZZ.2009.2029569

22. Wang, W., Vrbanek, J.: An evolving fuzzy predictor for industrial applications. IEEE Trans. Fuzzy Syst **16**(6), 1439–1449 (2008)

23. Williamson, J.: Gaussian ARTMAP: A neural network for past incremental learning of noisy multidimensional maps. Neural Netw. **9**(5), 881–897 (1996)

24. Yager, R.: A model of participatory learning. IEEE Trans. Syst. Man Cybern. B Cybern. **20**(5), 1229–1234 (1990)

Control: Advances on Fuzzy Model-Based Observers

Thierry-Marie Guerra and Miguel Angel Bernal

Abstract This chapter is concerned with recent developments on observer design for non-linear systems via Takagi-Sugeno models, both in continuous- and discrete-time. As traditionally done, the direct Lyapunov method is employed to derive conditions in the form of linear matrix inequalities. Novelties include a better handling of measurable and unmeasurable premises, the inclusion of multiple delays in non-quadratic Lyapunov functions, and the use of matrix properties to add slack variables; feasibility of conditions is thus enhanced while improving the quality of solutions.

1 Motivation

The development of control systems during the twentieth century followed two paths: one relied on the availability of a mathematical model of the plant, the other intended to use the difference between measured outputs and desired signals as the only guidance. Model-free methodologies are widely used in industry, the paradigmatic example being the PID; yet, they require tuning or heuristic considerations to properly work. Model-based theories, on the other hand, are countless and usually mathematically rigorous, but very often of little practical use. Using a model may help avoiding tuning for controller design, but many other steps require making choices, even at a mathematically analytical level. Moreover, once these choices are made, the designer is usually unaware of their suitability or optimality with respect to a number of performance goals. This led to an increasing interest for controller synthesis methods that guarantee a prescribed performance while being computationally efficient, i.e., that

T.-M. Guerra (✉)
Polytechnic University Hauts-de-France, CNRS, UMR 8201, LAMIH, 59313 Valenciennes, France
e-mail: Thierry.Guerra@univ-valenciennes.fr; guerra@univ-valenciennes.fr

M. A. Bernal
Sonora Institute of Technology, 5 de Febrero 818, Sur, Col. Centro, 85000 Ciudad Obregón, Mexico
e-mail: miguel.bernal@itson.edu.mx

© The Editor(s) (if applicable) and The Author(s), under exclusive license to Springer Nature Switzerland AG 2021
M.-J. Lesot and C. Marsala (eds.), *Fuzzy Approaches for Soft Computing and Approximate Reasoning: Theories and Applications*, Studies in Fuzziness and Soft Computing 394, https://doi.org/10.1007/978-3-030-54341-9_16

are able to find a solution or determine it does not exist in polynomial time. Among these methods, those able to cast a control design problem in the form of linear matrix inequalities, which belong to the class of convex optimization problems, are the most appreciated. Once a control problem has been put in such a way, the designer is only concerned to invoke a numerical routine to solve for gains and other variables of interest.

This is the case of up-to-date fuzzy control, which began as a model-free area that intended to implement expert knowledge through fuzzy logic inferences, but has become a full model-based methodology. Thanks to exact convex rewriting of nonlinearities and the direct Lyapunov method, derivation of controller design conditions in the form of linear matrix inequalities for a variety of systems with delays, uncertainties, and so on, has become possible. The dual problem of observation, which is critical for real-time applications where unavailable information might be needed for control tasks or diagnosis, has of course benefited from these developments: its recent advances are the matter of this chapter.

2   Introduction

This chapter discusses recent contributions on state estimation for continuous and discrete-time nonlinear systems via Takagi-Sugeno (TS) models, which originally appeared in the context of model-free fuzzy heuristic systems [22] and gradually moved towards model-based quasi-linear parameter varying (quasi-LPV) systems of the form [23]:

$$\delta x = \sum_{i=1}^{r} h_i(z)(A_i x + B_i u) = A_h x + B_h u$$

$$y = \sum_{i=1}^{r} h_i(z)C_i x = C_h x \tag{1}$$

where δx stands for $\dot{x}$ or $x(t+1)$ in continuous or discrete-time systems, respectively, $x \in \mathbb{R}^{n_x}$ represents the system state vector, $u \in \mathbb{R}^{n_u}$ the input vector, $y \in \mathbb{R}^{n_y}$ the measured output vector; $h_i(\cdot), i \in \{1, 2, \ldots, r\}$ are the membership functions (MFs) which depend on the vector of premise variables $z \in \mathbb{R}^p$; and matrices $A_i \in \mathbb{R}^{n_x \times n_x}$, $B_i \in \mathbb{R}^{n_x \times n_u}$, $C_i \in \mathbb{R}^{n_y \times n_x}$, result from modeling a given nonlinear system, a task usually done via the sector nonlinearity approach [24]. More general configurations may include perturbations, but they are intentionally left out of this presentation.

Generally, when using state space control techniques, not all the states are available (via measurement); therefore, an estimate of the state must be realized using only the inputs and output measured variables. This dynamical estimation is done via an observer that generally is based on model (1). Let us also notice that some conditions are required, so-called observability conditions, to be able to fully reconstruct the system state. The study of these properties is beyond the scope of this chapter, but they are assumed to be satisfied, at least locally.

Two cases can be distinguished for observation: the particular case where the premise variables in z depend on measured variables and the general case where they may depend on unmeasured variables. As with other control problems in the field, observer design with a parallel distributed compensation (PDC) structure [23] for TS models is usually achieved via conditions in the form of linear matrix inequalities (LMIs) which can be efficiently solved in polynomial time [2] via commercially available software [4, 21]. These conditions are derived via the direct Lyapunov method; as with other nonlinear systems, the simplest choice is a quadratic one $V(e) = e^T P e$ with $P = P^T > 0$, with e being the observation error and $P \in \mathbb{R}^{n_x \times n_x}$.

Despite a lot of efforts, in continuous time, there are no real proposition to break away with the quadratic case, but observer design for unmeasured premises remains challenging as the observer structure should handle membership functions of the form $h_i(\hat{z})$, where $\hat{z}$ stands for the premise estimate. Former results on the subject considered the membership function error $h_i(z) - h_i(\hat{z})$ altogether with classical Lipschitz constants [1, 13]; others such as [11, 12] are based on the differential mean value theorem.

In contrast, the discrete case has produced a wide variety of approaches with a lot of interesting proposals for Lyapunov functions such as k-sample ones [14], with multiple previous instants or delayed ones [8].

2.1 Notations and Preliminaries

Notations: As shown in (1), subscripts with h (formerly z) are customary within this field to denote convex matrix sums whose membership functions are h; this is particularly useful to denote delays in time as from $A_h = \sum_{i=1}^r h_i(z(t))A_i$ to $A_{h(t-1)} = \sum_{i=1}^r h_i(z(t-1))A_i$ and $A_{h(t-1)h(t)} = \sum_{i=1}^r \sum_{j=1}^r h_i(z(t-1))h_j(z(t))A_{ij}$. As LMIs involve symmetric matrix expressions, an asterisk $(*)$ is employed to denote a transpose quantity in expressions such as $A_{h(t)}^T P_{h(t)}(*) - P_{h(t-1)} < 0$ instead of

$$A_h^T P_h A_h - P_{h(t-1)} < 0 \text{ or } \begin{bmatrix} -P_{h(t-1)} & (*) \\ A_{h(t)} & -P_{h(t)} \end{bmatrix} < 0 \text{ instead of } \begin{bmatrix} -P_{h(t-1)} & A_{h(t)}^T \\ A_{h(t)} & -P_{h(t)} \end{bmatrix} < 0.$$

Properties: There are several matrix properties that are employed to derive LMI conditions for observer design in the TS context; they range from the very basic congruence or Schur complement to the more complex of Finsler's Lemma or Peaucelle's transformation [19]. The interested reader can find them in the literature and especially in the appendixes of comprehensive books such as [15–17, 23]. Within this chapter, some key properties are explicitly stated if considered relevant instead of providing them beforehand.

Relaxations: Whatever the choice of the Lyapunov function is, the analysis of stability/stabilization often leads to a co-positivity problem, i.e. finding the best way to drop the MFs from nested convex sums in order to get a finite set of LMIs guaranteeing the positive- or negative-definiteness of the original expression. Several sufficient and asymptotically sufficient and necessary conditions can be found for the classical double convex sum $\sum_{i=1}^{r} \sum_{j=1}^{r} h_i h_j x^T \Upsilon_{ij} x > 0$; the interested reader is referred to [20] and the references therein. In more complex schemes the number of nested convex sums may be larger, involving another set of MFs or, even more, convex sums of delayed MFs at different samples (discrete case). Whenever a specific relaxation is employed, it will be briefly presented if pertinent.

3   Measured Premise Variables

In this case, in order to estimate the states which are not available, the following observer, mimicking the TS structure in (1), is proposed considering $\hat{z} = z$:

$$\delta \hat{x} = A_h \hat{x} + B_h u + \mathcal{H}^{-1}(z) \mathcal{K}(z)(y - \hat{y})$$
$$\hat{y} = C_h \hat{x} \tag{2}$$

$\hat{x} \in \mathbb{R}^{n_x}$ is the observer state, $\hat{y} \in \mathbb{R}^{n_y}$ the estimated measured output, $e = x - \hat{x}$ the estimation error, and $\mathcal{H}(z) \in \mathbb{R}^{n_x \times n_x}$, $\mathcal{K}(z) \in \mathbb{R}^{n_x \times n_y}$ are matrix functions of the premise vector z to be designed in the sequel to guarantee that $\lim_{t \to \infty} e(t) = 0$. To ease notation, arguments of these matrix functions will be omitted wherever possible.

As it will be seen in detail, observer matrices $\mathcal{H}$ and $\mathcal{K}$ can be chosen as the corresponding approach requires, for instance, $\mathcal{H}(z) = H_{hh \cdots h}$ and $\mathcal{K}(z) = K_{hh \cdots h}$ for the observer gains with multiple nested summations or $\mathcal{H}(z) = H_h$ and $\mathcal{K}(z) = K_h$ for single sums. The link between Lyapunov function and the observer design can be removed when $\mathcal{H}(z) = H_h$, for instance; this sort of "decoupling" implies more flexibility in the conditions to be satisfied. The notation for "q" multiple nested convex sums, following the one provided in the previous section, will be

$$\Upsilon_{\tilde{h}} = \Upsilon_{\underbrace{hh \cdots h}_{q}} = \sum_{i_1=1}^{r} \sum_{i_2=1}^{r} \cdots \sum_{i_q=1}^{r} h_{i_1}(z) h_{i_2}(z) \cdots h_{i_q}(z) \Upsilon_{i_1 i_2 \cdots i_q}.$$

According to the notation above, the estimation error dynamics is described as:

$$\delta e = \left(A_h - \mathcal{H}^{-1} \mathcal{K} C_h \right) e. \tag{3}$$

3.1 The Continuous Case

Consider the TS model (1) in continuous time, i.e. for (2) $\delta\hat{x} = \dot{\hat{x}}$ and for (3) $\delta e = \dot{e}$ and the following quadratic Lyapunov function (QLF) candidate with $P = P^T > 0$:

$$V(e) = e^T P e. \tag{4}$$

Condition $\dot{V}(e) < 0$ is satisfied if:

$$e^T P \dot{e} + \dot{e}^T P e < 0. \tag{5}$$

The previous condition is equivalent to:

$$P A_h - P \mathcal{H}^{-1} \mathcal{K} C_h + \left(P A_h - P \mathcal{H}^{-1} \mathcal{K} C_h \right)^T < 0, \tag{6}$$

which leads to the next theorem [18]:

Theorem *The estimation error model* (3) *with* $w = 0$ *is asymptotically stable if there exist matrices* $P = P^T > 0$, $R_{i_1 i_2 \cdots i_q}$, $H_{i_1 i_2 \cdots i_q}$, *and* $K_{i_1 i_2 \cdots i_q}$, $i_1, i_2, \ldots, i_q \in \{1, 2, \ldots, r\}$ *of proper dimensions such that the following conditions hold:*

$$\sum_{i_0 i_1 i_2 \cdots i_q \in \rho(i_0 i_1 i_2 \cdots i_q)} \Upsilon_{i_0 i_1 i_2 \cdots i_q} < 0, \quad \forall (i_0, i_1, i_2, \ldots, i_q) \in \{1, 2, \ldots, r\}^{q+1} \tag{7}$$

where $\rho(i_0 i_1 i_2 \cdots i_q)$ *is the set of all terms similar to* $h_{i_1} h_{i_2} \cdots h_{i_q}$ *with*

$$\Upsilon_{i_0 i_1 i_2 \cdots i_q} = \begin{bmatrix} H_{i_1 i_2 \cdots i_q} A_{i_0} - K_{i_1 i_2 \cdots i_q} C_{i_0} + \left(H_{i_1 i_2 \cdots i_q} A_{i_0} - K_{i_1 i_2 \cdots i_q} C_{i_0} \right)^T & (*) \\ P - H_{i_1 i_2 \cdots i_q}^T + R_{i_1 i_2 \cdots i_q}^T A_{i_0} & -R_{i_1 i_2 \cdots i_q} - R_{i_1 i_2 \cdots i_q}^T \end{bmatrix}.$$

Proof The Peaucelle's transformation [19] states that $A^T \mathcal{P} + \mathcal{P}^T A + \mathcal{Q} < 0$ is equivalent to

$$\exists \mathcal{R}, \mathcal{L} : \begin{bmatrix} A^T \mathcal{L}^T + \mathcal{L} A + \mathcal{Q} & \mathcal{P}^T - \mathcal{L} + A^T \mathcal{R} \\ \mathcal{P} - \mathcal{L}^T + \mathcal{R}^T A & -\mathcal{R} - \mathcal{R}^T \end{bmatrix} < 0.$$

Therefore, taking $\mathcal{H}(z) = P$, $\mathcal{K}(z) = K_{\bar{h}}$, and applying this property with $A = A_h$, $\mathcal{L} = H_{\bar{h}}$, $\mathcal{R} = R_{\bar{h}}$, $\mathcal{Q} = -K_{\bar{h}} C_h - C_h^T K_{\bar{h}}^T$ to (6) yields:

$$\begin{bmatrix} H_{\bar{h}} A_h - K_{\bar{h}} C_h + \left(H_{\bar{h}} A_h - K_{\bar{h}} C_h \right)^T & (*) \\ P - H_{\bar{h}}^T + R_{\bar{h}}^T A_h & -R_{\bar{h}} - R_{\bar{h}}^T \end{bmatrix} < 0, \tag{8}$$

after which the relaxation scheme groups similar terms to guarantee the q nested convex sum expression above, thus concluding the proof.

Note that with $K_{\tilde{h}} = K_h$ and pre- and post-multiplying (8) by $\begin{bmatrix} I & A_h \end{bmatrix}$ and its transpose, respectively, the classical conditions are recovered:

$$PA_h - K_hC_h + (PA_h - K_hC_h)^T < 0. \tag{9}$$

3.2 The Discrete Case

We are now assuming the plant has the form (1) with $\delta x = x(t+1)$ while the observer has the structure (2) with $\delta\hat{x} = \hat{x}(t+1)$; the premise variables are still considered measurable. Again, the choice of matrices $\mathcal{H}(z)$ and $\mathcal{K}(z)$ is going to be established once the proper developments are made. The dynamic of the estimation error with $e(t) = x(t) - \hat{x}(t)$ is:

$$e(t+1) = \left(A_h - \mathcal{H}^{-1}(z)\mathcal{K}(z)C_h\right)e(t). \tag{10}$$

The stability analysis is based on the following non-quadratic Lyapunov function:

$$V(e, z) = e^T(t) \sum_{i=1}^{r} h_i(z(t))P_i \, e(t) = e^T(t)P_h e(t). \tag{11}$$

Classical results with membership function based approaches [7] arise directly choosing $\mathcal{H}(z) = H_h$ an $\mathcal{K}(z) = K_h$:

$$\begin{bmatrix} -P_{h(t)} & (*) \\ H_{h(t)}A_{h(t)} - K_{h(t)}C_{h(t)} & -H_{h(t)} - H_{h(t)}^T + P_{h(t+1)} \end{bmatrix} < 0 \tag{12}$$

Direct generalizations involving more complex Lyapunov functions are possible [8]. For instance, $V(e, z) = e^T(t)P_{h(t-1)}e(t)$ $\mathcal{H}(z) = H_{h(t-1)h(t)}$ and $\mathcal{K}(z) = K_{h(t-1)h(t)}$ lead to:

$$\begin{bmatrix} -P_{h(t-1)} & (*) \\ H_{h(t-1)h(t)}A_{h(t)} - K_{h(t-1)h(t)}C_{h(t)} & -H_{h(t-1)h(t)} - H_{h(t-1)h(t)}^T + P_{h(t)} \end{bmatrix} < 0 \tag{13}$$

Notice that the last term of (12) and (13) prove that the inverse $H_{(\cdot)}^{-1}$ is well-defined. Less conservative results can be found in (Guerra et al. [8]).

4 Unmeasured Premise Variables

When not all premise variables are available for measurement, the observer needs to deal with the difference between the estimate $\hat{z}$ and the real premise value z. In [1] sufficient LMI conditions for this case were found which depend on the knowledge of some bound for the difference between the system under the estimated premises $\hat{z}$ and under the real ones z. In [16, 17] a local observer with non-PDC structure was proposed whereas following similar arguments, in [12] the problem was extended to attenuate energy-bounded unknown inputs (disturbances). The following is a more recent development taken from [10] for the undistorted case.

Consider the TS model (1) in continuous time, i.e., $\delta x = \dot{x}$. Following [9] it is convenient to separate the expressions of the functions $h_i(\cdot)$ which depend exclusively on measured premise variables (z_α) and the ones depending on non-measured premise variable (z_β); therefore, (1) writes:

$$\dot{x} = \sum_{i=1}^{r_\alpha} \sum_{j=1}^{r_\beta} \alpha_i(z_\alpha)\beta_j(z_\beta)\big(A_{ij}x + B_{ij}u\big) = A_{\alpha\beta}x + B_{\alpha\beta}u$$

$$y = \sum_{i=1}^{r_\alpha} \sum_{j=1}^{r_\beta} \alpha_i(z_\alpha)\beta_j(z_\beta)C_{ij}x = C_{\alpha\beta}x \qquad (14)$$

$\alpha_i(z_\alpha) \geq 0, i \in \{1, \ldots, 2^{p_\alpha}\}$ depend on measured premise variables $r_\alpha = 2^{p_\alpha} \in \mathbb{N}$, and $\beta_j(z_\beta) \geq 0, j \in \{1, \ldots, 2^{p_\beta}\}$, depend on non-measured premise variable, $r_\beta = 2^{p_\beta} \in \mathbb{N}$. Of course, if the premises are fully measurable we have the correspondence $\alpha_i(z_\alpha) = h_i(z), \beta_j(z_\beta) = 1$; whereas, when none of them is measurable, $\alpha_i(z_\alpha) = 1, \beta_j(z_\beta) = h_j(z)$. Notice also that convex sums are preserved, i.e.: $\sum_{i=1}^{r_\alpha} \sum_{j=1}^{r_\beta} \alpha_i(z_\alpha)\beta_j(z_\beta) = 1, \sum_{i=1}^{r_\alpha} \alpha_i(z_\alpha) = 1$ and $\sum_{i=1}^{r_\beta} \beta_j(z_\beta) = 1$.

Let $\mathcal{O}_{xu}$ denote the operating set of the TS observer. This set covers the region where the TS model (14) perfectly matches the non-linear model and there exist known scalars λ_x and λ_u such that $\|x\| \leq \lambda_x$ and $\|u\| \leq \lambda_u$ hold. From these bounds, scalars σ_j are deduced such that $\|\partial\beta_j/\partial z_\beta\| \leq \sigma_j$ in $\mathcal{O}_{xu}$.

The interest of separating z_α and z_β is to recover the part z_α in the observer and the estimated $\hat{z}_\beta$ as:

$$\dot{\hat{x}} = \sum_{i=1}^{r_\alpha} \sum_{j=1}^{r_\beta} \alpha_i(z_\alpha)\beta_j(\hat{z}_\beta)\big(A_{ij}\hat{x} + B_{ij}u + P^{-1}K_{ij}(y - \hat{y})\big)$$

$$= A_{\alpha\hat{\beta}}\hat{x} + B_{\alpha\hat{\beta}}u + P^{-1}K_{\alpha\hat{\beta}}(y - \hat{y})$$

$$\hat{y} = \sum_{i=1}^{r_\alpha} \sum_{j=1}^{r_\beta} \alpha_i(z_\alpha)\beta_j(\hat{z}_\beta)C_{ij}\hat{x} = C_{\alpha\hat{\beta}}\hat{x}, \qquad (15)$$

with $\hat{x} \in \mathbb{R}^{n_x}$ as the observer state and $K_{\hat{h}} = K_{\alpha\hat{\beta}} \in \mathbb{R}^{n_x \times n_y}$ the matrix functions to be designed in the sequel. For the correspondence between both notations, it should be understood that $h = \alpha\beta$ and $\hat{h} = \alpha\hat{\beta}$, where $\hat{\beta}_j = \beta_j(\hat{z})$.

The following results are obtained using a quadratic Lyapunov function candidate of the form already used before, i.e., $V(e) = e^T P e$, $P = P^T > 0$.

Theorem *The error $e = x - \hat{x}$ defined from* (14) *and* (15) *goes asymptotically to zero for given scalars λ_x, λ_u, and σ_j, if there exist matrices $P = P^T > 0$, K_{kl}, Y_{ijkl}, Z_{ijkl} of proper dimensions and scalars μ_{ijkl}, $i, k \in \{1, \dots, r_\alpha\}$, $j, l \in \{1, \dots, r_\beta\}$ such that the following LMI conditions hold:*

$$\Upsilon_{ijij} < 0,$$

$$\frac{2}{r-1}\Upsilon_{ijij} + \Upsilon_{ijkl} + \Upsilon_{klij} < 0, \quad i \neq k, \ j \neq l \tag{16}$$

With

$$\Upsilon_{ijkl} = \begin{bmatrix} \begin{matrix} (PA_{ij} - K_{kl}C_{ij}) + (PA_{ij} - K_{kl}C_{ij})^T + \mu_{ijkl}\eta^2 T^T T \\ (PA_{i1} - K_{kl}C_{i1} + Y_{ijkl})^T \\ \vdots \\ (PA_{ir_\beta} - K_{kl}C_{ir_\beta} + Y_{ijkl})^T \\ (PB_{i1} + Z_{ijkl})^T \\ \vdots \\ (PB_{ir_\beta} + Z_{ijkl})^T \end{matrix} & \begin{matrix} (*) \\ \\ \\ -\mu_{ijkl}I_{r_\beta(n+m)} \end{matrix} \end{bmatrix},$$

$$\eta \geq \sqrt{(\lambda_x^2 + \lambda_u^2)\sum_{j=1}^{r_\beta}\sigma_j^2}.$$

Proof See [10].

The result above reduces to a linear mapping between the non-measured premises and the state vectors, i.e. $z_\beta = Tx$, $T \in \mathbb{R}^{p_\beta \times n_x}$. Extending the results to a class C^1 nonlinear mapping $z_\beta : \mathbb{R}^{n_x} \to \mathbb{R}^{p_\beta}$ is direct; see [10].

The main advantage of this solution is to provide conditions avoiding the inclusion of Lipschitz constants: that is the main drawback of classical results. Moreover, it has been shown to solve much more difficult cases than previous methods.

5 Example

Consider the following nonlinear model taken from [10]:

$$\dot{x} = \begin{bmatrix} -1.5 - 0.5\xi_1(x) & 0.5a - 1.5\,\xi_2(x) \\ 2 & b - 2 + \xi_2(x) \end{bmatrix} x + \begin{bmatrix} 0 \\ 1 \end{bmatrix} u$$

$$y = \begin{bmatrix} 1 & 0 \end{bmatrix} x,$$

with $\xi_1(x) = x_1^2$ and $\xi_2(x) = \sin(x_1^2 + x_2)$, where x_2 is an unmeasured state; the premise vector with measured (unmeasured) state variables has entries $z_\alpha = x_1^2$ ($z_\beta = x_1^2 + x_2$). Then, a TS model (14) in the compact set $\mathcal{C}_x = \{x \in \mathbb{R}^2 : |x_1| \le 1\}$ can be obtained with:

$$A_{11} = \begin{bmatrix} -1.5 & 0.5a + 1.5 \\ 2 & b - 3 \end{bmatrix}, \quad A_{21} = \begin{bmatrix} -2 & 0.5a + 1.5 \\ 2 & b - 3 \end{bmatrix}, \quad A_{12} =$$

$$\begin{bmatrix} -1.5 & 0.5a - 1.5 \\ 2 & b - 1 \end{bmatrix}, \quad A_{22} = \begin{bmatrix} -2 & 0.5a - 1.5 \\ 2 & b - 1 \end{bmatrix}, \quad B_{11} = B_{12} = B_{21} = B_{22} = \begin{bmatrix} 0 \\ 1 \end{bmatrix},$$

$C = \begin{bmatrix} 1 & 0 \end{bmatrix}$, $\beta_1(z_\beta) = \frac{1 - \sin(z_\beta)}{2}$, $\beta_2(z_\beta) = \frac{1 + \sin(z_\beta)}{2}$, $\alpha_1(z_\alpha) = 1 - z_\alpha$, $\alpha_2(z_\alpha) = z_\alpha$, $h_1(z) = \alpha_1(z_\alpha)\beta_1(z_\beta)$, $h_2(z) = \alpha_2(z_\alpha)\beta_1(z_\beta)$, $h_3(z) = \alpha_1(z_\alpha)\beta_2(z_\beta)$, $h_4(z) = \alpha_2(z_\alpha)\beta_2(z_\beta)$.

The following parameters are calculated:

$$\nabla\beta_1(c) = \left.\frac{\partial\beta_1(z_\beta)}{\partial z_\beta}\right|_{z_\beta = c} = \frac{\partial}{\partial z_\beta}\left(\frac{1 - \sin(z_\beta)}{2}\right) = \left|-\frac{\cos(c)}{2}\right| \le \frac{1}{2}, \quad \nabla\beta_2(c) =$$

$-\nabla\beta_1(c) \le \frac{1}{2}$, with bounds $\lambda_x^2 = 1 + \pi^2$ ($x^T x \le \lambda_x^2$) and $\sigma_1^2 = \sigma_2^2 = 0.25$ (so $\nabla\beta_i\nabla\beta_i^T \le \sigma_i^2$), which gives $\eta = \sqrt{\frac{1+\pi^2}{2}}$. Also, $e_z = z_\beta - \hat{z}_\beta = x_1^2 + x_2 - \hat{x}_1^2 - \hat{x}_2$

and because x_1 is a measured variable in this example, then $e_z = \begin{bmatrix} 0 & 1 \end{bmatrix}\begin{bmatrix} e_1 \\ e_2 \end{bmatrix} = Te$.

Now, consider the particular case of $a = 1$ and $b = -1$. Using conditions (16) for state estimation with unmeasured variables, a feasible solution is obtained with $\mu_{11} = \mu_{21} = 1.0236$ and $\mu_{12} = \mu_{22} = 0.9616$; on the other hand, conditions in Theorem 2 of [12] are rendered unfeasible. The corresponding Lyapunov matrix and observer gains are given by:

$$P = \begin{bmatrix} 0.5703 & 0.7258 \\ 0.7258 & 1.1844 \end{bmatrix}, \quad K_{11} = \begin{bmatrix} 1.1491 \\ -0.5241 \end{bmatrix}, \quad K_{12} = \begin{bmatrix} 1.1558 \\ -0.7917 \end{bmatrix}, \quad K_{21} =$$

$$\begin{bmatrix} 0.8640 \\ -0.8870 \end{bmatrix}, \quad K_{22} = \begin{bmatrix} 0.8707 \\ -1.1545 \end{bmatrix}.$$

In order to simulate this example, a PDC control law $u(t) = F_{\alpha\beta}x(t) + 0.4\sin(10t)$ will be applied to stabilize the nonlinear system; the following controller gains are used: $F_{11} = \begin{bmatrix} -16.8059 & 3.8224 \end{bmatrix}$, $F_{12} = \begin{bmatrix} 5.4607 & 1.7553 \end{bmatrix}$, $F_{21} = \begin{bmatrix} -16.7947 & 3.8223 \end{bmatrix}$, $F_{22} = \begin{bmatrix} 5.4719 & 1.7552 \end{bmatrix}$.

Time evolution of the estimation errors under initial conditions $x(0) = \begin{bmatrix} 0.7 & -1 \end{bmatrix}^T$ and $\hat{x}^T(0) = \begin{bmatrix} 0 & 0 \end{bmatrix}^T$ is shown Fig. 1.

Fig. 1 Time evolution of the estimation error: e_1 (solid line) and e_2 (dashed line)

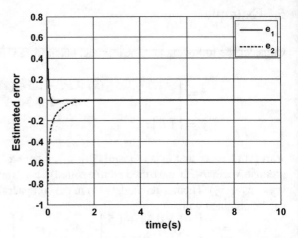

6 Concluding Remarks

Novel approaches for observer design of continuous-time nonlinear models that overcome up-to-date results in the state of the art, have been reported. A first set of results focuses on observer design based on measurable premises: taking advantage of a convex rewriting of the model (TS form) as well as from several matrix transformations such as the Finsler's Lemma [2], the observer is decoupled from their corresponding Lyapunov function; additionally, the proposed decoupling allows introducing progressively better results thanks to a nested convex structure. A second set of solutions in this chapter considers the unmeasured-premise case: the state estimation problem is expressed as a convex one using a quadratic Lyapunov function and the differential mean value theorem.

To go further, but beyond the scope of this chapter, there are many other developments based on the aforementioned techniques. To name few, for mechanical systems where the inertia matrix is nonlinear but always invertible, descriptor based techniques are to be used [5, 6] that corresponds to the extension of (1) to the following model:

$$\sum_{i=1}^{r} h_i(z)E_i \times \delta x = \sum_{i=1}^{r} h_i(z)(A_i x + B_i u) \Leftrightarrow E_h \delta x = A_h x + B_h u$$

Another important class is the so-called Unknown Input Observer (UIO) that is generally used to decouple state estimates from disturbance and/or input signals [3]. This variety of observer is generally used in fault detection scheme for certain class of actuator faults and uses extended forms of (2) such as:
$$\begin{cases} \delta x = N_h x(t) + G_h u(t) + L_h y(t) \\ \hat{x}(t) = z(t) - E y(t) \end{cases}$$
, the variables $N_i, G_i, L_i \ i \in \{1, \ldots, r\}$ being related to the model matrices of (1) via equality constraints.

References

1. Bergsten, P., Palm, R., Driankov, D.: Fuzzy observers. In: 2001 The 10th IEEE International Conference on Fuzzy Systems, pp. 700–703 (2001)
2. Boyd, S., El Ghaoui, L., Feron, E., Balakrishnan, V.: Linear matrix inequalities in system and control theory. Soc. Ind. Math. (1994)
3. Chadli, M., Karimi, H.R.: Robust observer design for unknown inputs Takagi–Sugeno models. IEEE Trans. Fuzzy Syst. **21**(1), 158–164 (2013)
4. Gahinet, P., Nemirovski, A., Laub, A.J., Chilali, M.: LMI Control Toolbox. The Math Works Inc., Massachusetts (1995)
5. Guelton, K., Delprat, S., Guerra, T.M.: An alternative to inverse dynamics joint torques estimation in human stance based on a Takagi-Sugeno unknown-inputs observer in the descriptor form. Control Eng. Practice **16**, 1414–1426 (2008)
6. Guerra, T.M., Estrada-Menzo, V., Lendek, Z.: Observer design for Takagi-Sugeno descriptor models: an LMI approach. Automatica **52**, 154–159 (2015)
7. Guerra, T.M., Vermeiren, L.: LMI-based relaxed nonquadratic stabilization conditions for nonlinear systems in the Takagi–Sugeno's form. Automatica **40**(5), 823–829 (2004)
8. Guerra, T.M., Kerkeni, H., Lauber, J., Vermeiren, L.: An efficient Lyapunov function for discrete TS models: observer design. IEEE Trans. Fuzzy Syst. **20**(1), 187–192 (2012)
9. Guerra, T.M., Kruszewski, A., Vermeiren, L., Tirmant, H.: Conditions of output stabilization for nonlinear models in the Takagi-Sugeno's form. Fuzzy Sets Syst. **157**(9), 1248–1259 (2006)
10. Guerra, T.M., Marquez, R., Kruszewski, A., Bernal, M.: H∞ LMI-based observer design for nonlinear systems via TS models with unmeasured premise variables. IEEE Trans. Fuzzy Syst. **26**(3), 1498–1509 (2017)
11. Ichalal, D., Marx, B., Maquin, D., Ragot, J.: Observer design and fault tolerant control of Takagi-Sugeno nonlinear systems with unmeasurable premise variables. In: Fault Diagnosis in Robotic and Industrial Systems, pp. 1–21 (2012)
12. Ichalal, D., Marx, B., Maquin, D., Ragot, J.: On observer design for nonlinear Takagi-Sugeno systems with unmeasurable premise variable. In: International Symposium on Advanced Control of Industrial Processes, pp. 353–358. IEEE (2011)
13. Ichalal, D., Marx, B., Ragot, J., Maquin, D.: Conception d'observateurs pour un modèle de Takagi-Sugeno à variables de décision non mesurables. In: 8ème Conférence Internationale Sciences et Techniques de l'Automatique (2007)
14. Kruszewski, A., Wang, R., Guerra, T.M.: Nonquadratic stabilization conditions for a class of uncertain nonlinear discrete time TS fuzzy models: a new approach. IEEE Trans. Autom. Control **53**(2), 606–611 (2008)
15. Lam, H.K., Keung, A.: Stability Analysis of Fuzzy-Model-Based Control Systems, vol. 264, 226 pp. Springer-Verlag, Berlin, Heidelberg (2010)
16. Lendek, Z., Guerra, T.M., Babuska, R., De Schutter, B.: Stability Analysis and Nonlinear Observer Design Using Takagi-Sugeno Fuzzy Models, vol. 262, 196 pp. Springer, Berlin, Heidelberg (2010)
17. Lendek, Z., Lauber, J., Guerra, T.-M., Babuska, R., De Schutter, B.: Adaptive observers for TS fuzzy systems with unknown polynomial inputs. Fuzzy Sets Syst. **161**, 2043–2065 (2010)
18. Márquez, R., Guerra, T.M., Bernal, M., Kruszewski, A.: Asymptotically necessary and sufficient conditions for Takagi-Sugeno models using generalized non-quadratic parameter-dependent controller design. Fuzzy Sets Syst. **306**, 48–62 (2017)
19. Peaucelle, D., Arzelier, D., Bachelier, O., Bernussou, J.: A new robust D-stability condition for real convex polytopic uncertainty. Syst. Control Lett. **40**, 21–30 (2000)
20. Sala, A., Ariño, C.: Asymptotically necessary and sufficient conditions for stability and performance in fuzzy control: applications of Polya's theorem. Fuzzy Sets Syst. **158**(24), 2671–2686 (2007)
21. Sturm, J.F.: Using SeDuMi 1.02, a MATLAB toolbox for optimization over symmetric cones. Optim. Methods Softw. **11**(1–4), 625–653 (1999)

22. Takagi, T., Sugeno, M.: Fuzzy identification of systems and its applications to modeling and control. IEEE Trans. Syst. Man Cybern. **1**, 116–132 (1985)
23. Tanaka, K., Wang, H.O.: Fuzzy Control Systems Design and Analysis: A Linear Matrix Inequality Approach. Wiley (2001)
24. Taniguchi, T., Tanaka, K., Ohtake, H., Wang, H.O.: Model construction, rule reduction, and robust compensation for generalized form of Takagi-Sugeno fuzzy systems. IEEE Trans. Fuzzy Syst. **9**(4), 525–538 (2001)

Fuzzy Extensions of Databases

Olivier Pivert and Grégory Smits

Abstract   In this chapter, we provide a brief overview of recent research works based on fuzzy set theory that aim to make data management systems more flexible. Three aspects are considered: (i) preference queries to classical databases, (ii) modeling and handling of uncertain data, and (iii) cooperative query-answering techniques.

1   Introduction

Since Lotfi Zadeh introduced fuzzy set theory in 1965, many applications of fuzzy logic to various domains of computer science have been achieved. As far as databases are concerned, the potential interest of fuzzy sets in this area has been identified as early as 1977, by V. Tahani [34]—then a Ph.D. student supervised by L. A. Zadeh— who proposed a simple fuzzy query language extending SEQUEL. This first attempt was then followed by many researchers who strove to exploit fuzzy logic for giving database languages more expressiveness and flexibility. Then, in 1978, Zadeh coined possibility theory [36], a model for dealing with uncertain information in a qualitative way, which also opened new perspectives in the area of uncertain databases. The pioneering work by Prade and Testemale [28] has had a rich posterity and the issue of modeling/querying uncertain databases in the framework of possibility theory is still an active topic of research nowadays. Unfortunately, the term "fuzzy databases" which is still widely used in the fuzzy research community is somewhat confusing since it is applied to both research topics (fuzzy querying of classical databases on the one hand, handling of uncertain values on the other hand) whereas they have not much in common and raise very specific problems. We hope that the present chapter contributes to clarify this point. Beside these two main research lines, several

O. Pivert (✉) · G. Smits
University of Rennes 1/IRISA, Lannion, France
e-mail: olivier.pivert@irisa.fr

G. Smits
e-mail: gregory.smits@irisa.fr

© The Editor(s) (if applicable) and The Author(s), under exclusive license   191
to Springer Nature Switzerland AG 2021
M.-J. Lesot and C. Marsala (eds.), *Fuzzy Approaches for Soft Computing and Approximate Reasoning: Theories and Applications*, Studies in Fuzziness and Soft Computing 394,
https://doi.org/10.1007/978-3-030-54341-9_17

other ways of exploiting fuzzy logic have been proposed along the years for dealing with various other aspects of data management, for instance *fuzzy data summaries* [3, 18, 29, 32]. More recently, fuzzy logic has also been applied to model and query non-relational databases such as RDF databases [26] or graph databases [8, 27]. In this chapter, due to space limitation, we mainly focus on works that use the relational framework. The remainder of the chapter is organized as follows. Section 2 presents important results related to fuzzy querying of classical databases. Section 3 is devoted to possibilistic databases. Section 4 deals with fuzzy cooperative query-answering approaches. Finally Sect. 5 outlines a few research perspectives.

2 Fuzzy Querying of Classical Databases

As mentioned above, the first work that aimed to use fuzzy logic to enrich database query languages is that by Tahani. The approach he proposed in [34] also constitutes the first attempt to define a *database preference query model*. Preference queries remained a rather confidential subarea of database research up to the end of the 1990s when it suddenly became a "hot topic" and many approaches were proposed (Skyline queries, top-k queries, Preference SQL, CP-nets, etc). The motivations for introducing user preferences into queries are manifold. A first goal is to offer more expressive query languages that can be more faithful to what a user intends to express. Second, preferences provide a basis for rank-ordering the retrieved items, which is especially valuable in case of large sets of items satisfying a query. Third, a classical query may also have an empty set of answers, while a flexible (thus less restrictive) version of the query might be matched by some items. Compared to the other approaches of the literature, the fuzzy-set-based one has the advantages of providing a *complete pre-order* over the answers and to offer a very wide range of operators for combining conditions (making it possible for instance to express different *trade-off effects*).

2.1 Fuzzy Extension of SQL

After the seminal paper by Tahani where a relatively basic fuzzy extension of SEQUEL was proposed (that made it possible to use fuzzy predicates in the *select* clause of a query), different researchers strove to enrich this framework. Kacprzyk and Zadrożny [17] proposed to use *fuzzy quantified statements* for combining the atomic fuzzy predicates involved in a selection condition, in order to express require-ments such as "find the elements that satisfy *most of* the conditions among $\{P_1, \ldots, P_n\}$". Then, Bosc and Pivert [5, 22], building on a fuzzy extension of relational alge-bra, defined a wide-scope generalization of the SQL language, their objective being to fuzzify (i.e. to make gradual) whatever could be in the language. Their proposal, called SQLf, thus includes the possibility to express:

- fuzzy selection conditions in the *where* clause, using a wide range of connectives, including mean operators, non-commutative connectives (e.g., *and if possible*, *or else*) and fuzzy quantifiers;
- fuzzy joins, using gradual comparison operators such as "approximately equal", "much greater than", etc;
- fuzzy nesting operators extending the classical SQL ones, namely θ {ALL | ANY}, IN and EXISTS, where θ is a comparison operator;
- fuzzy quantified conditions on groups of tuples in the *having* clause;
- a division of fuzzy relations, as in "find the customers who have ordered *all of* the expensive products of the catalog in high quantity";
- fuzzy relation partitioning by means of a fuzzy generalization of the *group by* clause that uses a fuzzy partition of the attribute domain considered;
- a quantitative or qualitative thresholding of the result.

Using SQLf, one may express for instance a query that looks for the 10 best companies that have a *rather high* annual growth and in which *most* employees have a *medium or high* education level. The authors of [5] showed that almost all of the query equivalences that are valid in SQL remain so in SQLf, thus allowing the user to choose the formulation that he/she finds the most intuitive for expressing his/her information need.

Obviously, as any formal language, SQLf [5, 22] is not so easy to use for novice users. This is why it is necessary to propose intuitive user interfaces to help them formulate their queries. Smits et al. [30] describes such an interface, based on graphical tools.

2.2 Fuzzy Query Optimization

Fuzzy query evaluation [4, 12, 13, 17] raises different types of difficulties, among which the main ones are listed hereafter:

- commercial RDBMSs do not offer any tools primarily aimed at defining membership functions, fuzzy connectives and so on;
- it is not possible to directly use existing indexes while evaluating fuzzy selection conditions;
- a step devoted to the computation of the degrees and the calibration of the result (top-k or qualitative thresholding) is needed, which induces an extra cost.

Then, implementing a fuzzy querying system can be done according to three main types of architectures [35]:

- *loose coupling*: the new features are integrated through a software layer on top of the RDBMS. The main advantage of this type of architecture is its portability, which allows connecting to any RDBMS. Its weakness resides in its poor scalability and performances.

- *mild coupling*: the new features can be integrated through stored procedures using a procedural language for relational databases such as Oracle PL/SQL; or through external function calls. With this type of solution, the data are directly managed from the RDBMS, which entails better performances.
- *tight coupling*: the new features are incorporated into the RDBMS inner core. This solution, which is of course the most efficient in terms of query evaluation, implies to entirely rewrite the query engine (including the parser and query optimizer), which is a very heavy task, all the more as it has to be reconsidered when the DBMS evolves.

Since commercial RDBMSs are not able to natively interpret fuzzy queries, the loose coupling type of approach presented in [6] suggested to perform a so-called *derivation step* in order to generate a regular Boolean query used to prefilter the relation (or Cartesian product of relations) concerned. The idea is to restrict the degree computation phase only to those tuples that somewhat satisfy the fuzzy selection (or join) condition. With this type of evaluation strategy, fuzzy query processing involves three steps:

1. derivation of a Boolean query from the fuzzy one,
2. processing of this Boolean query, which produces a classical relation,
3. computation of the satisfaction degree attached to each tuple of the relation returned (accompanied by a tuple filtering step based on the degree if necessary), which yields the final fuzzy relation that constitutes the result.

In terms of performances, the interest of using a mild coupling architecture lies in the fact that the fuzzy resulting relation is directly computed during the tuple selection phase (no external program needs to be called to perform step 3).

The type of implementation presented in [30] is at the junction between mild coupling and tight coupling inasmuch as i) the membership functions corresponding to the user-specified fuzzy predicates are defined as stored procedures and ii) the gradual extensions of the operators (triangular norm for the conjunction, triangular co-norm for the disjunction, fuzzy quantifiers) are implemented in C and integrated on demand in the query processing engine of the RDBMS PostgreSQL.

3   Possibilistic Databases

Many authors have made proposals to model and handle databases involving uncertain data. In particular, the last two decades have witnessed a blossoming of researches on this topic (cf. [33] for a survey). Even though most of the literature about uncertain databases uses probability theory as the underlying uncertainty model, some approaches rather rest on possibility theory [36]. The initial idea of applying possibility theory to this issue goes back to the early 80s [28]. The possibilistic framework constitutes an interesting alternative to the probabilistic one inasmuch as it captures a different kind of uncertainty (of a qualitative nature). An example is that of a person who witnesses a car accident and is not sure about the model of the car involved. In such a case, it seems reasonable to model the uncertain value by means of a possibility distribution, e.g., {1/Mazda, 1/Toyota, 0.7/Honda}—where 0.7 is a numerical encoding in a usually finite possibility scale—rather than with a probability distribution which would be artificially normalized.

In this section, we first give general notions about uncertain databases, then we provide a brief overview of the recent works that use possibility theory for handling uncertain data. More details about their positioning one with respect to the other can be found in [25].

3.1 About Uncertain Databases and Possible Worlds

Classically, in an uncertain database model, an attribute value is represented as a disjunctive weighted set, which can be interpreted as a probability distribution or a possibility distribution depending on the underlying uncertainty model considered. From a semantic point of view, an uncertain database D can be interpreted as a set of usual databases, called possible worlds $W_1, ..., W_p$, and the set of all interpretations of D is denoted by $rep(D) = \{W_1, ..., W_p\}$. Any world W_i is obtained by choosing a value in each disjunctive set appearing in D. One of these (regular) databases is supposed to correspond to the actual state of the universe modeled. The assumption of independence between the sets of candidates is usually made and then any world W_i corresponds to a conjunction of independent choices, thus the (probability or possibility) degree associated with a world is computed using a conjunction operator, e.g., "product" or "min".

When processing a query, a naive way of doing would be to make explicit all the interpretations of D in order to query each of them. Such an approach is intractable in practice and it is of prime importance to find a more realistic alternative. To this end, the notion of a representation system has been introduced by Imielinski and Lipski [16]. The basic idea is to represent both initial tables and those resulting from queries in such a way that the representation of the result of a query q against any database D denoted by $q(D)$, is equivalent (in terms of worlds) to the set of results obtained by applying q to every interpretation of D, i.e.: $rep(q(D)) = q(rep(D))$ where $q(rep(D)) = \{q(W) \mid W \in rep(D)\}$. If this property holds for a representation system ρ and a subset σ of the relational algebra, ρ is called a *strong representation system* for σ. From a querying point of view, this property enables a direct (or compact) calculus of a query q, which then applies to D itself without making the worlds explicit.

3.2 Possibilistic Database Models

Full possibilistic model In the "full possibilistic model" defined in [7], any attribute value can be a possibility distribution. Besides, there is a need for expressing that some tuples may not be represented in some worlds. Thus every relation includes an attribute denoted by N which states whether or not it is legal to build worlds where the corresponding tuple has no representative, and, if so, the influence of this choice in terms of possibility degree. A second aspect is related to the fact that it is sometimes necessary to express dependencies between candidate values of different attributes of a same tuple. For instance, let A and B be two attributes whose respective candidates in a given tuple t are $\{a_1, a_2\}$ and $\{b_1, b_2, b_3\}$. If, according to a given selection criterion, the only legal associations are (a_1, b_1) and (a_2, b_3), one cannot call on a Cartesian product of subsets of $t.A$ and $t.B$. In other words, A and B values cannot be kept separate (which would mean that they are independent) and the correct associations have to be explicitly represented as weighted tuples of a *nested relation* (with a disjunctive meaning). In [7], the authors show that this model is a strong representation system for selection, projection and foreign key join.

Certainty-based model   In [23], the authors consider relational databases containing uncertain attribute values, when some knowledge is available about the more or less *certain* value (or disjunction of values) that a given attribute in a tuple may take. As that described above, the model they propose is based on possibility theory, but it only represents the values that are *more or less certain* instead of all those which are *more or less possible*. This corresponds to the most important part of information (a possibility distribution is "summarized" by keeping its most plausible elements). This choice obviously implies a certain loss of information as only the candidate values or disjunctions of candidate values that are *somewhat certain* are kept. However, this loss is compensated by a very important gain in terms of query evaluation cost and simplicity/intelligibility of the model. Moreover, the information lost is quite weak, indeed when a statement is *just* possible, its negation is possible as well (otherwise, the statement would be somewhat certain). The basic idea is to attach a certainty level to each piece of data (by default, a piece of data has certainty 1). It is shown in [23] that this certainty-based model is a representation system for the whole relational algebra. An important result is that the data complexity associated with the extended operators in this context is the same as in the classical database case, which makes the approach highly scalable.

Layered possibilistic database model   In [20], the authors model uncertainty in a qualitative way by assigning to tuples a degree of possibility with which they occur, and assigning to functional dependencies a degree of certainty which says to which tuples they apply. A design theory is developed for possibilistic functional dependencies. In this context, the possibility degrees of tuples result in a scale of different degrees of data redundancy. Scaled versions of the classical syntactic Boyce-Codd and Third Normal Forms are established and semantically justified in terms of avoiding data redundancy of different degrees.

Possibilistic c-tables   In [16], Imielinski and Lipski introduced a very powerful database model for representing uncertain information, called *c*-tables (where *c* stands for conditional). A *c*-table is a table that may contain variables as attribute values, as well as conditions (constraints) involving these variables. In the original definition, conditions may only be attached to individual tuples. An extension is to consider that a condition may also be attached to the c-table globally, see [1]. This makes it possible to represent uncertain databases involving complex constraints. For instance, let us consider a relation representing the associations between some students and the courses they take. With a c-table, one may represent information such as: Sally is taking math or computer science (but not both) and another course; Alice takes biology if Sally takes math, and math or physics (but not both) if Sally takes physics.

A probabilistic extension of c-tables was proposed in [15] where the candidate values associated with a given variable may be assigned a probability degree. This makes it possible to represent, for instance that "Alice takes a course that is math with probability 0.3, or physics with probability 0.3, or chem with probability 0.4; Bob takes the same course as Alice provided that the course is physics or chem; Theo takes math with probability 0.85." In [24], the authors proposed a possibilistic extension of c-tables and showed that it generalizes the certainty-based model from [23]. Possibilistic c-tables constitute a general setting that appears quite appropriate for handling uncertainty in a qualitative way. The qualitative nature of possibility theory makes simpler the elicitation of the possibility and certainty degrees, and leads to a modeling less sensitive to modifications of the values of the degrees than in a probabilitistic framework.

4 Cooperative Query-Answering

The paradigm of cooperative answering is originated in the work by Kaplan [19] in the context of natural language question-answering. The general aim of cooperative answering techniques is to prevent users from being faced with an uninformative answer (for instance, "there is no result" when a query fails). Cooperative systems should rather correct any false presupposition of the user, anticipate follow-up queries and provide information not explicitly requested by the user [10]. With respect to Boolean queries, fuzzy queries reduce the risk of obtaining an empty set of answers since the use of a finer discrimination scale—[0, 1] instead of {0, 1}—increases the chance for an element to be considered somewhat satisfactory. Nevertheless, the situation may occur where none of the elements of the target database satisfies the query even to a low degree: this is the Empty Answer Set (EAS) problem. Symmetrically, it may happen that a fuzzy query produces a very large answer, when many elements satisfy the fuzzy condition to a high degree: this is the Plethoric Answer Set (PAS) problem.

When a fuzzy conjunctive selection query leads to an empty result, several techniques can be used. A first option is to weaken one or several fuzzy predicates by means of an appropriate modifier in order to make the condition more tolerant, see e.g. [2]. In [31], the authors propose an alternative approach that relies on a fuzzy cardinality-based summary of the database and show how this summary can be efficiently used to explain failing queries (or "almost failing" ones, i.e., producing insufficiently satisfactory answers) or to revise queries returning a plethoric answer set. In this latter case, the idea is to augment the user query Q with predefined predicates which are semantically correlated with those present in Q, in order to reduce the initial answer set.

Fuzzy sets can also play an interesting role regarding another issue related to cooperative answering: that of *answers intelligibility*. An example of explanation needs is when the set of answers obtained can be clustered in clearly distinct subsets of similar or close answers. Then, it may be interesting for the user to know what meaningful differences exist between the tuples leading to the answers that may explain the discrepancy in the result. For instance, if one looks for possible prices for houses to let obeying some (possibly fuzzy) specifications, and that two clusters of prices are found, one may discover, e.g., that this is due to two categories of houses having, or not, some additional valuable equipment. This issue is dealt with in [9] where the authors describe a case-based reasoning type of method. In [21], Moreau *et al.* propose an alternative approach that first uses a clustering algorithm to detect groups of answers (a group corresponds to elements that have similar values on the attributes from the projection clause of the query) described by means of a fuzzy expert vocabulary. Then, the explanation step looks for common properties between the elements of each cluster (and specific to the cluster) for the other attributes.

5   Conclusion

In this chapter, we have presented a brief overview of fuzzy approaches aimed to make database systems more flexible and cooperative. Due to space limitation, it is of course far from exhaustive. In any case, the ways fuzzy set theory can contribute to data management is still an open (and active) field of research, and much remains to do regarding the three axes presented above. When it comes to fuzzy querying, the most important challenge is certainly to make the current techniques *scalable* so that it becomes realistic to query *big data* in a fuzzy way. Concerning uncertain databases, it would be worth studying whether a common general framework could "unify" the probabilistic and the possibilistic c-table models. The notion of *provenance semirings* discussed in [14] looks like a promising theoretical tool in this respect. As to cooperative answering, much effort is still needed to transform DBMSs from "black boxes" into more transparent ones, which means being able to *explain* every aspect of the querying process.

More generally, we believe that coupling data mining techniques and database technology is a very promising perspective in the area of flexible data management, the objective being to benefit from the power of machine learning algorithms on the one hand, and from the efficient data access methods that the DBMSs provide on the other hand. Data analysis is a crucial task at the center of many professional activities and now constitutes a support for decision making, communicating and reporting. However, domain experts (insurers, data journalists, communication managers, decision makers, etc.) are not, most of the time, data or computer scientists, and need efficient tools that help them turn data into useful knowledge. This explains the recent growing interest for so-called Agile Business Intelligence (ABI) systems that reconsider classical data integration processes to favor pragmatic approaches that make domain experts self-reliant in the analysis of raw data. In this respect, the challenge is to define efficient strategies able to generate meaningful, condensed and human-interpretable representations of a dataset. Several approaches based on fuzzy set theory have been proposed for meeting this goal. Much attention has been paid by the soft computing community to data summarization using personalized linguistic explanations, see e.g. [18, 29]. However, ABI raises new issues that have not been addressed by existing approaches of data summarization and exploration, as e.g. the interpretability and visualization of the linguistic summaries [11] and the scalability of the linguistic rewriting process. A first step towards an efficient fuzzy approach to ABI is outlined in [32], which relies on a rewriting of the data using a fuzzy vocabulary, and the construction of condensed graphical and personalized views that can be used both as a knowledge visualization tool and a data exploration mechanism.

References

1. Abiteboul, S., Hull, R., Vianu, V.: Foundations of Databases. Addison-Wesley (1995)
2. Bosc, P., HadjAli, A., Pivert, O.: Incremental controlled relaxation of failing flexible queries. J. Intell. Inf. Syst. **33**(3), 261–283 (2009)
3. Bosc, P., Lietard, L., Pivert, O.: Extended functional dependencies as a basis for linguistic summaries. In: Proceedings of PKDD'98, pp. 255–263 (1998)
4. Bosc, P., Pivert, O.: On the evaluation of simple fuzzy relational queries: principles and measures. In: Lowen, R., Roubens, M. (eds.) Fuzzy logic—State of the Art, pp. 355–364. Kluwer Academic Publishers, Dordrecht, The Netherlands (1993)
5. Bosc, P., Pivert, O.: SQLf: a relational database language for fuzzy querying. IEEE Trans. Fuzzy Syst. **3**, 1–17 (1995)
6. Bosc, P., Pivert, O.: SQLf query functionality on top of a regular relational database management system. In: Pons, O., Vila, M.A., Kacprzyk, J. (eds.) Knowledge Management in Fuzzy Databases, pp. 171–190. Physica-Verlag, Heidelberg, Germany (2000)
7. Bosc, P., Pivert, O.: About projection-selection-join queries addressed to possibilistic relational databases. IEEE Trans. Fuzzy Syst. **13**(1), 124–139 (2005)
8. Castelltort, A., Laurent A.: Fuzzy queries over nosql graph databases: perspectives for extending the cypher language. In: Proceedings of the Information Processing and Management of Uncertainty in Knowledge-Based Systems—15th International Conference, IPMU'14, pp. 384–395 (2014)
9. de Calmès, M., Dubois, D., Hüllermeier, E., Prade, H., Sedes, F.: Flexibility and fuzzy case-based evaluation in querying: An illustration in an experimental setting. Int. J. Uncertain. Fuzziness Knowl.-Based Syst. **11**(1), 43–66 (2003)
10. Gaasterland, T., Godfrey, P., Minker, J.: An overview of cooperative answering. J. Intell. Inf. Syst. **1**(2), 123–157 (1992)
11. Gacto, M.J., Alcalá, R., Herrera, F.: Interpretability of linguistic fuzzy rule-based systems: an overview of interpretability measures. Inf. Sci. **181**(20), 4340–4360 (2011)
12. Galindo, J., Medina J.M., Pons O., Cubero J.C.: A server for fuzzy SQL queries. In Proceedings of FQAS'98, pp. 164–174 (1998)

13. Goncalves, M., Tineo, L.: SQLf3: an extension of SQLf with SQL features. In: Proceedings of FUZZ-IEEE'01, pp. 477–480 (2001)
14. Green, T.J., Karvounarakis, G., Tannen, V.: Provenance semirings. In: Proceedings of PODS'07, pp. 31–40 (2007)
15. Green, T.J., Tannen, V.: Models for incomplete and probabilistic information. In: Proceedings of the IIDB'06 Workshop, pp. 278–296 (2006)
16. Imielinski, T., Lipski, W.: Incomplete information in relational databases. J. ACM **31**(4), 761–791 (1984)
17. Kacprzyk, J., Zadrożny, S.: FQUERY for ACCESS: fuzzy querying for a Windows-based DBMS. In Bosc, P., Kacprzyk, J. (eds.) Fuzziness in Database Management Systems, pp. 415–433. Physica Verlag (1995)
18. Kacprzyk, J., Zadrożny, S.: Linguistic database summaries and their protoforms: towards natural language based knowledge discovery tools. Inf. Sci. **173**(4), 281–304 (2005)
19. Kaplan, S.-J.: Cooperative responses from a portable natural language query system. Artif. Intell. **19**, 165–187 (1982)
20. Link, S., Prade, H.: Relational database schema design for uncertain data. In Proceedings of CIKM'16, pp. 1211–1220 (2016)
21. Moreau, A., Pivert, O., Smits, G.: A fuzzy approach to the characterization of database query answers. In: Proceedings of IPMU'16, Part II, pp. 329–3402 (2016)
22. Pivert, O., Bosc, P.: Fuzzy Preference Queries to Relational Databases. Imperial College Press, London, UK (2012)
23. Pivert, O., Prade, H.: A certainty-based model for uncertain databases. IEEE Trans. Fuzzy Syst. **23**(4), 1181–1196 (2015)
24. Pivert, O., Prade, H.: Possibilistic conditional tables. In: Proceedings of FoIKS'16, pp. 42–61 (2016)
25. Pivert, O., Prade, H.: Handling uncertainty in relational databases with possibility theory—a survey of different modelings. In: Proceedings of SUM'18, pp. 1–9 (2018)
26. Pivert, O., Slama, O., Thion, V.: An extension of SPARQL with fuzzy navigational capabilities for querying fuzzy RDF data. In Proceedings of the 2016 IEEE International Conference on Fuzzy Systems, (FUZZ-IEEE'16), pp. 2409–2416 (2016)
27. Pivert, O., Smits, G., Thion, V.: Expression and efficient processing of fuzzy queries in a graph database context. In: 2015 IEEE International Conference on Fuzzy Systems, (FUZZ-IEEE'15), pp. 1–8 (2015)
28. Prade, H., Testemale, C.: Generalizing database relational algebra for the treatment of incompleteuncertain information and vague queries. Inf. Sci. **34**, 115–143 (1984)
29. Saint-Paul, R., Raschia, G., Mouaddib, N.: Database summarization: the SaintEtiq system. In: 2007 IEEE 23rd International Conference on Data Engineering, pp. 1475–1476 (2007)
30. Smits, G., Pivert, O., Girault, T.: Towards reconciling expressivity, efficiency and user-friendliness in database flexible querying. In: Proceedings of the IEEE International Conference on Fuzzy Systems (FUZZ-IEEE'13) (2013)
31. Smits, G., Pivert, O., HadjAli, A.: Fuzzy cardinalities as a basis to cooperative answering. In: Pivert, O., Zadrozny, S. (eds.) Flexible Approaches in Data, Information and Knowledge Management. Studies in Computational Intelligence, vol. 497, pp. 261–289. Springer (2013)
32. Smits, G., Yager, R.R., Pivert, O.: Interactive data exploration on top of linguistic summaries. In: Proceedings of FUZZ-IEEE'17 (2017)
33. Suciu, D., Olteanu, D., Ré, C., Koch, C.: Probabilistic Databases. Synthesis Lectures on Data Management. Morgan & Claypool Publishers (2011)
34. Tahani, V.: A conceptual framework for fuzzy query processing—a step toward very intelligent database systems. Inf. Process. Manag. **13**(5), 289–303 (1977)
35. Urrutia, A., Tineo, L., Gonzalez, C.: FSQL and SQLf: towards a standard in fuzzy databases. In: Galindo, J. (ed) Fuzzy Databases. Handbook of Research on Fuzzy Information Processing in Databases, pp. 270–298. Information Science Reference, Hershey, PA, USA (2008)
36. Zadeh, L.A.: Fuzzy sets as a basis for a theory of possibility. Fuzzy Sets Syst. **1**, 3–28 (1978)

On Maxitive Image Processing

Olivier Strauss, Kevin Loquin, and Florentin Kucharczak

Abstract Most digital processing nowadays is in digital format. Many transformations, such as contrast enhancement, restoration, color correction, etc. can be achieved through very simple and versatile algorithms However, a major branch of image processing algorithms makes extensive use of spatial transformations such as rotation, translation, zoom, anamorphosis, homography, distortion, derivation, etc. that are only defined in the analog domain. A digital image processing algorithm that mimics a spatial transformation is usually designed by using the so-called *kernel based approach*. This involves two kernels to ensure the continuous to discrete interplay: a sampling kernel and a reconstruction kernel, whose choice is highly arbitrary. The maxitive kernel based approach can be seen as an extension of the conventional kernel based approach and it reduces the impact of this arbitrary choice. It consists in replacing at least one of the kernels by a normalized fuzzy subset of the image plane. This replacement in the digital image spatial transformation framework leads to computing the convex set of all the images that would be obtained by using a (continuous convex) set of conventional kernels. Use of this set generates a kind of robustness that can reduce the risk of false interpretation. Medical imaging, for example, would be an application that could benefit from such an approach.

O. Strauss (✉) · F. Kucharczak
LIRMM – Université de Montpellier, Montpellier, France
e-mail: olivier.strauss@lirmm.fr

F. Kucharczak
e-mail: florentin.kucharczak@lirmm.fr

K. Loquin
DMS Imaging, Montpellier, France
e-mail: kloquin@dms-imaging.com

201

M.-J. Lesot and C. Marsala (eds.), *Fuzzy Approaches for Soft Computing and Approximate Reasoning: Theories and Applications*, Studies in Fuzziness and Soft Computing 394, https://doi.org/10.1007/978-3-030-54341-9_18

1 Introduction

In digital image processing, fuzzy subsets have been used since the outset for representing image information at different levels (see e.g. [2] for a nice overview on fuzzy set based image processing). Though, high level image processing is the focus of most previous studies [16]. The maxitive approach we present in this paper is a low-level processing strategy where a membership function is used to represent a local neighborhood around each point of the image plane.

Digital image processing refers to the set of algorithms used to transform, filter, enhance, modify, analyze, distort, fuse, etc., digital images. Many of these algorithms are designed to mimic an underlying physical operation defined in the continuous illumination domain and formerly achieved via optical or electronic filters or through manipulations, including painting, cutting, moving or pasting of image patches. Spatial domain operations like derivation [24], morphing, filtering, geometric and perspective transformations [12], super-resolution [17], etc. are usually derived using a kernel based approach [26]. Among all spatial domain operations, we are interested here in geometric transformations like affine and projective transformations or, more generally, diffeomorphisms or homeomorphisms. In kernel-based approaches, the choice of a particular kernel shape or spread is usually prompted more by practical aspects than by any theoretical thrust. Unfortunately, this choice can highly impact the output of the obtained discrete operator. Figure 2 illustrates this by highlighting the noise introduced by a digital approximation of a continuous rigid transformation operation. Two different interpolation methods have been used to rotate a detail of Fig. 1 (see Fig. 2a, b). The obtained images are not identical, as illustrated by enhancing their absolute difference (Fig. 2c).

This dependance is not a real problem when the considered operations are dedicated to artistic modifications of an image. Photographs have their own rule to choose among the three main interpolation methods (nearest neighbor, bilinear, bicubic), while some dedicated software packages have developed their own interpolation (or more generally reconstruction) method. It is more problematic when the information within by an image is quantitative, e.g. in medical applications where quantization is sought.

For example, to study lung tumor growth, the patient is subjected to a hybrid PET-CT scan, where the CT[1] provides the anatomical structure information and PET[2] gives quantitative information about the tumor metabolism. After image acquisition, it is necessary to bring the image of either of one the modality w.r.t. other (PET w.r.t. CT or vice versa). This operation involves geometrical transformations. If details in the images are comparable to the image resolution, the choice of registration algorithm can be critical.

---

[1] Computed Tomography scan is a computational method for reconstructing cross-sectional images from sets of X-ray measurements taken from different angles.

[2] Positron-Emission Tomography is a nuclear medicine functional imaging technique used to observe metabolic processes in the body.

Fig. 1 A digitized version of an illustration of Ivan Bilibin for Fairytale of the Tsar Saltan (1905)

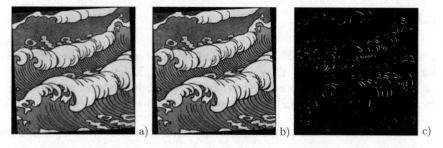

Fig. 2 Rotation (3°) of a detail of Fig. 1 using bilinear interpolation (**a**) bicubic interpolation (**b**) and an enhanced view of the difference between the two rotated images (**c**)

A more careful approach would be to compute not a single transformed image but rather the set of all images that can be obtained by using different kernels. In medical diagnosis, this can help confirm a diagnosis (if all the images lead to the same diagnosis) or highlight the need for a complementary medical investigation (if different images lead to different diagnoses). This is what the maxitive approach proposes.

Maxitive image processing takes advantage of the obvious analogy between probability density functions (pdf) and positive kernels to extend the signal processing theory to the case where the modeling is imprecisely known [13]. Within this technique, the possibilistic interpretation [5] of fuzzy subsets is used to define *maxitive kernels* that can be seen as convex sets of conventional positive kernels. These convex sets aim to represent scant knowledge on the appropriate kernel to be used in a given application. Maxitive-based signal processing extensions lead to an interval-valued signal that includes the set of all signals that would have been obtained by the corresponding conventional method using a positive kernel that belongs to the maxitive kernel core [19]. This approach has been extended to two dimensions while being involved in image processing applications (see e.g. [6, 10, 14, 21]).

In this article, we propose a formalization of the maxitive approach for extending geometrical transformations in digital image processing, i.e. we show how this technique can be used to compute the convex set of all images that would be obtained by considering a convex set of possible kernels when applying geometrical transformations to digital images.

After this introduction, Sect. 2 provides some notations and necessary background knowledge. In Sect. 3 we present the kernel-based method to design discrete operators mimicking continuous geometrical transformations. We then propose the maxitive approach as a simple extension of the former method. We propose to use this extension to design a rigid transformation. We then conclude this article.

2 Preliminary Considerations, Definitions and Notations

2.1 Notations

Let $\mathbb{R}$ be the real line and $\mathbb{IR}$ be the set of all intervals of $\mathbb{R}$. Let Ω be the image plane, i.e. a box of $\mathbb{R}^2$: $\Omega = \Omega_1 \times \Omega_2$, where $\Omega_1, \Omega_2 \subseteq \mathbb{R}$. Let $\mathscr{P}(\Omega)$ be the set of all Lebesgue measurable subsets of Ω. With N being a positive integer, we define Θ_N by $\Theta_N = \{1, \dots, N\} \subset \mathbb{N}$. Let $\mathscr{P}(\Theta_N) = 2^{\Theta_N}$ be the power set of Θ_N.

2.2 Digital Images

A continuous (real) image is usually obtained by projecting on a plane, via an optical device, the light reflected by objects placed in front of the camera. This projection is often referred to as *the illumination function*.

A digital image is a numeric representation of a continuous image. It is composed of a finite set of digital values called pixels, with each value being associated to the measurement of the illumination function at a specific location on the continuous image plane.

2.3   Capacities and Expectations

A capacity is a confidence measure that is more general than a probability measure [8]. It can be defined on both continuous and discrete domains. A capacity defined on a continuous reference set is called a *continuous capacity*, while a capacity defined on a discrete reference set is called a *discrete capacity*. Let Φ be either Ω or Θ_N.

Definition 18.1 A (continuous or discrete) capacity ν is a set function $\nu : \mathscr{P}(\Phi) \to [0, 1]$ such that $\nu(\varnothing) = 0$, $\nu(\Phi) = 1$ and $\forall A, B \in \mathscr{P}(\Phi), A \subseteq B \Rightarrow \nu(A) \leq \nu(B)$.

Given a capacity ν, its conjugate ν^c is defined as: $\nu^c(A) = 1 - \nu(A^c)$ for any subset $A \in \mathscr{P}(\Phi)$, with A^c being the complementary set of A in Φ. A capacity ν such that for all A, B in $\mathscr{P}(\Phi)$, $\nu(A \cup B) + \nu(A \cap B) \leq \nu(A) + \nu(B)$ is said to be concave. Here we only consider this kind of capacity. The core of a concave capacity ν, denoted $\mathscr{M}(\nu)$, is the set of probabilities P on $\mathscr{P}(\Phi)$ such that $\nu(A) \geq P(A)$ for all subsets $A \in \mathscr{P}(\Phi)$. A probability measure is a capacity that equals its conjugate. Thus, if ν is a probability, $\nu(A \cup B) + \nu(A \cap B) = \nu(A) + \nu(B)$ [4].

The concept of expected value associated with a probability measure has been extended to concave capacities through a Choquet integral (see e.g. [20]). Let ν be a concave capacity defined on Φ and let $f : \Phi \to \mathbb{R}$ be a L_1 bounded function.

The (imprecise) expectation of f w.r.t. ν is the real interval $\overline{\mathbb{E}}_\nu(f)$ defined by:

$$\overline{\mathbb{E}}_\nu(f) = \left[\underline{\mathbb{E}}_\nu(f), \overline{\mathbb{E}}_\nu(f) \right] = \left[\check{\mathbb{C}}_{\nu^c}(f), \check{\mathbb{C}}_\nu(f) \right], \tag{1}$$

where $\check{\mathbb{C}}$ denotes the asymmetric Choquet integral (see e.g. [23]). One of the important properties of this extension is that $\overline{\mathbb{E}}_\nu(f)$ is an interval that contains all of the $\mathbb{E}_P(f)$ with $P \in \mathscr{M}(\nu)$. Conversely, any value of this interval corresponds to an expected value of the form $\mathbb{E}_P(f)$ with $P \in \mathscr{M}(\nu)$ [4].

Remark 18.1 Upper expectations and concave capacities coincide when considering the characteristic function. Let ν be a concave capacity, $\forall A \in \mathscr{P}(\Phi)$, $\nu(A) = \overline{\mathbb{E}}_\nu(\chi_A)$, with χ_A being the characteristic function of A.

2.4   Summative and Maxitive Kernels

In discrete image processing, the role of a kernel is to define a weighted neighborhood of spatial locations in the image plane, with those weights being used in an aggregation process. Among different kernels, summative kernels play an important role since they define normalized positive weighted neighborhoods. They are intensively used to establish discrete operators defined in the continuous domain [13]. Let $N \in \mathbb{N}$ and Φ be either Ω or Θ_N.

A **summative kernel** is a positive function $\kappa : \Phi \longrightarrow \mathbb{R}^+$ complying with the summative property, i.e. $\int_\Omega \kappa(x)dx = 1$ (if $\Phi = \Omega$) or $\sum_{n \in \Theta_N} \kappa_n = 1$ (if $\Phi = \Theta_N$).

Such a function defines a probability measure P_κ on Φ by: $\forall A \in \mathscr{P}(\Phi)$,
$P_\kappa(A) = \int_A \kappa(x)\, dx$ (if $\Phi = \Omega$) or $P_\kappa(A) = \sum_{n \in A} \kappa_n$ (if $\Phi = \Theta_N$).
$\mathscr{K}(\Phi)$ is the set of all summative kernels defined on Φ.

A **maxitive kernel** [13] is a function $\pi : \Phi \longrightarrow [0, 1]$ complying with the maxitive property, i.e. $\sup_{x \in \Phi} \pi(x) = 1$. Such a function defines a concave capacity on Φ, called a possibility measure Π_π, by: $\Pi_\pi(A) = \sup_{x \in A} \pi(x)$. A maxitive kernel π defines a convex subset of $\mathscr{K}(\Phi)$ as follows:
$\mathscr{M}(\pi) = \{\kappa \in \mathscr{K}(\Phi) \mid \forall A \in \mathscr{P}(\Phi),\ P_\kappa(A) \leq \Pi_\pi(A)\}$ called its *core*.

2.5 Crisp and Fuzzy Partitions

In image processing, partitioning is mandatory to define the relation between the continuous domain, where the illumination function is defined, and the discrete domain, where the measured illumination is depicted. Traditional image processing is based on crisp partitions, while more advanced image processing has been based on fuzzy partitions (see e.g. [6, 15]).

An image partition of Ω is a set of N subsets $\{C_n\}_{n \in \Theta_N}$ such that (i) $\forall (n, m) \in \Theta_N$, $C_n \cap C_m \neq \emptyset \iff n = m$, (ii) $\forall \omega \in \Omega$, $\exists n \in \Theta_N$ such that $\omega \in C_n$.

An image partition is said to be uniform if it can be generated by a simple generic subset E: let χ_E be the characteristic function of E, $\forall n \in \Theta_N$, $\exists \omega_n \in \Omega$ such that $\forall \omega \in \Omega$, $\chi_{C_n}(\omega) = \chi_E(\omega - \omega_n)$.

A fuzzy image partition of Ω is a set of N fuzzy subsets $\{C_n\}_{n \in \Theta_N}$ such that $\forall \omega \in \Omega$: (i) $\sum_{n=1}^{N} \mu_{C_n}(\omega) = 1$ and (ii) μ_{C_n} is continuous, with μ_{C_n} being the membership function of C_n ($n \in \Theta_N$). A fuzzy partition is said to be uniform if it can be generated by a simple generic fuzzy subset E: let $\{\omega_n\}_{n \in \Theta_N}$ be a set of N regularly spaced locations of Ω, then C_n is generated by E i.e. $\forall \omega \in \Omega$, $\mu_{C_n}(\omega) = \mu_E(\omega_n - \omega)$. A fuzzy partition is said to be normalized if $\forall n \in \Theta_N$, $\exists \omega \in \Omega$ such that $\mu_{C_n}(\omega) = 1$ [22]. In image processing, partitions are generally uniform to comply with the geometry of the image sensors.

Fuzzy partitions are instrumental for performing reconstructions [18]. Let $\{F_n\}_{n \in \Theta_N}$ be a discrete function, a reconstructed continuous function $\hat{F} : \Omega \to \mathbb{R}$ can be defined by $\forall \omega \in \Omega$, $\hat{F}(\omega) = \sum_{n \in \Theta_N} F_n \mu_{C_n}(\omega)$. The following definition allows us to extend this instrumentality to link the continuous space Ω to the discrete space Θ_N.

Definition 18.2 Let $A \subseteq \Theta_N$. We define Υ_A as being the membership function of $\bigcup_{n \in A} C_n$, where the union is defined by the Łukasiewicz T-conorm: $\forall \omega \in \Omega$, $\Upsilon_A(\omega) = min(1, \sum_{n \in A} \mu_{C_n}(\omega)) = \sum_{n \in A} \mu_{C_n}(\omega)$ due to the fact that $\sum_{n=1}^{N} \mu_{C_n}(\omega) = 1$.

Remark 18.2 Note that a uniform crisp partition is a special case of fuzzy partition where the generic fuzzy subset E is a crisp subset.

3   From Continuous to Digital Image Processing

3.1   *Continuous Image/Digital Image*

In the continuous domain, an image can be seen as a measurable physical illumination phenomenon, i.e. the projection, via an optical device, of real-world light information in a particular direction. It is generally modeled by a bounded positive integrable function $\mathscr{I}$ defined on $\mathbb{R}^2$. More precisely, $\mathscr{I}$ is a $L_1(\mathbb{R}^2)^+$ function defined on a compact subset Ω (e.g. a closed rectangle) of $\mathbb{R}^2$. This function is usually extended throughout the continuous domain $\mathbb{R}^2$ by assigning an arbitrary value (usually 0) to Ω^c (the complementary set of Ω in $\mathbb{R}^2$).

Some optical systems allow image processing to be performed in the continuous domain. Yet, nowadays image processing is mainly performed on computers or smartphones, i.e. on images stored in the computer memory as discrete quantities. From a signal processing point of view, a sampled image can be considered as being obtained by measuring the continuous illumination function $\mathscr{I}$ defined on $\mathbb{R}^2$ projected by an optical device on a matrix of sensors called the retina (see Fig. 3). The sensors are usually regularly spaced along each axis at a limited number N of locations, called the *sampling locations*. Those measurement values or locations are usually referred to as *pixel values or locations*. Let $\Theta_N = \{1, \ldots, N\} \subset \mathbb{N}$ be the set of indices of the sampling locations and $\{\omega_n\}_{n \in \Theta_N}$ be the set of sampling locations also referred to as the *sampling grid*.

Ideally, each measure I_n can be modeled by an integral of the illumination function $\mathscr{I}$ in a crisp neighborhood around ω_n. Let $\phi^{\omega_n} \subset \mathbb{R}^2$ be this neighborhood, then the relation between I and $\mathscr{I}$ can be expressed by: $\forall n \in \Theta_N,\ I_n = \int_{\phi^{\omega_n}} \mathscr{I}(\omega)d\omega$.

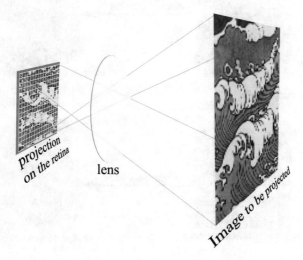

projection on the retina   lens

Image to be projected

Fig. 3   Measure of a continuous image

When $\chi_{\phi^{\omega_n}}$ is the characteristic function of the subset ϕ^{ω_n}, it can be rewritten as:

$$\forall n \in \Theta_N, \ I_n = \int_{\mathbb{R}^2} \mathscr{I}(\omega) \chi_{\phi^{\omega_n}}(\omega) d\omega. \tag{2}$$

Finally, the measured pixel values are quantized to obtain the digital image. It should be kept in mind that what we have at hand is not the image but rather discrete measures of it.

There are many operations for which it is mandatory to account for the underlying continuous nature of the image: derivation, morphing, filtering, geometric and perspective transformations, etc. For those operations, the aim is to define a discrete operator that can mimic the equivalent operation in the continuous domain. This idea is illustrated in Fig. 4 when considering a rotation of a detail of the image depicted in Fig. 1 around the optical axis. Let us consider the input discrete image as being obtained by sampling a continuous image. The image we would like to obtain by using the discrete rotation operator is the image that would be obtained by rotating and then sampling the original continuous image. Such an operation is not possible due to the loss of information induced by the sampling. It has to be approximated in a way that preserves at best the original (discrete) information. For example, a particularly desirable property would be the reversibility of a digital operation. However, continuous based discrete operations always lead to information loss [3]. Therefore, the original information cannot be reconstructed from the processed image.

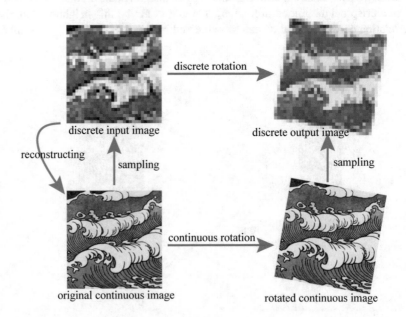

Fig. 4 How does going from continuous to discrete image processing work?

Different methods have been proposed in the relevant literature to counteract this information loss (see e.g. [9] for a lossless rotation that uses permutations of pixel values). Instead, the method we propose aims at quantifying the information loss induced by a geometrical transformation.

3.2 Kernel-Based Image Processing

Kernel-based image processing, as illustrated in Fig. 4, consists of defining discrete operations on digital images that are analogous to operations defined on continuous images in the continuous domain. Kernels are used for defining weighted neighborhoods of a location in the image plane, with the aim of reconstructing a continuous image from a discrete image, or sampling a continuous image to build a discrete image.

3.2.1 Kernel-Based Image Sampling and Reconstruction

Let $\{\omega_n\}_{n \in \Theta_N}$ be the N sampling locations. In Sect. 3.1, sampling a continuous image was very straightforwardly modeled by integrating the illumination function $\mathscr{I}$ in a crisp neighborhood around each sampling location ω_n (see Eq. 2). This modeling supposes that the sensor impulse response is uniform. To account for a known non-uniform sensor impulse response, a more general link between the sampled pixel values I and the continuous illumination $\mathscr{I}$ can be obtained by replacing the neighborhood function χ_ϕ by a sampling kernel κ [6]:

$$\forall n \in \Theta_N, \ I_n = \int_{\mathbb{R}^2} \mathscr{I}(\omega)\kappa(\omega - \omega_n)d\omega = \int_{\mathbb{R}^2} \mathscr{I}(\omega)\kappa^n(\omega)d\omega, \tag{3}$$

with κ^n being the kernel κ translated in ω_n.

A certain degree of consistency has to be kept between the continuous and discrete domains. As the range of the real measure of the illumination value is generally unknown, it can be supposed, without loss of generality, that the correspondance between digital gray-scale and real value is linear, i.e. quantifying a real illumination value amounts to replacing this value by its nearest integer. That comes down to assuming that the original illumination measurement scale is the real valued counterpart of the digital available grayscale (e.g. if the digital grayscale is $\{0, \ldots 255\}$, then the real illumination scale is $[0, 255]$). The consistency of both continuous and discrete images can be expressed as follows: "if $\mathscr{I}$ is a constant image such that $\forall \omega \in \Omega$, $\mathscr{I}(\omega) = a$, then $\forall n \in \Theta_N, I_n = a$". Considering Eq. 3, $a = \int_{\mathbb{R}^2} a\kappa(\omega - \omega_n)d\omega$, i.e. $\int_{\mathbb{R}^2} \kappa(\omega)d\omega = 1$, thus κ is a summative kernel.

Reconstruction can be considered as the converse sampling procedure. However, since sampling induces information loss, the recomposed image usually cannot be seen as a perfect reconstruction of the original continuous illumination $\mathscr{I}$, but rather as an estimate $\hat{\mathscr{I}}$ of the continuous function $\mathscr{I}$. This estimate is obtained by a finite weighted sum of the pixel values I_n ($n \in \Theta_N$):

$$\hat{\mathscr{I}}(\omega) = \sum_{n \in \Theta_N} I_n \eta(\omega_n - \omega) = \sum_{n \in \Theta_N} I_n \eta_n^\omega, \tag{4}$$

where η is a continuous reconstruction kernel and η_n^ω is the discrete kernel induced by sampling η translated in ω on the sampling grid.

The same consistency between the continuous and discrete domains evoked above implies that $\forall \omega \in \Omega$, $\sum_{n \in \Theta_N} \eta(\omega - \omega_n) = \sum_{n \in \Theta_N} \eta_n^\omega = 1$: sampling a reconstruction kernel translated at any location $\omega \in \Omega$ leads to a discrete summative kernel. Moreover, if the sampling is uniform, then η is even [25].

3.2.2 From Summative to Maxitive Kernel-Based Image Processing

Let φ be a geometric transformation (i.e. a mapping form $\mathbb{R}^2$ to $\mathbb{R}^2$) transforming a continuous image $\mathscr{I}$ into another continuous image $\mathscr{I}'$ such that $\forall \omega \in \Omega$, $\mathscr{I}'(\omega) = \mathscr{I}(\varphi(\omega))$. Let $\{\omega_n\}_{n \in \Theta_N}$ be the sampling locations and $\{I_n\}_{n \in \Theta_N}$ be the pixel values of the original discrete image—supposedly obtained by sampling $\mathscr{I}$.

Kernel based image processing consists of deriving a discrete operation that is equivalent to sampling $\mathscr{I}'$ on the sampling grid.

Following Eq. 3, sampling the continuous image $\mathscr{I}'$ leads to the sampled image I' such that:

$$\forall k \in \Theta_N, I_k' = \int_{\mathbb{R}^2} \mathscr{I}'(\omega) \kappa(\omega - \omega_k) d\omega, \tag{5}$$

where κ is the sampling kernel. Considering $\mathscr{I}'(\omega) = \mathscr{I}(\varphi(\omega))$, Eq. 5 becomes:

$$\forall k \in \Theta_N, I_k' = \int_{\mathbb{R}^2} \mathscr{I}(\varphi(\omega)) \kappa(\omega - \omega_k) d\omega. \tag{6}$$

Now, let η be a reconstruction kernel, i.e. $\forall \omega \in \Omega$, $\mathscr{I}(\omega) = \sum_{n \in \Theta_N} I_n \eta(\omega_n - \omega)$, then $\forall k \in \Theta_N$:

$$
\begin{aligned}
I_k' &= \int_{\mathbb{R}^2} \sum_{n \in \Theta_N} I_n \eta(\varphi(\omega_n - \omega)) \kappa(\omega - \omega_k) d\omega \\
&= \sum_{n \in \Theta_N} I_n \int_{\mathbb{R}^2} \eta(\varphi(\omega_n - \omega)) \kappa(\omega - \omega_k) d\omega = \sum_{n \in \Theta_N} I_n \rho_n^k,
\end{aligned}
\tag{7}
$$

with $\rho_n^k = \int_{\mathbb{R}^2} \eta(\varphi(\omega_n - \omega))\kappa(\omega - \omega_k)d\omega$.

Thus, estimating the discrete values of I' based on I comes down to defining, for each location k, a positive discrete kernel ρ^k by sampling the continuous kernel κ^k, defined by $\forall \omega \in \Omega$, $\kappa^k(\omega) = \kappa(\omega - \omega_k)$, with the continuous kernel η^φ defined by: $\forall \omega \in \mathbb{R}$, $\eta^\varphi(\omega) = \eta(\varphi(\omega))$. The obtained weights define discrete summative kernels (for each k) if $\sum_{n \in \Theta_N} \rho_n^k = 1$ which is not guaranteed for every transformation φ. This is due to the fact that φ is not an equiareal mapping. Within this approach, a way to cope with this problem is to normalize ρ by replacing ρ_n^k by $\frac{\rho_n^k}{\sum_{i \in \Theta_N} \rho_i^k}$.

Since we consider a uniform partition, η is a positive even kernel. Let C_n be the (possibly non-normalized) fuzzy subset whose membership function is defined by: $\forall \omega \in \Omega$, $\mu_{C_n}(\omega) = \eta(\omega - \omega_n) = \eta(\omega_n - \omega)$. By construction $\forall \omega \in \Omega$, $\sum_{n \in \Theta_N} \mu_{C_n}(\omega) = 1$. Therefore the subsets $\{C_n\}_{n \in \Theta_N}$ form a fuzzy partition. Now, let C_n^φ ($n \in \Theta_N$) be the fuzzy subsets defined by $\forall \omega \in \Omega$, $\mu_{C_n^\varphi}(\omega) = \eta(\varphi(\omega - \omega_n)) = \eta(\varphi(\omega_n - \omega))$. Since φ is a mapping, the property $\forall \omega \in \Omega^\varphi$, $\sum_{n \in \Theta_N} \mu_{C_n^\varphi}(\omega) = 1$ is kept (with Ω^φ being defined by: $\Omega^\varphi = \{\omega | \exists u \in \Omega, \omega = \varphi(u)\}$). $\{C_n^\varphi\}_{n \in \Theta_N}$ is a fuzzy partition of Ω^φ.

Now, let Q_κ^k be the measure defined by: $\forall A \subseteq \Theta_N$, $Q_\kappa^k(A) = \sum_{n \in A} \rho_n^k$, then the value I_k' can be seen as being the estimate of I w.r.t. Q_κ^k:

$$I_k' = \mathbb{E}_{Q_\kappa^k}(I). \tag{8}$$

The measure Q_κ^k can also be seen as an estimate:

$$\forall A \subseteq \Theta_N, \quad Q_\kappa^k(A) = \sum_{n \in A} \int_{\mathbb{R}^2} \eta^\varphi(\omega_n - \omega)\kappa(\omega - \omega_k)d\omega$$

$$= \int_{\mathbb{R}^2} \kappa^k(\omega)\Big(\sum_{n \in A} \mu_{C_n^\varphi}(\omega)\Big)d\omega \tag{9}$$

$$= \int_{\mathbb{R}^2} \kappa^k(\omega)\Upsilon_A^\varphi(\omega)d\omega,$$

with Υ_A^φ being defined as in Definition 18.2: $\forall \omega \in \mathbb{R}$, $\Upsilon_A^\varphi(\omega) = \sum_{n \in A} \mu_{C_n^\varphi}(\omega)$. Thus Eq. 9 becomes:

$$\forall A \subseteq \Theta_N, Q_\kappa^k(A) = \mathbb{E}_{P_{\kappa^k}}(\Upsilon_A), \tag{10}$$

where P_{κ^k} is the set measure whose density is defined by the kernel κ^k.

As an example, let φ be a perspective transformation, then $\varphi(\omega) = \frac{A.\omega + \alpha}{B.\omega + \beta}$, where A and B are a 2×2 matrix and α and β translation vectors. Then, $\forall \omega \in \Omega$, $\mu_{C_n^\varphi}(\omega) = \eta(\varphi(\omega - \omega_n)) = \eta(\frac{A.\omega + \alpha - A.\omega_n}{B.\omega + \beta - B.\omega_n})$. This is illustrated in Fig. 5 where the interpolation kernel is considered as being the nearest-neighbor kernel $\eta^\square$ defined by: $\forall \omega \in \mathbb{R}^2$, $\eta^\square(\omega) = 1$, if $\omega \in \Xi$ and 0 else, with $\Xi =]-0.5, 0.5] \times]-0.5, 0.5]$.

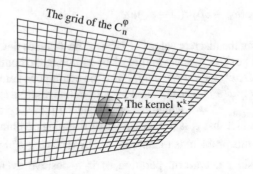

Fig. 5 Computing the ρ_n^k by sampling the kernel κ^k on the transformed interpolation partition when φ is a perspective transformation and $\eta = \eta^{\square}$

3.2.3 Maxitive Kernel-Based Image Processing

The maxitive extension of the kernel-based image processing simply involves replacing the summative sampling kernel by a maxitive sampling kernel. This replacement aims at representing imprecise knowledge on the sampling kernel, because the original sampling kernel is unknown, or because the kernel that would lead to the least distortion in the transformed image is unknown.

Let π be a maxitive kernel. Replacing κ by π in Eq. 3 leads to an interval-valued discrete image $\overline{I}$ such that $\forall n \in \Theta_N$, $\overline{I}_n = \overline{\mathbb{E}}_{\Pi_{\pi^k}}(\mathscr{I})$, with Π_{π^k} being the possibility measure induced by π^k, the maxitive kernel π translated in ω_k. This interval valued discrete image represents the convex set of all discrete images that could be obtained by sampling $\mathscr{I}$ with a kernel $\kappa \in \mathscr{M}(\pi)$ (see [13]). Following [7], Expression 10 can be extended to define the capacity v^k, for each location ω_k:

$$\forall A \subseteq \Theta_N, v_\pi^k(A) = \overline{\mathbb{E}}_{\Pi_{\pi^k}}(\Upsilon_A^\varphi). \tag{11}$$

By construction, v_π^k is a concave capacity such that $\forall \kappa \in \mathscr{M}(\pi)$ and $\forall A \subseteq \Theta_N$, $Q_\kappa^k(A) \leq v_\pi^k(A)$.

The final step of this extension consists of estimating the interval valued image $\overline{I}'$ by replacing, in Eq. 8, the discrete additive set measure Q_κ^k by the discrete non-additive set measure v_π^k. This leads to:

$$\forall k \in \Theta_N, \overline{I}_k' = \overline{\mathbb{E}}_{v_\pi^k}(I). \tag{12}$$

The estimation operator $\overline{\mathbb{E}}$ propagates imprecise knowledge of the sampling kernel to the interval-valued transformed image.

3.3 Example: Rigid Transformations

As illustrated bellow, the computation can be very easy. But at first we should answer the question: "which maxitive kernel has to be chosen?". As in image processing mostly separable kernels or radial kernels are considered, a very interesting kernel would be one that dominates any sampling kernel whose support is bounded. This problem was addressed in [6] by considering separable kernels and separable estimations. As mentioned by the authors, such an approach is not suitable for transformations like rotations. A nice answer to this problem is proposed in [1] where different $2D$ maxitive kernels are given depending on which summative kernels are to be considered. For example, the continuous maxitive kernel $\check{\pi}(\omega) = \max(0, 1 - \|\omega\|^2)$ dominates every bell-shaped radial continuous summative kernel whose support is included in $[-1, 1]$. This is very convenient because it can represent the fact that the support of the sampling kernel is at most the distance between two pixels but all that is known about the shape is that it is bell-shaped and radial. Note that the Dirac impulse is included in $\mathcal{M}(\check{\pi})$, ensuring a kind of guaranteed preservation of the information borne by the digital original image.

Thus now let $\check{\pi}$ be the sampling maxitive kernel and $\eta^{\square}$ the interpolation kernel. Then the partition $\{C_n\}_{n \in \Theta_N}$ is a crisp partition, with each C_n being a box centered in ω_n. Then Eq. 11 leads to:

$$\forall A \subseteq \Theta_N, v_\pi^k(A) = \overline{\mathbb{E}}_{\Pi_{\pi^k}}(\Upsilon_A) = \sup_{\omega \in \bigcup_{n \in A} C_n^\varphi} \check{\pi}(\omega - \omega_k).$$

In that case, since $\{C_n^\varphi\}_{n \in \Theta_N}$ forms a crisp partition, v_π^k is a possibility measure associated with the discrete possibility distribution γ^k defined by:

$$\forall n \in \Theta_N, \gamma_n^k = \sup_{\omega \in C_n^\varphi} \check{\pi}(\omega - \omega_k). \tag{13}$$

where φ is a rigid transform, it is defined by: $\forall \omega \in \mathbb{R}^2, \varphi(\omega) = R.\omega + \tau$, where R is a 2×2 orthonormal matrix and $\tau \in \mathbb{R}^2$ is a translation vector.

C_n^φ can be computed very easily based on using geometric considerations. If the partition is not crisp, then v_π^k is not a possibility measure but can be computed analytically (see [7] for an example of this kind of computation).

To illustrate this approach, we propose to take another look at the experiment illustrated in Fig. 2. The same detail of Fig. 1 has been rotated by 3° using the approach proposed in Sect. 3.2.3 by considering a crisp partition of the original image and the maxitive kernel $\check{\pi}(\omega)$ described above. The discrete capacity v_n^k is a possibility measure defined by the distribution computed via Eq. 13. By construction, the images in Figs. 2.a and 2.b are included in the interval-valued image represented by its lower bound Fig. 6a and its upper bound Fig. 6b. Figure 6c represents the imprecision of the interval valued image. Note also that the most important imprecision is concentrated near the edges of the drawing, i.e. the regions where two different interpolations yield two different values.

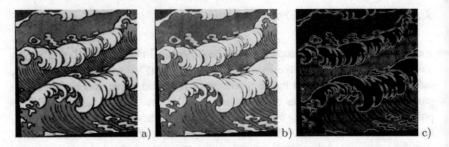

Fig. 6 Imprecise rotation (3°) of the same detail of Fig. 1. The interval-valued image is represented by its lower bound (**a**) and its upper bound (**b**). (**c**) is an enhanced view of the difference between the upper and lower image, i.e. the imprecision of the interval-valued images

4 Conclusion and Discussion

In many applications, images are subject to geometric distortions introduced by perspective, misalignment, optical aberrations, imaging sensor movements etc., which need to be corrected for further interpretation. Digital geometrical transformations are essential for reversing these distortions, aligning different images in a common frame or simply enabling image analysis or interpretation. Many such transformations aim a mimicking a physical operation defined in the continuous illumination domain. The kernel-based approach is a convenient way for designing digital transformations that may ensure a certain degree of preservation of digital image topological and illumination properties. However, this method relies on modeling the interplay between the continuous and discrete domains via two arbitrarily chosen kernels that model the sampling (to go from the continuous to discrete domain) and the reconstruction (to go from the discrete to continuous domain) operations. The arbitrariness of this choice can have serious consequences in applications where details have to be preserved with comparable image resolution.

Maxitive image processing can be an interesting solution to preserve information borne by the original image. It allows us to compute the (convex) set of all images that would have been obtained by considering a (convex) set of kernels that could be appropriate for computing this transformation. This computation can be achieved with a very minor increase in computation complexity (see e.g. [6]).

This article is focussed on modeling imprecise knowledge on the sampling kernel. Imprecise knowledge on the approximation kernel has to be taken into account in another way. Note however that, due to the interchangeable role of kernels in Expression 7, this modeling actually addresses imprecise knowledge on the convolution of both reconstruction and sampling kernels (see e.g. [7]). This needs to be further investigated. Other problems have to be investigated, including reversibility—how are both illumination and topological information preserved when an image is subject to a transformation and then its inverse transformation?—and selection, in the convex set of obtained images, of a representative image to be presented to experts (see e.g. [11]).

References

1. Akrout, H., Crouzet, J-F., Strauss, O.: Multidimensional possibility/probability domination for extending maxitive kernel based signal processing. Fuzzy Sets Syst. (to appear)
2. Bloch, I.: Fuzzy sets for image processing and understanding. Fuzzy Sets Syst. **281**, 280–291 (2015). Special Issue Celebrating the 50th Anniversary of Fuzzy Sets
3. Danielsson, P.-E., Hammerin, M.: High-accuracy rotation of images. CVGIP: Graph. Models Image Process. **54**(4), 340–344 (1992)
4. Denneberg, D.: Non-Additive Measure and Integral. Kluwer Academic Publishers (1994)
5. Dubois, D.: Possibility theory and statistical reasoning. Comput. Stat. Data Anal. **51**(1):47–69 (2006). The Fuzzy Approach to Statistical Analysis
6. Graba, F., Comby, F., Strauss, O.: Non-additive imprecise image super-resolution in a semi-blind context. IEEE Trans. Image Process. **26**(3), 1379–1392 (2017)
7. Graba, F., Strauss, O.: An interval-valued inversion of the non-additive interval-valued f-transform: Use for upsampling a signal. Fuzzy Sets Syst. **288**, 26–45 (2016)
8. Grabisch, M., Marichal, J.-L., Mesiar, R., Pap, E.: Aggregation Functions. Cambridge University Press (2009)
9. Komatsu, K., Sezaki, K.: Lossless rotation transformations with periodic structure. In: IEEE International Conference on Image Processing 2005, vol. 2, pp. II–273 (2005)
10. Kucharczak, F., Loquin, K., Buvat, I., Strauss, O., Mariano-Goulart, D.: Interval-based reconstruction for uncertainty quantification in pet. Phys. Med. Biol. **63**(3), 035014 (2018)
11. Kucharczak, F., Mory, C., Strauss, O., Comby, F., Mariano-Goulart, D.: Regularized selection: a new paradigm for inverse based regularized image reconstruction techniques. In: Proceedings of IEEE International Conference on Image Processing, Beijin, China (2017)
12. Lehmann, T.M., Gonner, C., Spitzer, K.: Survey: interpolation methods in medical image processing. IEEE Trans. Med. Imaging **18**(11), 1049–1075 (1999)
13. Loquin, K., Strauss, O.: On the granularity of summative kernels. Fuzzy Sets Syst. **159**(15), 1952–1972 (2008)
14. Loquin, K., Strauss,O.: Linear filtering and mathematical morphology on an image: a bridge. In: Proceedings of IEEE International Conference on Image Processing, pp. 3965–3968, Le Caire, Egypt (2009)
15. Nie, Y., Barner, K.E.: The fuzzy transformation and its applications in image processing. IEEE Trans. Image Process. **15**(4), 910–927 (2006)
16. Omhover, J., Detyniecki, M., Rifqi, M., Bouchon-Meunier, B.: Ranking invariance between fuzzy similarity measures applied to image retrieval. In: 2004 IEEE International Conference on Fuzzy Systems (IEEE Cat. No.04CH37542), vol. 3, pp. 1367–1372 (2004)
17. Park, S.C., Park, M.K., Kang, M.G.: Super-resolution image reconstruction: a technical overview. Signal Process. Mag. IEEE **20**(3), 21–36 (2003)
18. Perfilieva, I., Novák, V., Dvořák, A.: Fuzzy transform in the analysis of data. Int. J. Approx. Reason. **48**(18), 36–46 (2008)
19. Rico, A., Strauss, O.: Imprecise expectations for imprecise linear filtering. Int. J. Approx. Reason. **51**(8), 933–947 (2010)
20. Rico, A., Strauss, O.: Imprecise expectations for imprecise linear filtering. Int. J. Approx. Reason. **51**(8), 933–947 (2010)
21. Rico, A., Strauss, O., Mariano-Goulart, D.: Choquet integrals as projection operators for quantified tomographic reconstruction. Fuzzy Sets Syst. **160**(2), 198–211 (2009)
22. Ruspini, E.: New experimental results in fuzzy. Inf. Sci. **6**, 273–284 (1973)
23. Schmeidler, D.: Integral representation without additivity. Process. Am. Math. Soc. **2** (1986)
24. Shen, J., Castan, S.: Towards the unification of band-limited derivative operators for edge detection. Signal Process. **31**(2), 103–119 (1993)
25. Strauss, O.: Non-additive interval-valued f-transform. Fuzzy Sets Syst., 1–24 (2015)
26. Unser, M.: Splines: a perfect fit for signal and image processing. IEEE Signal Process. Mag. **16**, 22–38 (1999)

F-Transform Representation of Nonlocal Operators with Applications to Image Restoration

Irina Perfilieva

Abstract In image processing, nonlocal operators define a new type of functionals that extend the ability of classical PDE-based algorithms in handling textures and repetitive structures. We showed that in the particular space with a fuzzy partition, the nonlocal Laplacean and partial derivatives can be represented by the F^0- and F^1-transforms. For the restoration problem specified by relatively large damaged areas, we propose a new total variation model with the F-transform-based nonlocal operators. We show that the proposed model together with the corresponding algorithm increase the quality of a (usually considered) patch-based searching algorithm.

1 Introduction

We are focused on new theoretical and application aspects of image processing in general and advantage of fuzzy (soft computing) techniques in this field in particular. Together with perpetual development of soft computing methods, their applications in analysis and representation of images grow in numbers and establish stable and developing approach in this field of research. The explanation of this phenomenon is that soft computing approach perceives images as meaningful entities and represents their immanent meaning formally. The purpose of relevant representation is to make important aspects of physical objects explicit and by this, accessible for processing.

There are many methods involved in image processing. A vast majority of them consider image as a two-dimensional signal, and process it accordingly. As a result, the technique of Fourier transform is widely used, but leads to the situation where the connection between a problem and a method of its solution is lost. Contrary to this, we propose the method of fuzzy (F-)transform that is based on a *fuzzy partition* of a universe of discourse. The method allows us to stay in the same space where a problem is formulated.

I. Perfilieva (✉)
University of Ostrava, Institute for Research and Applications of Fuzzy Modeling, NSC IT4Innovations, 30. dubna 22, 2270103 Ostrava 1, Czech Republic
e-mail: Irina.Perfilieva@osu.cz

217

M.-J. Lesot and C. Marsala (eds.), *Fuzzy Approaches for Soft Computing and Approximate Reasoning: Theories and Applications*, Studies in Fuzziness and Soft Computing 394, https://doi.org/10.1007/978-3-030-54341-9_19

A space with a fuzzy partition appears in the fuzzy literature using various requirements and consequently, under various names: granulation, rough sets, space with similitude, etc., see in [6–8, 13, 15]. In a certain sense, this notion is comparable with notions of metric, topological or other spaces where structure characterizes closeness between elements. In a space with a fuzzy partition, the atomic elements are particular fuzzy sets, represented by basic functions, providing a low-dimensional representation of original data lying in a high-dimensional space.

The F-transform technique maps the data from a high to low dimensional spaces. This makes it attractive and useful in many applications ranging from image processing to numerical methods for differential equations. The F-transform can be explained in the language of discrete convolutions, scale-space representations, integral transforms and generally, aggregation operators. This fact emphasizes its high position in the application field. Below, we discuss one typical problem of image processing—restoration and explain how it can be properly formulated and solved in the language of F-transform.

Image restoration (inpainting) is a process of filling in damaged regions of images using information from surrounding areas. Formally, image I should be reconstructed from its damaged version f. This problem is connected with the two well developed problems of image processing: upscaling and denoising.

From the technical point of view, restoration is a type of (extra-) interpolation. However, there are many other approaches where restoration is considered as a boundary valued problem, based on nonlinear partial differential equations [1], as a particular inverse problem [14], or as a modification of the variational image denoising or segmentation [2, 3]. The latter model (called ROF—short of Rudin-Osher-Fatemi) focuses on the minimization of the functional

$$\int_{\Omega} |\nabla I| + \lambda \|f - GI\|^2_{L^2(\Omega)}, \tag{1}$$

where Ω is a domain of f, ∇ stands for the gradient, and G—a damaging operator. In (1), the first summand (regularization term) is a *total variation* (TV) of I, and the second one is the fidelity term. Regularization aims at successful propagation of sharp edges into the damaged domain, while fidelity is responsible for closeness between a solution and the original image. Formally, this model is focused on the minimization of the functional.

In this contribution, we stem from the ROF-type model with nonlocal operators and modify it with the F-transform-based operators. Our purpose is to find a suitable model for the restoration problem specified by relatively large damaged areas. In this case, the best known results are achieved using patches.

Our contribution to this problem is two-fold: at first, we modify the functional in the underlying ROF model to make it suitable for the patch-based restoration; and at second, we show how the nonlocal operators can be easily represented with the help of higher degree F-transform components.

The structure of this contribution is as follows. The preliminaries and a short introduction into the theory of F-transforms are given in Sect. 2. The F-transform-based representation of nonlocal operations is considered in Sect. 3. Application of the presented theoretical constructions to the inpainting process is explained in Sect. 4. Experiments and comparison with the similar technique is given in Sect. 5.

2 Preliminaries

The F-transform (originally, *fuzzy transform*, [10]) is a particular integral transform whose peculiarity consists in using a *fuzzy partition* of a universe of discourse (usually, the set of reals $\mathbb{R}$). The F-transform maps a function onto a sequence of its F-transform components and by this, provides a compact (compressed) representation of this function.

By this we mean that every F-transform component is a "nonlocal" functional value at an element of a fuzzy partition of the function domain. To emphasize the difference with the conventional "local" representation of a function, let us draw a parallel with the representation of a smooth function f:

$$f(t) = \int_{-\infty}^{\infty} f(x)\delta(t - x)dx, \tag{2}$$

where δ is the Dirac's delta.[1] The collection $\{\delta_t, \ t \in \mathbb{R}\}$, where $\delta_t(x) = \delta(t - x)$ can be considered as a *partition* of $\mathbb{R}$ by one-element sets $\{t\}$ with "characteristic" functions δ_t. We say that the partition $\{\delta_t, \ t \in \mathbb{R}\}$ is generated by δ or δ is a "generating function" of this partition.

With this particular and extreme case in mind, let us give details of what we call a (uniform) fuzzy partition. In [12], we generalized (2) and defined the notion of a *generating function of a fuzzy partition*. We recall that a non-negative, continuous, even, bell-shaped function $a : \mathbb{R} \to [0, 1]$ is a generating function of a (uniform) fuzzy partition, if it vanishes outside $[-1, 1]$ and fulfills $\int_{-1}^{1} a(t) \, dt = 1$. The following are two examples of a generating function:

$$a^{cos}(t) = \begin{cases} (\cos^2(\frac{\pi t}{2})), & -1 \le t \le 1, \\ 0, & \text{otherwise,} \end{cases} \quad a^{tr}(t) = \begin{cases} (1 - |t|), & -1 \le t \le 1, \\ 0, & \text{otherwise.} \end{cases}$$

Generating function a produces infinitely many *rescaled* functions $a_H : \mathbb{R} \to [0, 1]$, where $H > 0$ is a *scale factor*, and

---

[1]Speaking formally, the Dirac's delta is not a function, but a generalized function or a linear functional. Therefore, it makes sense to use it only with respect to its action on some function.

$$a_H(t) = a\left(\frac{t}{H}\right).$$

A collection of translations $\mathscr{A}_{h,H} = \{a_H(k \cdot h - t), \ k \in \mathbb{Z}\}$ is called an (h, H)-*fuzzy partition*[2] of $\mathbb{R}$ generated by a, where H is a scale factor, h is a *step-value*, and $0 < h < 2H$. The latter condition guarantees that every point in $\mathbb{R}$ is covered by a certain partition element. We will be working with the two particular cases of an (h, H)-fuzzy partition, generated by a. The first case is specified by $h = H$ and will be referred to as h-fuzzy partition. The second case has no step-value and the corresponding fuzzy partition of $\mathbb{R}$ is given by

$$\mathscr{A}_H = \{a_H(x - t), \ t \in \mathbb{R}\}.$$

Contrary to $\delta(t - x)$, every fuzzy set $a_H(x - t)$ is a *nonlocal* element in the universe of discourse. Fuzzy partition $\mathscr{A}_H$ is connected with a symmetric *weight function* $w(x, t) = a_H(x - t)$, whose symmetry follows from the fact that the generating function a is even.

To summarize our discussion, we introduce two new spaces on the set of reals with *fuzzy and symmetry structures* on them:

- $(\mathbb{R}, \mathscr{A}_{h,H})$—space with an (h, H)-fuzzy partition;
- $(\mathbb{R}, w)$—space with symmetrical weights, where $w : \mathbb{R} \times \mathbb{R} \to [0, +\infty)$ and $w(x, y) = w(y, x)$.

Let us remark that the structure of each space determines a "closeness" between its elements, i.e.,

- x, y are *close in* $(\mathbb{R}, \mathscr{A}_{h,H})$, if there exists $k \in \mathbb{Z}$, such that $a_H(k \cdot h - t)(x) \cdot a_H(k \cdot h - t)(y) > 0$;
- x, y are *close in* $(\mathbb{R}, w)$, if the value $w(x, y)$ is large.

2.1 F^m-*transform*

Let us fix $[a, b] \subset \mathbb{R}$, generating function a, $n \geq 2$, and the h-uniform fuzzy partition $\{A_1, \ldots, A_n\}$, where $A_k(x) = a_h(x - x_k)|_{[a,b]}$, $x_k = a + h(k - 1)$, $k = 1, \ldots, n$, and $h = \frac{b-a}{n-1}$. Let us notice that

$$h = \int\limits_{x_{k-1}}^{x_{k+1}} A_k(x)dx, \ k = 2, \ldots, n - 1.$$

---

[2]Every function $a_H(k \cdot h - t)$ is considered as a membership function of a fuzzy set, so that the whole collection $\mathscr{A}_{h,H}$ is said to be a fuzzy partition. Each element in $\mathscr{A}_{h,H}$ is a fuzzy set with a bounded support.

Let k be a fixed integer from $\{1, \ldots, n\}$, $L^2(A_k)$ a set of square-integrable functions on $[x_{k-1}, x_{k+1}]$, and $L^2([a, b])$ a set of square-integrable functions on $[a, b]$. We consider $\int_{x_{k-1}}^{x_{k+1}} f(x)g(x)A_k(x)dx$ as a (weighted) inner product and denote it by $\langle f, g \rangle_k$. The functions $f, g \in L^2(A_k)$ are *orthogonal* in $L^2(A_k)$ if $\langle f, g \rangle_k = 0$.

Let us denote by $L_m^2(A_k)$ a linear subspace of $L^2(A_k)$ spanned by restrictions of orthogonal polynomials $P_k^0, P_k^1, P_k^2 \ldots, P_k^m$ to the interval $[x_{k-1}, x_{k+1}]$. The orthogonal projection of $f \in L^2(A_k)$ on $L_m^2(A_k)$ is as follows

$$F_k^m(f) = c_{k,0}P_k^0 + c_{k,1}P_k^1 + \cdots + c_{k,m}P_k^m, \tag{3}$$

where for all $i = 0, 1, \ldots, m,$

$$c_{k,i} = \frac{\langle f, P_k^i \rangle_k}{\langle P_k^i, P_k^i \rangle_k} = \frac{\int_{x_{k-1}}^{x_{k+1}} f(x)P_k^i(x)A_k(x)dx}{\int_{x_{k-1}}^{x_{k+1}} P_k^i(x)P_k^i(x)A_k(x)dx}. \tag{4}$$

The n-tuple $(F_1^m, \ldots, F_n^m)$ is an F^m-transform of f with respect to $\{A_1, \ldots, A_n\}$, or formally,

$$F^m[f] = (F_1^m(f), \ldots, F_n^m(f)).$$

$F_k^m(f)$ is called the kth F^m-transform component of f. We will omit the reference to f, if it is clear from the context.

In particular, the restrictions of orthogonal polynomials P_k^0 and P_k^1 to $[x_{k-1}, x_{k+1}]$ are: $1_k(x)$ and $x - x_k$, respectively. By $1_k(x)$ we denote a constant function with the value 1 on the domain $[x_{k-1}, x_{k+1}]$. Then, the F^0-transform of f or simply, the F-transform of f with respect to the partition $\{A_1, \ldots, A_n\}$ is given by the n-tuple

$$F^0[f] = (c_{1,0}1_1(x), \ldots, c_{n,0}1_n(x)),$$

of constant functions (0-degree polynomials) where for $k = 1, \ldots, n,$

$$c_{k,0} = \frac{\langle f, 1 \rangle_k}{\langle 1, 1 \rangle_k} = \frac{1}{h} \int_{x_{k-1}}^{x_{k+1}} f(x)A_k(x)dx. \tag{5}$$

In many cases, we identify the F^0-transform of f with the n-tuple $(c_{1,0}, \ldots, c_{n,0})$ of coefficients.

The F^1-transform of f with respect to $\{A_1, \ldots, A_n\}$ is given by the n-tuple

$$F^1[f] = (c_{1,0} + c_{1,1}(x - x_1), \ldots, c_{n,0} + c_{n,1}(x - x_n)),$$

of linear functions (1-degree polynomials), where instead of writing $c_{k,0}1_k(x)$ we simply wrote $c_{k,0}, k = 1, \ldots, n$.

$F^1[f]$ is fully represented by its vectorial coefficients $((c_{1,0}, c_{1,1})^T, \ldots, (c_{n,0}, c_{n,1})^T)$, which in addition to (5), have the following representation:

$$c_{k,1} = \frac{\langle f, x - x_k \rangle_k}{\langle (x - x_k), (x - x_k) \rangle_k} = \frac{\int_{x_{k-1}}^{x_{k+1}} f(x)(x - x_k)A_k(x)dx}{\int_{x_{k-1}}^{x_{k+1}} (x - x_k)^2 A_k(x)dx}.$$

The inverse F^m-transform ($iF^m zT$) is applied to a sequence of corresponding components and transforms it linearly into a function from $L_2(\mathbb{R})$. In general, it is different from the original function. In [10], it has been shown that the $iF^m zT$ can approximate a continuous function with an arbitrary precision.

2.2 F^0- and F^1-transforms for Multivariate Functions

In the view of the F-transform applications to image processing, we extend the F^0- and F^1-transform techniques for multivariate functions, considering (for the simplicity) the case of functions with 2 variables. Let $[\mathbf{a}, \mathbf{b}] \subset \mathbb{R}^2$, where $\mathbf{a} = (a_1, a_2)$, $\mathbf{b} = (b_1, b_2)$, and $[\mathbf{a}, \mathbf{b}] = [a_1, b_1] \times [a_2, b_2]$.

Below, we introduce one particular case of a fuzzy partition of $[\mathbf{a}, \mathbf{b}]$ with a single generating function a and a scale factor $h > 0$:

$$\mathscr{A}_h = \{a_h(x - t)a_h(y - s), \ t \in [a_1, b_1], s \in [a_2, b_2]\}, \tag{6}$$

where $a_h(x) = a\left(\frac{x}{h}\right)|_{[a_1,b_1]}$ and $a_h(y) = a\left(\frac{y}{h}\right)|_{[a_2,b_2]}$.

Let $f \in L^2([\mathbf{a}, \mathbf{b}])$. Then, the (t, s)-F^0-transform component of f with respect to the above given partition $\mathscr{A}_h$ is

$$F_{t,s}^0(x, y) = c_{t,s}^{00} 1_{t,s}(x, y), \tag{7}$$

where

$$c_{t,s}^{00} = \frac{1}{h^2} \int\limits_{a_1}^{b_1} \int\limits_{a_2}^{b_2} f(x, y)a_h(x - t)a_h(y - s)dxdy,$$

and function $1_{t,s}(x, y)$ is equal to 1 on $[t - h, t + h] \times [s - h, s + h]$, and 0 otherwise.

The (t, s)-F^1-transform component of f with respect to the partition $\mathscr{A}_h$ is

$$F_{t,s}^1(x, y) = c_{t,s}^{00} 1_{t,s}(x, y) + c_{t,s}^{10}(x - t)|_{[t-h,t+h]} + c_{t,s}^{01}(y - s)|_{[s-h,s+h]}, \tag{8}$$

where

$$c_{t,s}^{10} = \frac{\int_{a_1}^{b_1} \int_{a_2}^{b_2} f(x, y)(x - t)a_h(x - t)a_h(y - s)dxdy}{\int_{a_1}^{b_1} \int_{a_2}^{b_2} (x - t)^2 a_h(x - t)a_h(y - s)dxdy},$$

and

$$c_{t,s}^{01} = \frac{\int_{a_1}^{b_1} \int_{a_2}^{b_2} f(x, y)(y - s)a_h(x - t)a_h(y - s)dxdy}{\int_{a_1}^{b_1} \int_{a_2}^{b_2} (y - s)^2 a_h(x - t)a_h(y - s)dxdy}.$$

By the similar technique as in [11], it can be shown that F^1-transform coefficients $c_{t,s}^{00}$, $c_{t,s}^{10}$, and $c_{t,s}^{01}$ approximate respectively, the function value and its partial derivatives at the point (t, s) with the accuracy h^2.

3 F-transform-Based Representation of Nonlocal Operations

Nonlocal operators are introduced for data that has a corresponding structure and consequently, representation. Above, the two spaces with nonlocal structures were discussed: a space $(\mathbb{R}, \mathscr{A}_{h,H})$ with an (h, H)-fuzzy partition, and a space $(\mathbb{R}, w)$ with symmetrical weights. Historically, the nonlocal operators were introduced in the language of the space $(\mathbb{R}, w)$, where different criterion of closeness than that based on the Euclidean distance is used. In the space $(\mathbb{R}, w)$, closeness between x and y is measured by the weight $w(x, y)$, associated with a "distance" in such a way that

$$w(x, y) = \tilde{d}^{-2}(x, y), \tag{9}$$

where $\tilde{d}^3$ is assumed to be ranged in $(0, +\infty]$.

Close objects are assigned large values of w. This fact gives a possibility to introduce nonlocal operators by replacing small distances in definitions of local ones (partial derivatives, etc.) by weights.

Recently, nonlocal operators were introduced in [5] with the focus on image processing. Below we repeat some useful definitions and show how the nonlocal partial derivative and Laplacian can be represented by the corresponding F-transform components, using the language of a space with a fuzzy partition.

Let $(\mathbb{R}, w)$ be a space with symmetrical weights. For function $f : \Omega \to \mathbb{R}, \Omega \subset \mathbb{R}^s$, the *nonlocal derivative* is defined as follows:

$$\partial_y f(x) = \frac{f(y) - f(x)}{\tilde{d}(x, y)}, \tag{10}$$

where $x, y \in \mathbb{R}^s$. Using (9), we rewrite

$$\partial_y f(x) = (f(y) - f(x))\sqrt{w(x, y)}. \tag{11}$$

---

[3] Actually, $\tilde{d}$ is not a distance in the pure sense of this notion. It does not fulfill the basic requirements such as "identity of indiscernibles" or "triangle inequality".

By (11), any point x can "interact" directly with any other point y in the function domain.

The *nonlocal gradient* $\nabla_w f(x)$ is defined as a function $\nabla_w f : \Omega \times \Omega \to \mathbb{R}$:

$$\nabla_w f(x, y) = (f(y) - f(x))\sqrt{w(x, y)}, \ x, y \in \Omega, \tag{12}$$

and the *nonlocal Laplacian* $\Delta_w f : \Omega \to \mathbb{R}$, is defined by

$$\Delta_w f(x) = \int_\Omega (f(y) - f(x))w(x, y)dy. \tag{13}$$

Finally, the *nonlocal partial derivatives* are as follows

$$\left(\frac{\partial f}{\partial x_i}\right)_w (x) = \int_\Omega (f(y) - f(x))w(x, y)(y_i - x_i)dy, \tag{14}$$

where i corresponds to the i-th coordinate space in $\mathbb{R}^s$.

In the following Proposition, we show how the last two expressions can be rewritten for functions of two variables and for the case where the weight function w is determined by a generating function of a uniform fuzzy partition. In this case, we show that the nonlocal Laplacian (13) and the partial derivative (14) can be represented with the help of the F^0- and F^1-transform components.

Proposition 19.1 *Let* $[\mathbf{a}, \mathbf{b}] \subset \mathbb{R}^2$, *where* $\mathbf{a} = (a_1, a_2)$, $\mathbf{b} = (b_1, b_2)$, *and* $[\mathbf{a}, \mathbf{b}] = [a_1, b_1] \times [a_2, b_2]$. *Let* $h > 0$ *and* $\mathscr{A}_h$ *be a fuzzy partition of* $[\mathbf{a}, \mathbf{b}]$ *in accordance with (6). Let finally,* $f \in L^2([\mathbf{a}, \mathbf{b}])$ *and for every* $(t, s) \in [\mathbf{a}, \mathbf{b}]$, *the* (t, s)-*th* F^0-*transform component and the* (t, s)-*th* F^1-*transform component are given by (7) and (8), respectively, so that*

$$F^0_{t,s}(x, y) = c^{00}_{t,s} 1_{t,s}(x, y),$$
$$F^1_{t,s}(x, y) = c^{00}_{t,s} 1_{t,s}(x, y) + c^{10}_{t,s}(x - t)|_{[t-h,t+h]} + c^{01}_{t,s}(y - s)|_{[s-h,s+h]}.$$

We put

$$w(\mathbf{x}, \mathbf{y}) = \frac{1}{h^2} a_h(x_1 - x_2)a_h(y_1 - y_2), \tag{15}$$

where $\mathbf{x} = (x_1, y_1)$, $\mathbf{y} = (x_2, y_2)$. *Then, the nonlocal Laplacian* $\Delta_w f$ *and nonlocal partial derivatives* $\left(\frac{\partial f}{\partial x}\right)_w$ *and* $\left(\frac{\partial f}{\partial y}\right)_w$ *have the following F-transform-based representation*

$$\Delta_w f(t, s) = c^{00}_{t,s} - f(t, s), \ t, s \in [\mathbf{a}, \mathbf{b}], \tag{16}$$

$$\left(\frac{\partial f}{\partial x}\right)_w (t, s) = c_{t,s}^{10} \cdot \frac{\langle P_t^1, P_t^1 \rangle_t}{\langle P_t^0, P_t^0 \rangle_t}, \tag{17}$$

$$\left(\frac{\partial f}{\partial y}\right)_w (t, s) = c_{t,s}^{01} \cdot \frac{\langle P_s^1, P_s^1 \rangle_s}{\langle P_s^0, P_s^0 \rangle_s}, \tag{18}$$

where P_t^0, P_t^1 are the polynomials from the orthogonal basis of $L_1^2(a_t)$ with respect to x, and P_s^0, P_s^1 are the polynomials from the orthogonal basis of $L_1^2(a_s)$ with respect to y.

Remark 19.1 We remark that both coefficients $\frac{\langle P_t^1, P_t^1 \rangle_t}{\langle P_t^0, P_t^0 \rangle_t}$ and $\frac{\langle P_s^1, P_s^1 \rangle_s}{\langle P_s^0, P_s^0 \rangle_s}$ in the F-transform-based Laplacian do not depend on f.

Further on, we denote $\Delta_w^{FT} \left(\left(\frac{\partial f}{\partial x}\right)_w^{FT}, \left(\frac{\partial f}{\partial y}\right)_w^{FT} \right)$ the F-transform-based representation of the nonlocal Laplacian (nonlocal partial derivatives) and refer to it as to the FT-Laplacian (FT-partial derivatives).

4   Image Restoration by the Patch-Based-Inpainting

We propose to use the ROF-type model with nonlocal operators, specified for a space with a fuzzy partition. We consider the problem of restoration for relatively large damaged areas where the best known results are achieved using patches. Our goal is to increase the quality of a patch-based algorithm by improving the underlying searching mechanism. Another concern is restoration with a lowest complexity. Therefore, we propose a new ROF model with nonlocal F-transform-based Laplacian (further, FT-Laplacian) in the regularization term, where the minimization is considered over a searching space restricted to a finite set of possible reconstructions, each of them is a result of a patch-based-inpainting.

With this purpose on mind, we make the following changes in the functional (1). Assume that $f(\bar{x})$, where $\bar{x} = (x, y)$, is a given image in the domain Ω with an h-uniform fuzzy partition over it, determined by the generating function a with the scale factor h. Let $D \subset \Omega$ be the inpainting domain. Let $I(\bar{x})$ evolve in time to become an inpainted version of $f(\bar{x})$, i.e. I minimizes the following functional:

$$\int_\Omega |\Delta_w^{FT} I| + \|\lambda(f - I)\|_{\tilde{W}^{1,2}(\Omega)}^2, \tag{19}$$

where the weight function w is given by (15), and

$$\lambda(\bar{x}) = \begin{cases} 0, & \text{if } \bar{x} \in D, \\ 1, & \text{if } \bar{x} \in \Omega \setminus D. \end{cases}$$

Our proposal in (19) relates to both regularization and fidelity terms. Firstly, the regularization is proposed to be based of the nonlocal FT-Laplacian $\Delta_w^{FT} I$, given by (16). Secondly, the fidelity is proposed to be estimated by the norm in a Sobolev-like space, such that

$$\|g\|_{\tilde{W}^{1,2}}^2 = \sum_{0 \leq \alpha \leq 1} \|D^\alpha g\|_{L^2},$$

and for $\alpha_1 + \alpha_2 = 1$,

$$D^0 g = \Delta_w^{FT} g, \quad D^\alpha g = \left(\frac{\partial^\alpha g}{\partial x^{\alpha_1} \partial y^{\alpha_2}} \right)_w^{FT},$$

where $\Delta_w^{FT} g$ is in accordance with (16) and $D^\alpha g$ is in accordance with (17) and (18).

Furthermore, we make use of the fact that f and I coincide on $\Omega \setminus D$ so that $D^0(f - I) = F_{\bar{x}}^0(f) - F_{\bar{x}}^0(I)$. Moreover by Remark 19.1, we identify $D^\alpha g$ with $\left(\frac{\partial f}{\partial x} \right)_w^{FT}$, if $\alpha_1 = 1$, $\alpha_2 = 0$, or with $\left(\frac{\partial f}{\partial y} \right)_w^{FT}$, if $\alpha_1 = 0$, $\alpha_2 = 1$. Finally, we introduce a searching space $\mathscr{I}$ such that the minimization of (19) would be restricted to $\mathscr{I}$.

Let parameter γ characterizes a size (width or height) of a rectangular patch $P_{\bar{x}}^\gamma$, centered at pixel $\bar{x}$ (all patches are of the same size and shape) and such that $h \leq \gamma$. The latter means that a patch is covered by more than one basic function from the chosen h-uniform fuzzy partition of Ω. Let

$$\Omega^\gamma = \{\bar{x} \in \Omega \mid P_{\bar{x}}^\gamma \subseteq \Omega\}.$$

With each pixel $\bar{x} \in \Omega^\gamma$, we correspond the following subset

$$S_{\bar{x}} = \{\bar{y} \mid P_{\bar{y}}^\gamma \subseteq \Omega \setminus D \,\&\, \bar{y} = \arg\min \|\lambda(\bar{x})(f|_{P_{\bar{y}}^\gamma} - f|_{P_{\bar{x}}^\gamma})\|_{\tilde{W}^{1,2}(\Omega)}^2\}.$$

It is easy to see that $\bar{x} \in S_{\bar{x}}$, if and only if $P_{\bar{x}}^\gamma \subseteq \Omega \setminus D$.

Let D^c be a set of pixels such that

$$D^c = D \cup (\delta\Omega \cap D),$$

where $\delta\Omega$ is a border of Ω.

Then, the searching space $\mathscr{I} \subseteq \mathbb{R}^\Omega$ consists of all functions I such that for all $\bar{x} \in D^c$,

$$I|_{P_{\bar{x}}^\gamma} = f|_{P_{\bar{y}}^\gamma}, \text{ for some } \bar{y} \in S_{\bar{x}}.$$

Thus, the minimization of the functional in (19) is considered over $\mathscr{I}$.

Remark 19.2 It is worth noticing that the estimation of the fidelity term is based on the closeness between F^0- and F^1-transform components. We know that both of them are not sensitive to noise [12]. Therefore, the proposed selection of a searching space is not influenced by noise.

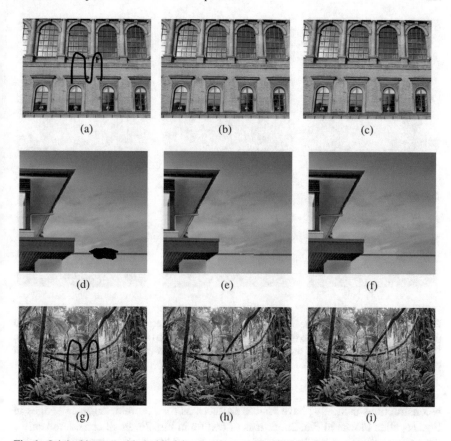

Fig. 1 Original images with the black inpainted area are in the left column and marked by (**a**), (**d**), (**g**). Reconstruction according to [4] with the visible artifact in (**e**), is in the middle column, marked by (**b**), (**e**), (**h**). Reconstruction on the basis of the proposed ROF-type model with the FT-Laplacian in the right column, marked by (**c**), (**f**), (**i**)

5 Examples

Below, we show several examples of the proposed ROF model with the FT-Laplacian and compare its performance with the results from [4] where the similar filling order strategy was realized. The images are taken from the database used in [9]. They have different characteristics as for example, repetitive patterns, strong gradients or a single dominant color. In all of them, the black area represents the inpainting region. It is of various sizes and shapes. As it is seen from the examples, some

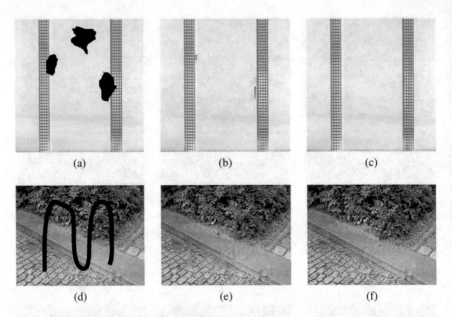

Fig. 2 Original images with the black inpainted area are in the left column and marked by (**a**) and (**d**). Reconstruction according to [4] with the visible artifacts in (**b**) and (**e**) is in the middle column. Reconstruction on the basis of the proposed ROF-type model with the FT-Laplacian in the right column, marked by (**c**), (**f**), (**i**)

reconstructions using [4] have visible artifacts, as for example, "zig-zag" edges in Fig. 1e, alien blocks in Fig. 2b, or loss of texture in Fig. 2e. In all considered cases, the proposed method has better performance than that of [4], especially when the size of a damaged area is big enough.[4]

6   Conclusion

Aiming at the problem of restoration for relatively large damaged areas, we proposed a new ROF model with nonlocal FT-Laplacian in the regularization term, where the minimization is considered over a searching space restricted to a finite set of possible patches. The fidelity term in the proposed ROF model is estimated by the norm in a Sobolev-like space, which increases the overall quality of reconstruction.

Due to the finiteness of the searching space of patches, we did not apply the technique of the calculus of variations (normally required for the ROF-type models) and by this, significantly increased the efficiency of a computation.

---

[4]The term "big enough" means that a damaged area of this size cannot be reconstructed by any noise-removing technique.

The proposed algorithm is demonstrated on a representative set of images with different characteristics. In all considered cases, the proposed method has better performance than that of [4], especially when the size of a damaged area is big enough.

Acknowledgements The SW and original idea to connect the filling order strategy with the technique of F-transforms belong to Dr. Pavel Vlašánek. This work was partially supported by the project LQ1602 IT4Innovations excellence in science. The implementation of the F-transform technique is available as a part of the OpenCV framework. The module `fuzzy` is included in `opencv_contrib` and available at https://github.com/itseez/opencv_contrib.

References

1. Bertalmio, M., Sapiro, G., Caselles, V., Ballester, C.: Image inpainting. In: Proceedings of the 27th Annual Conference on Computer Graphics and Interactive Techniques, pp. 417–424. ACM Press/Addison-Wesley Publishing Co. (2000)
2. Chan, T., Shen, J., Kang, S.: Euler's elastica and curvature-based image inpainting. SIAM J. Appl. Math. **63**(2), 564–592 (2002)
3. Chan, T.F., Shen, J.: Mathematical models of local non-texture inpaintings. SIAM J. Appl. Math. **62**(3), 1019–1043 (2001)
4. Criminisi, A., Pérez, P., Toyama, K.: Region filling and object removal by exemplar-based image inpainting. IEEE Trans. Image Process. **13**(9), 1200–1212 (2004)
5. Gilboa, G., Osher, S.: Nonlocal operators with applications to image processing. Multiscale Model. Simul. **7**(3), 1005–1028 (2008)
6. Holčapek, M., Perfilieva, I., Novák, V., Kreinovich, V.: Necessary and sufficient conditions for generalized uniform fuzzy partitions. Fuzzy Sets Syst. **277**, 97–121 (2015)
7. Mellouli, N., Bouchon-Meunier, B.: Abductive reasoning and measures of similitude in the presence of fuzzy rules. Fuzzy Sets Syst. **137**, 177–188 (2003)
8. Močkoř, J., Holčapek, M.: Spaces with fuzzy partitions and fuzzy sets. Oxf. J. Intell. Decis. Data Sci. 25–33 (2017)
9. Merget, D., Tiefenbacher, P., Bogischef, V., Rigoll G.: Subjective and objective evaluation of image inpainting quality. In: IEEE International Conference on Image Processing (ICIP), pp. 447–451. IEEE (2015)
10. Perfilieva, I.: Fuzzy transforms: theory and applications. Fuzzy Sets Syst. **157**(8), 993–1023 (2006)
11. Perfilieva, I., Danková M.: Towards F-transform of a higher degree. In: IFSA/EUSFLAT Conference, Citeseer, pp. 585–588 (2009)
12. Perfilieva, I., Holčapek, M., Kreinovich, V.: A new reconstruction from the F-transform components. Fuzzy Sets Syst. **288**, 3–25 (2016)
13. Perfilieva, I., Singh, A., Tiwari, S.: On the relationship among F-transform, fuzzy rough set and fuzzy topology. Soft Comput. **21**, 3513–3523 (2017)
14. Rudin, L., Osher, S., Fatemi, E.: Non linear total variation based noise removal algorithms. Physica **60**, 259–268 (1992)
15. Zadeh, L.A.: Toward a theory of fuzzy information granulation and its centrality in human reasoning and fuzzy logic. Fuzzy Sets Syst. **90**, 111–127 (1997)

Forensic Identification by Craniofacial Superimposition Using Fuzzy Set Theory

Oscar Ibáñez, Carmen Campomanes-Álvarez,
B. Rosario Campomanes-Álvarez, Rubén Martos, Inmaculada Alemán,
Sergio Damas, and Oscar Cordón

Abstract Skeleton-based forensic identification techniques involve the assessment of human osseous remains to identify the deceased person's identity and cause of death. Craniofacial superimposition (CFS) is one of the most extended techniques of such kind. It involves the superimposition of an image of a skull with a number of ante-mortem face images of an individual and the analysis of their morphological correspondence. Designing automatic methods to address CFS and support the forensic anthropologist remains a challenge. Our research group has a long-term collaboration track with the University of Granada's Physical Anthropology Lab and some other international forensic labs to automate the whole CFS identification process. The procedure is affected by different sources of uncertainty and thus fuzzy set theory becomes an appealing approach to automate this task. The current contribution reviews these developments specifically focusing on the fuzzy set-based solutions applied to deal with each of the uncertainty sources inherent to the process.

1 Introduction

Forensic identification techniques share the search for equal/compatible patterns in two (or more) different images/models/data/etc. [1]. Within the forensic identification techniques, craniofacial superimposition (CFS) [2] is one of the most relevant

O. Ibáñez · S. Damas · O. Cordón (✉)
Andalusian Research Institute DaSCI, University of Granada, Granada, Spain
e-mail: ocordon@decsai.ugr.es

B. R. Campomanes-Álvarez
Data for Value Unit, CTIC Technological Centre, Gijón, Asturias, Spain

C. Campomanes-Álvarez
Vision Technologies Unit, CTIC Technological Centre, Gijón, Asturias, Spain

R. Martos · I. Alemán
Doctoral program in biomedicine, Department of Legal Medicine, Toxicology and Physical Anthropology, University of Granada, Granada, Spain

© The Editor(s) (if applicable) and The Author(s), under exclusive license to Springer Nature Switzerland AG 2021
M.-J. Lesot and C. Marsala (eds.), *Fuzzy Approaches for Soft Computing and Approximate Reasoning: Theories and Applications*, Studies in Fuzziness and Soft Computing 394, https://doi.org/10.1007/978-3-030-54341-9_20

skeleton-based approaches. It involves the process of overlaying a skull with one or more photographs of missing persons and the analysis of their morphological correspondence. This identification technique has a great application potentiality since nowadays the wide majority of the people have photographs, ante-mortem (AM) material, where their faces are clearly visible. The counterpart, the skull, post-mortem (PM) material, is a bone that hardly degrades with the effect of fire, humidity, high or low temperatures, time lapse, etc.

Despite some authors consider this technique as a valid identification method [1], many others only attribute CFS the ability to serve as an exclusion method [3]. The fact that two different objects have to be compared (contrary to the majority of the forensic identification techniques such as DNA, fingerprints, face recognition, dental identification) involves a number of sources of uncertainty. This makes CFS identification a subjective technique, highly dependent on the forensic anthropologist's experience and knowledge: acquisition parameters of the photographs [4, 5], mandible articulation [24], landmark location imprecision [6], soft tissue depth variability [7], and ultimately morphological comparison. Thus, a systematic, unbiased, accurate, and automatic method properly considering and modeling all sources of uncertainty is a need [8].

The following Sect. 2 introduces the sources of uncertainty experts and automatic systems have to deal with in each of the four CFS stages. Then, in Sect. 3 we summarize our previous proposals to model these sources, with a special focus on those based on fuzzy sets. Finally, in Sect. 4 we compile future research directions and demands.

2   Sources of Uncertainty

Although the most recent publications identified three different stages within CFS [2, 9, 10], they all omit an important stage in any forensic identification process: the description of available AM and PM data. Thus, the CFS is composed of four stages. Each of the following subsections analyzes the different CFS stages together with their sources of uncertainty.

2.1   Stage 1: Acquisition and Processing of AM-PM Materials

The original materials used in the CFS process include photographs of the person and the proper skull. The international consortium for CFS standardization, MEPROCS, recommends to use more than one AM images, in different poses and with a good quality, as well as to avoid images with obscuring objects, e.g. spectacles and beards [7]. Skull 2D images, skull live images (video superimposition), and skull 3D models can be used in CFS but a 3D model should be preferred because it is definitively a more accurate representation.

Despite the technical procedure followed by the experts, cranial and facial land-marks are often used in one o more of the CFS stages. The selected landmarks are located in those parts where the thickness of the soft tissue is low (see Fig. 1). The goal is to facilitate their location when the anthropologist must deal with changes in age, weight, and facial expressions.

Finally, within this first stage we can include the delineation (or segmentation) of certain anatomically corresponding structures in both the face and the skull. They are employed by novel approaches [3, 11, 12] to automatically measure their anatomical correspondence and consistency.

Uncertainty related to the quality and quantity of the materials: Although nowadays is becoming easier to access to a larger number of digital photographs, there are still scenarios where these ideal conditions are not fulfilled (as for example old cases from dictatorship periods or wars). Regarding the skull, it can be found fragmented, eroded, with missing parts, and at different states of conservation.

Landmark location uncertainty: The uncertainty related to the location of facial landmarks refers to the difficult task of precisely and invariably placing landmarks on a photograph [6, 13]. The ambiguity of placing points in a photograph may also arise from landmark definition and reasons such as variation in shade distribution depending on light condition during photographing, unsuitable camera focusing, poor image quality, face pose and expression, partial landmarks occlusion, etc.

Anatomical region delineation uncertainty: The delineation/segmentation of anatomically corresponding structures is influenced by the quality of the materials (photographs and skulls) but it is subjective by nature since in most of these regions there are not clear/objective borders delimiting each of them.

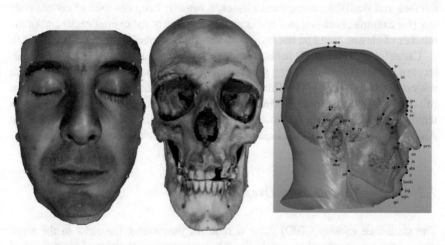

Fig. 1 Anthropometric landmarks commonly employed in craniofacial identification: cephalomet-ric in the face (left) and craniometric in the skull (center). CT scan showing the soft tissue irregular presence among landmarks (right)

Mandible positioning uncertainty: The Temporomandibular Joint connects the mandibular condyle to the temporal bone in the cranium, which are separated by a small oval fibrocartilage called articular disc. This cartilage is not present in skeletonized corpses and therefore it has to been estimated with the corresponding uncertainty. In addition, with no access to the occlusion as it was in life, it is not possible to determine the exact position of the mandible.

2.2 Stage 2: Face and Skull Examination

This stage is very important as a source of information allowing anthropologists to filter a large number of potential candidates while providing data for subsequent stages of the identification process. It involves morphological and morphometric description of different anatomical regions on the face and the skull, and the subsequent quantification and comparison. The morphological description is intended to classify each anatomical region within a given set of categories. For example, the nasal root could be narrow, medium, or broad; the cranial vault shape can be categorized as semisphere, pentagonoid, oval, or rectangular, etc. On the other hand, numerous anthropometric facial and skull proportions have been proposed, and researchers have collected, recorded and analyzed their values on several human populations. Within anthropometry, dimensions and proportionality indices of the face/skull are measured using landmarks.

Uncertainty related to morphological categories: The type of information to be analyzed and compared is mostly qualitative (morphology of anatomical regions on the face and skull). As categorical classes, it oftently becomes hard to choose only one (for example, between oval and semisphere shape of the cranial vault) and/or the number of classes could be not enough to accurately describe the data (granularity).

Uncertainty related 2D measurements: Intrinsic and extrinsic camera parameters are responsible of changing proportions and measurements in a 2D plane with respect to the real (3D) world. Thus forensic standard guidelines only recommend the use of photo-anthropometric techniques under controlled conditions rarely encountered in forensic case work. Another problem of measuring in 2D, i.e. measuring in AM facial photographs, is the landmark location uncertainty described in Stage 1.

2.3 Stage 3: Skull-Face Overlay

The skull-face overlay (SFO) process requires positioning the skull in the same pose as the face in the photograph. From a computer vision point of view, the AM image is the result of the 2D projection of a real (3D) scene that was acquired by a particular (unknown) camera. In such a scene, the living person was somewhere inside the camera field of view in a given pose [9]. The most natural way to deal with the SFO problem is to replicate that original scenario. The goal is to adjust its size and its orientation with respect to the head in the photograph. In addition, the specific characteristics of the camera must also be replicated to reproduce the original situation as much as possible [9]. To do this, the skull 3D model is positioned in the camera coordinate system through geometric transformations, i.e. translation, rotation, and scaling which corresponds to the adjustment of the skull size and its orientation at the same angle as the face in the image. Then, a perspective projection of the skull 3D model is performed onto the facial photograph. In the state-of-the-art automatic SFO procedures [2], this process is guided by a set of cranial and facial landmarks. Once the location of these landmarks is provided by the anthropologist, the SFO procedure is based on automatically searching for the skull orientation leading to the best matching of the two sets of landmarks.

Uncertainty related to landmark cooplanarity: It makes the equation system (the objective function to be minimized) undetermined (or near–undetermined) since most of the facial landmarks are located in the same plane [14, 15]. I.e., there is an uncertainty source (there is not enough information or it is imprecise) regarding which of the possible solutions is the best.

Landmark matching uncertainty: The correspondence between facial and cranial anthropometric landmarks is not always symmetrical and perpendicular to the skin surface and to the underlying bone. In addition, the facial soft tissue depth varies for each cephalometric landmark, as well as for different populations.

Skull-face overlay evaluation uncertainty: Although modelling SFO as a 3D-2D IR problem removes the subjectivity of the manual process traditionally performed by the anthropologist, there is incomplete and vague information guiding the process (landmark location and matching uncertainties). Thus, there is no precisely quantifiable way to determine when an accurate overlay has been achieved.

Mandible articulation uncertainty: Once the mandible has been positioned (see Stage 1) it must reproduce the facial expression of the subject in each available photograph. This is time-consuming and prone to errors due to the impossibility to exactly know the mandible articulation from the photograph.

2.4   Stage 4: Decision Making

Once one or several SFOs have been achieved for the same identification case, the main goal is to determine the degree of support that the skull and the face of the photograph(s) belong to the same person or not. This decision is guided by different criteria studying the anatomical relationship between the skull and the face [16].

Morphological correspondence uncertainty: The morphological correspondence of face and skull structures is never precise due to the presence of soft tissue which at the same time is affected (to a lower or higher degree) by the age, sex, ancestry (biological profile), and/or body mass index.

Information aggregation uncertainty: Decisions about the skull and face relationships are traditionally made in a subjective way where there is a complete absence of measurements and countable aggregation of positive and negative factors. There is a need of considering multiple criteria together with their corresponding uncertainty.

Degrees of confidence as a result: Uncertainty and degrees of confidence are inherent to the final identification result. In fact, a scale for a craniofacial matching evaluation has been recently defined by some of the most representative experts in craniofacial identification in [7]. Accordingly, the final decision is provided in terms of strong, moderate, and limited support.

3   State of the Art Approaches Modeling Craniofacial Superimposition Uncertainties

Soft computing can be extremely useful for modeling the said sources of uncertainty. It is aimed for the design of intelligent systems to process uncertain, imprecise, and incomplete information. Its main techniques include fuzzy logic and fuzzy set theory, and evolutionary algorithms (EAs), which are powerful bio-inspired search and optimization tools to automate problem solving in areas such as modeling and global optimization. In this section we will give the most relevant details of those approaches, which based on fuzzy set theory, tackle one or more of the sources of uncertainty identified in Sect. 2. The remainder are out of the scope of this publication. Nevertheless, they are briefly introduced in the discussion.

3.1   Modeling the Uncertainty Related to the Location of Facial Landmarks

Ibáñez et al. [14] developed a fuzzy approach to deal with the uncertainty related to facial landmark location. This proposal is based on imprecise landmarks, i.e., the forensic anthropologist can mark the approximate location of any landmark using an ellipse. This ellipse delimitates a region where the experts can actually assure the anatomical location of the landmark. The size of the ellipse will be directly related with the imprecision in the landmark location. This way, the spray pen metaphor defined by Prof. Zadeh is considered [17]. Those additional landmarks are essential to deal with the coplanarity problem in the automatic search of the best SFO. Fuzzy landmarks were defined as a fuzzy convex set of points having a nonempty core and a bounded support. That is, all its α-levels are nonempty bounded and convex sets. In the SFO case, since we are dealing with 2D photographs with an $x \times y$ resolution, fuzzy landmarks were defined as 2D masks represented as a matrix M with $m_x \times m_y$ points (i.e., a discrete fuzzy set of pixels). Each fuzzy landmark will have a different size depending on the imprecision on its localization but at least one pixel (i.e. crisp point related to a matrix cell) will have membership with degree one. These masks are easily built starting from two triangular fuzzy sets $\widetilde{V}$ and $\widetilde{H}$

modeling the approximate vertical and horizontal position of the ellipse representing the location of the landmark, thus becoming 2D fuzzy sets.

3.2   Modeling the Landmark Matching Uncertainty

The fuzzy sets-based approach developed in [18] models the imprecision related to the facial soft tissue depth between corresponding pairs of cranial and facial land-marks. To do so, the population-based statistical minimum (min), mean (mean) and maximum (max) distances between a pair of cranial and facial landmarks are represented by fuzzy sets assuming a certain degree of perpendicularity between cranial and facial landmarks as most soft tissue studies do. To do so, authors consider the normal vector $\widetilde{v}$ on the surface of the skull 3D model at each cranial landmark. In order to estimate the position of the facial landmarks, the unit corresponding vector $\widetilde{u}$ (same direction that normal vector $\widetilde{v}$ but magnitude of the unit) is multiplied by the specific distance (minimum, mean or maximum). In addition, different inclination angles can be applied to the unit vector $\widetilde{u}$ in order to define the volume where the facial landmark will be likely located. The landmark matching uncertainty is defined using 3D masks. Hence, a fuzzy set $\widetilde{B}_p$ with $p \in \{x, y, z\}$ is determined by its center $c \in \{c_x, c_y, c_z\}$ (the 3D coordinates of the cranial landmark), the normal vector coordinates $u \in \{u_x, u_y, u_z\}$, and the min, mean, and max soft tissue distances. Therefore, the original definition of our evolutionary SFO technique's fitness function in [4, 14] was modified in [3] to take into account distances between two fuzzy sets, Fuzzy Mean Error (FME). In [19] the authors studied the performance and influence of the most significant and suitable fuzzy distances proposed in the specialized literature, eight in total. They concluded that the distance metric has a crucial impact on the performance of the method with the weighted mean distance presenting the best performance in most of the cases and that the results are both more accurate and robust than the other studied metrics. The FME considering this distance is:

$$FME = \frac{\sum_{i=1}^{N_{crisp}} (d'(x_i, f(\widetilde{C}^i))) + \sum_{j=1}^{N_{fuzzy}} (d''(\widetilde{F}^j, f(\widetilde{C}^j)))}{N} \qquad (1)$$

where N_{crisp} is the number of 2D facial landmarks precisely located (crisp points), N_{fuzzy} is the number of 2D facial landmarks imprecisely located and defined as 2D fuzzy sets, N is the total number of landmarks ($N = N_{crisp} + N_{fuzzy}$), x_i corresponds to a 2D facial landmark defined as a crisp point ($x_i \in F$), $\widetilde{C}^i$ and $\widetilde{C}^j$ are fuzzy sets modeling each 3D cranial landmark and the soft tissue distance to the corresponding 3D facial landmark i or j; f is the function that determines the 3D-2D perspective transformation that properly projects every 3D skull point onto the 2D photograph; $f(\widetilde{C}^i)$ and $f(\widetilde{C}^j)$ are two fuzzy sets corresponding to the result of applying the perspective transformation f to the 3D volume that model the landmark matching uncertainty; $\widetilde{F}^j$ represents the fuzzy set of points of the imprecise 2D facial landmark; $d'(x_i, f(\widetilde{C}^i))$ is the distance between a point and a fuzzy set of points, and $d''(\widetilde{F}^j, f(\widetilde{C}^j))$ is the distance between two fuzzy sets.

3.3 Fuzzy Hierarchical Decision Support Framework

With the objective of automating the whole decision making process while modeling all the sources of uncertainty identified in Sect. 2.4, a complete framework for a Decision Support System (DSS) Framework in CFS was proposed in [12].

The system develops fusion of information concerning skull-face anatomical correspondence at three different levels: criterion evaluation, SFO evaluation, and CFS evaluation. Following this structure, different Computer Vision (CV) methods for evaluating a specific criterion are aggregated considering the accuracy of each one at level 3. CV-based methods were used [11, 12, 20, 21] to measure each craniofacial correspondence criterion and fuzzy integrals [22] were used to obtain the final matching degree from them. The result of this aggregation is the matching degree C_m.

The next level (level 2) corresponds to the SFO matching degree. In order to obtain this value, the craniofacial anatomical correspondence of several criteria of the overall face has to be studied. For each criterion, the skull-face matching degree is obtained applying a specific CV method (level 3). The quality of the materials (PQ_m and BQ_m), the biological profile variability (BP_m), and the discriminative power of each isolated region are taken into consideration to compute this SFO degree. Three different sublevels are distinguished in level 2. The first one aggregates, using either the minimum (min), the product (prod) or the arithmetic Mean (mean), the sources of uncertainty. The second one integrates this uncertainty with the matching degree of the criterion (level 3) using either the weighted arithmetic mean (wam) or the weighted geometric mean (wgm). The third sublevel aggregates the different previous values, using either wam, wgm, Choquet or Sugeno integrals, for all the regions weighting them by the discriminative power of the isolated region.

Finally, at the CFS evaluation level (level 1), the authors [3] analyzed the aggregation of the different SFO degrees using the following four operators: the mean, the maximum, the minimum, and a weighting function based on the number of regions analyzed in each SFO. In view of the results obtained in the experiments, they concluded that the best current performance is obtained using the combination of aggregation functions mean-wgm-wgm (at level 2) and weighting function based on the number of regions (at level 1). The system performance was validated in 14 identification scenarios involving real forensic cases and its results compared with those achieved by 26 experts from 17 different institutions dealing with exactly the same identification scenarios. The overall performance (correct decision rate) of the system was better than the mean of the experts with respect to the correct decisions rate (90% vs 79%) and better than 25 out of the 26 forensic experts who participated in the study (see Fig. 2).

Method	Correct Decisions	Ground Truth	Decision		Decision(%)	
			Positive	Negative	Positive	Negative
Experts Mean	78.99%	Positive	100	90	52.63%	47.37%
		Negative	152	810	15.80%	84.20%
Best Expert 1	93.33%	Positive	8	2	80.00%	20.00%
		Negative	2	48	4.00%	96.00%
Best Expert 2	88.14%	Positive	6	3	66.67%	33.33%
		Negative	4	46	8.00%	92.00%
Best Expert 3	86.21%	Positive	5	3	62.50%	37.50%
		Negative	5	45	10.00%	90.00%
DSS-0.86	90.00%	Positive	6	4	60.00%	40.00%
		Negative	2	48	4.00%	96.00%

Fig. 2 Performance of the different CFS approaches. DSS-0.86 refers to our Fuzzy Hierarchical Decision Support System thresholding its output by 0.86)

4 Discussion and Conclusions

In this chapter we have summarized for the first time all the sources of uncertainty that have a significant impact in the CFS identification technique. We have also introduced the main fuzzy set–based approaches to deal with some of those sources of uncertainty. The following lines will be devoted to analyze the work done and to be done regarding the whole list of sources of uncertainty.

Identification, i.e. individualization, always involves examination, description, and ultimately comparison of two or more objects. In the case of CFS, the skull and the face. The challenges of this technique are numerous but of all of them share the same origin, the comparison of two different objects (skull-face) in two different formats (3D-2D). Different sources of uncertainty of different nature arise and the tools to model them also have to be also different.

The use of imprecise landmarks (modeled using fuzzy sets) allows forensic experts to locate a larger number of landmarks. This has an important impact in stages 2 and 3 as it permits calculating a larger number of facial dimensions and proportionality indices. In addition, more pairs of facial-cranial landmarks will be available to guide automatic the SFO process. In [23] the authors developed a methodology to allow the direct comparison of 2D measures (pixels)-3D (millimetres) by studying the variation of cephalometric indices measured on 3D facial models, and their correlation with 2D indices obtained in 2D projections of the previous ones. However, neither this group of authors nor others dealt with the estimation of facial dimensions and indices when the location of the landmarks employed could be imprecise (something already demonstrated [6, 13] although precise points are employed). Other sources of uncertainty not tackled in the literature are the positing of the mandible and the delineation of anatomical regions. Inter- and intra-observer variability should be at least quantified together with the impact they have in subsequent stages (mandible positioning in SFO and region delineation in decision making).

The morphometric description of a 3D object (the skull) through dimensions and indices is straight forward once landmarks have been located. However, in the 2D case (facial photograph) a 3D estimation (in the form of an interval) is possible [23]. However, landmark location uncertainty in this scenario has not been modeled yet. Still in stage 2, the morphological description and comparison of anatomical traits in the skull and the face present an opportunity for fuzzy sets. It seems appropriate to use this tool to model the fact that a particular trait could be included within two different categories at a certain degree. This way we could reduce the effect of intra-observer variability while giving the opportunity to be more discriminating.

SFO is probably the most challenging stage in CFS. It is affected by landmark location and matching uncertainties to a point that an objective and properly correlated evaluation of a given overlay has not been possible yet. Contrary to this unsolved issue, mandible articulation has been recently modelled within the automatic SFO process [24] by extending the state–of–the–art EA-based method with the ability to allocate the mandible in the right position according to an anatomical model. Based on a dataset of simulated AM images with different mandible apertures and facial poses, the authors proved experimentally that the proposed method is able to effectively tackle cases displaying a much larger range of mandible positions. In fact, thanks to the new genetic design, it is able to outperform the original method, even when the mandible aperture is very small.

The influence of the SFO achieved on the last stage is out of any doubt. An important criteria for analysing anatomical correspondence is by studying positional relationship [11] which strongly depends on the skull-face relative position. A part from the importance of achieving a reliable SFO, or at least a measurement of the quality of the achieved SFO, the remaining sources of uncertainty affecting decision making has been modelled [3, 11, 12]. At this point the possibility of computing two separate values for the matching degree and the uncertainty around it seems to be an interesting research line to be explored. In the state–of–the–art proposal a unique value is computed. While this facilitates ranking candidates and binary decision (positive vs negative identification) for a given threshold, it is not aligned with anthropologist's standards involving a decision (positive, negative or undetermined) together with a degree of support of such a decision: strong, moderated, or limited.

Acknowledgements This work was supported by the Spanish Ministry of Science, Innovation and Universities, and European Regional Development Funds (ERDF) under grant EXASOCO (PGC2018-101216-B-I00), and by the Regional Government of Andalusia under grant EXA-ISFI (P18-FR-4262). Dr. Ibáñez's work is funded by Spanish Ministry of Science, Innovation and Universities-CDTI, Neotec program 2019 [reference EXP-00122609/SNEO-20191236]. The results of this work are part of the doctoral thesis in Biomedicine by the University of Granada of Rubén Martos Fernández. On a personal basis, the authors would like to acknowledge the strong positive influence Prof. Lotfi A. Zadeh has played on their research careers. They are very proud of having met and learned from Lotfi in different occassions, especially at the European Centre for Soft Computing and during the research stays some of the authors carried out in the BISC lab at Berkeley.

References

1. Jayaprakash, P.T.: Practical relevance of pattern uniqueness in forensic science. Forensic Science International **231**(1), 403–e1 (2013)
2. Huete, M.I., Ibáñez, O., Wilkinson, C., Kahana, T.: Past, present, and future of craniofacial superimposition: literature and international surveys. Legal Med. **17**, 267–278 (2015)
3. Campomanes-Álvarez, C., Martos-Fernández, R., Wilkinson, C., Ibáñez, O., Cordón, O.: Modeling skull-face anatomical/morphological correspondence for craniofacial superimposition-based identification. IEEE Trans. Inf. Forensics Secur. **13**, 1481–1494 (2018)
4. Ibáñez, O., Cordón, O., Damas, S., Santamaría, J.: An experimental study on the applicability of evolutionary algorithms to craniofacial superimposition in forensic identification. Inf. Sci. **79**, 3998–4028 (2009)
5. Stephan, C.N.: Perspective distortion in craniofacial superimposition: logarithmic decay curves mapped mathematically and by practical experiment. Forensic Sci. Int. **257**, 520–e1 (2015)
6. Campomanes-Álvarez, B.R., Ibáñez, O., Navarro, F., Alemán, I., Cordón, O., Damas, S.: Dispersion assessment in the location of facial landmarks on photographs. Int. J. Legal Med. **129**(1), 227–236 (2015)
7. Damas, S., Wilkinson, C., Kahana, T., Veselovskaya, E., Abramov, A., Jankauskas, R., Jayaprakash, P., Ruiz, E., Navarro, F., Huete, M., Cunha, E., Cavalli, F., Clement, J., Leston, P., Molinero, F., Briers, T., Viegas, F., Imaizumi, K., Humpire, D., Ibáñez, O.: Study on the performance of different craniofacial superimposition approaches (ii): best practices proposal. Forensic Sci. Int. **257**, 504–508 (2015)
8. Damas, S., Ibáñez, O., Cordón, O.: Handbook on craniofacial superimposition. Springer (2020). ISBN: 978-3-319-11136-0
9. Campomanes-Álvarez, B.R., Ibáñez, O., Navarro, F., Botella, M., Damas, S., Cordón, O.: Computer vision and soft computing for automatic skull-face overlay in craniofacial superimposition. Forensic Sci. Int. **245**, 77–86 (2014)
10. Damas, S., Cordón, O., Ibáñez, O., Santamaría, J., Alemán, I., Botella, M., Navarro, F.: Forensic identification by computer-aided craniofacial superimposition: a survey. ACM Comput. Surv. **43**(4), 27 (2011)
11. Campomanes-Alvarez, C., Ibáñez, O., Cordón, O.: Design of criteria to assess craniofacial correspondence in forensic identification based on computer vision and fuzzy integrals. Appl. Soft Comput. **46**, 596–612 (2016)
12. Campomanes-Alvarez, C., Ibáñez, O., Cordón, O., Wilkinson, C.: Hierarchical information fusion for decision making in craniofacial superimposition. Inf. Fusion **39**, 25–40 (2018)
13. Cummaudo, M., Guerzoni, M., Marasciuolo, L., Gibelli, D., Cigada, A., Obertová, Z., Ratnayake, M., Poppa, P., Gabriel, P., Rizt-Timme, S., Cattaneo, C.: Pitfalls at the root of facial assessment on photographs: a quantitative study of accuracy in positioning facial landmarks. Int. J. Legal Med. **127**, 699–706 (2013)
14. Ibáñez, O., Cordón, O., Damas, S., Santamaría, J.: Modeling the skull-face overlay uncertainty using fuzzy sets. IEEE Trans. Fuzzy Syst. **16**, 946–959 (2011)
15. Santamaría, J., Cordón, O., Damas, S., Ibáñez, O.: Tackling the coplanarity problem in 3D camera calibration by means of fuzzy landmarks: a performance study in forensic craniofacial superimposition. In: 2009 IEEE 12th International Conference on Computer Vision Workshops (ICCV Workshops), pp. 1686–1693. IEEE (2009)
16. Jayaprakash, P.T., Srinivasan, G.J., Amravaneswaran, M.G.: Cranio-facial morphanalysis: a new method for enhancing reliability while identifying skulls by photo superimposition. Forensic Sci. Int. **117**(1), 121–143 (2001)
17. Zadeh, L.A.: Toward extended fuzzy logic a first step. Fuzzy Sets Syst. **160**, 3175–3181 (2009)
18. Campomanes-Álvarez, B.R., Ibáñez, O., Campomanes-Álvarez, C., Damas, S., Cordón, O.: Modeling the facial soft tissue thickness for automatic skull-face overlay. IEEE Trans. Inf. Forensics Secur. **10**, 2057–2070 (2015)

19. Campomanes-Álvarez, C., Campomanes-Álvarez, B.R., Guadarrama, S., Ibáñez, O., Cordón, O.: An experimental study on fuzzy distances for skull-face overlay in craniofacial superimposition. Fuzzy Sets Syst. **318**, 100–119 (2017)

20. Campomanes-Alvarez, C., Ibáñez, O., Cordón, O.: Modeling the consistency between the bony and facial chin outline in craniofacial superimposition. In: 16th World Congress of the International Fuzzy Systems Association (IFSA). pp. 1612–1619 (2015)

21. Campomanes-Alvarez, C., Ibáñez, O., Cordón, O.: Experimental study of different aggregation functions for modeling craniofacial correspondence in craniofacial superimposition. In: The 2016 IEEE International Conference on Fuzzy Systems (FUZZ-IEEE 2016). pp. 437–444 (2016)

22. Sugeno, M.: Fuzzy measures and fuzzy integrals: a survey. Fuzzy Automata Dec. Process. **78**(33), 89–102 (1977)

23. Martos, R., Valsecchi, A., Ibáñez, O., Alemán, I.: Estimation of 2D to 3D dimensions and proportionality indices for facial examination. Forensic Sci. Int. **287**, 142–152 (2018)

24. Bermejo, E., Campomanes-Álvarez, C., Valsecchi, A., Ibáñez, O., Cordón, O.: Genetic algorithms for skull-face overlay including mandible articulation. Inf. Sci. **420**, 200–217 (2017)

On the Applicability of Fuzzy Rule Interpolation and Wavelet Analysis in Colorectal Image Segment Classification

Szilvia Nagy, Ferenc Lilik, Brigita Sziová, Melinda Kovács, and László T. Kóczy

Abstract The automatic detection of colorectal polyps could serve as a visual aid for gastroenterologists when screening the population for colorectal cancer. A fuzzy inference based method was developed for determining whether a segment of an image has polyps. Its antecedent dimensions were the mean pixel intensity, the intensity's standard deviation, the edge density, the structural entropies and the gradients, not only for the original image segments, but for its wavelet transformed versions. The method performed moderately well, even though the number of the input parameters was very large. In the present contribution we studied, that based on the necessary and usually applied conditions of the applicability of fuzzy rule interpolation, which antecedent dimensions should remain, and how omitting the other input parameters influences the results of the method.

S. Nagy · F. Lilik · B. Sziová · M. Kovács · L. T. Kóczy (✉)
Széchenyi István University, Egyetem ter 1, 9026 Gyor, Hungary
e-mail: koczy@tmit.bme.hu; koczy@sze.hu

S. Nagy
e-mail: nagysz@sze.hu

F. Lilik
e-mail: lilikf@sze.hu

B. Sziová
e-mail: szi.brigitta@sze.hu

M. Kovács
e-mail: kovacsmeli@gmail.com

L. T. Kóczy
Budapest University of Technology and Economics, Magyar tudosok krt. 2, 1111 Budapest, Hungary

M.-J. Lesot and C. Marsala (eds.), *Fuzzy Approaches for Soft Computing and Approximate Reasoning: Theories and Applications*, Studies in Fuzziness and Soft Computing 394, https://doi.org/10.1007/978-3-030-54341-9_21

243

1   Introduction

Colonoscope is one of the most efficient equipments in studying the colorectal part of the bowel. As over 50 years of age the screening for colorectal polyps is recommended for the whole population, theoretically much more checks should be carried out in order to find and diagnose every possible colorectal cancer in the very early stage. The medical staff operating the colonoscopes are trained well enough to detect the colorectal polyps, however, they are still humans, and a visual aid for drawing their attention to a part of the image can help with their work significantly. If polyps are found, then, based on their surface pattern or shape [1–3], the classification of the polyp can be made, as most of the polyps are benign, and it is not necessary to remove them. A malign polyp missed can develop into cancer, and healing a colorectal cancer is still not always possible, and even if it is possible, it is very expensive and painful.

Thus, detecting colorectal polyps by computer is an important topic; training databases and challenges for this task can be found [4–7]. In our previous studies we found, that database [7] has the largest sized, least blurred images with the least image processing artifacts, thus the method presented here uses this database, which consists of 196 images of size 1225×966 pixels, together with the masks of the polyps.

In our previous studies [8], as well as in this work, we cut the images to segments of $N \times N$ pixels. After determining, whether a small enough segment contains polyp or not, the edges of the polyp can later be found by e.g., an active contour method [9, 10]. To determine, if such a segment contains polyp, we used the mask provided by database [7], and tried to reproduce these already given results by a fuzzy inference system with a large number of inputs, and two possible outputs. The inputs are mainly statistical properties of the image and the consequent can be yes or no. As many of the statistical parameters had only slightly different membership functions in the two possible rules, the investigation, whether omitting such antecedent dimensions induce a large decrease in the precision of finding a polyp will be presented in this work.

First we summarize the fuzzy inference methods, focusing on fuzzy rule interpolation in Sect. 2. In Sect. 2.2 we give the antecedent dimensions, while in Sect. 3, we determine, whether antecedent dimensions can be omitted because of the large overlap of the membership functions. Using a KH rule interpolation [11] based fuzzy inference method we compare the results of several antecedent choices in Sect. 4. As a last step, we conclude the results in Sect. 5.

2   Fuzzy Sets and Reasoning

The human mind can draw consequences based on simple rules using general concepts of human languages. For a computer the same procedure can be extremely complicated if conventional methods based on Boolean algebra are used.

Zadeh suggested to use fuzzy sets and operators, instead of the crisp Boolean ones in 1965 [12, 13]. In 1973 he proposed the compositional rule of inference on "if...then" type fuzzy production rules [14], and in 1974 Mamdani proposed [15] a simplified version with orthogonal projections,—later known as Mamdani controller. The operators "and" and "or" in the composition rules can be implemented by fuzzy operators t-norms and s-norms, respectively. For example, Mamdani [15] used minimum for "and" and maximum for "or". Usually the membership functions representing the ith antecedent dimension in the rules belonging to the jth consequent, i.e., in the jth rule are denoted by A_{ij}, whereas the ith consequent by B_i. The observation (input) vector is usually denoted by $(A_1^*, \ldots, A_n^*)$, while the conclusion by B^*, or in the Takagi-Sugeno type systems, by b^*. For translating the conclusion into an interpretable value, it might need defuzzification [16].

2.1 Fuzzy Rule Interpolation

Sparse rule bases are rule bases, where in at least one of the dimensions the antecedent sets do not entirely cover the possible set of input values to a reasonable depth, e.g., at least to 0.5. Thus, Mamdani [15] or Sugeno-type [17–19] inference methods are not always suitable for automatic decision making. If one of the dimensions of the observation vector contains a value from an area uncovered by any of the antecedent sets, then its membership value is 0 for all or some rules in this particular dimension. This results in zero weight for all or of the rules, thus there is no firing rule in the rule base.

This problem can be solved by using fuzzy rule interpolation, based on the rule antecedents of the rule base and the observation vector. The first member of the interpolation methodology was proposed by Koczy and Hirota (usually referred to as KH rule interpolation) [11, 20]. The fuzzy rule interpolation approaches are well summarized in [21]. Bernadette Bouchon-Meunier and her coworkers gave a summary of the mathematics of the interpolative reasoning [22], and she has contributed to the understanding and foundation of fuzzy rule interpolation, especially similarity relations [23, 24] and graduality [25] based interpolation methods.

The main idea of the KH rule interpolation method is that a fuzzy set itself is the collection of its α-cuts, and the interpolation in the predefined (e.g., linear) sense is done for each α-cuts separately. Theoretically the interpolation has to be carried out on all α-cuts, however in real life cases—e.g., for piecewise linear membership functions—it is sufficient to interpolate the characteristic alpha cuts. In the case of fuzzy sets with triangular or trapezoidal membership functions, it is sufficient to use interpolation at α 0 and 1. The necessary condition for using KH interpolation is that the fuzzy sets are convex, normal (CNF sets), and at least partially ordered [26]. The definition of two fuzzy sets being ordered is defined by the following [27]

$$A \prec B \text{ ha } \forall \alpha \in [0, 1] : \inf\{A_\alpha\} \leq \inf\{B_\alpha\} \text{ and } \sup\{A_\alpha\} \leq \sup\{B_\alpha\}. \quad (1)$$

The principle of the rule interpolation is the ratio of the distances between the consequents of the rules and the result, should be the same as the ratio between the distances of the antecedents and the observation [28]; according to the fundamental equation of rule interpolation (FERI)

$$d(A^*, A_{i1}) : d(A_{i2}, A^*) = d(B^*, B_{i1}) : d(B_{i2}, B^*). \tag{2}$$

In the KH interpolation the distance of the α-cuts is determined by the following

$$\begin{aligned} d_{\alpha L}(A_1, A_2) &= \inf\{A_{2\alpha}\} - \inf\{A_{1\alpha}\} \\ \text{and} & \\ d_{\alpha U}(A_1, A_2) &= \sup\{A_{2\alpha}\} - \sup\{A_{1\alpha}\}, \end{aligned} \tag{3}$$

and the α-cuts of the of the conclusion are [26],

$$\min\{B_\alpha^*\} = \frac{\frac{\inf\{B_{i1\alpha}\}}{d_{\mathrm{L}}(A_\alpha^*, A_{i1\alpha})} + \frac{\inf\{B_{i2\alpha}\}}{d_{\mathrm{L}}(A_\alpha^*, A_{i2\alpha})}}{\frac{1}{d_{\mathrm{L}}(A_\alpha^*, A_{i1\alpha})} + \frac{1}{d_{\mathrm{L}}(A_\alpha^*, A_{i2\alpha})}}$$

and

$$\max\{B_\alpha^*\} = \frac{\frac{\inf\{B_{i1\alpha}\}}{d_{\mathrm{U}}(A_\alpha^*, A_{i1\alpha})} + \frac{\inf\{B_{i2\alpha}\}}{d_{\mathrm{U}}(A_\alpha^*, A_{i2\alpha})}}{\frac{1}{d_{\mathrm{U}}(A_\alpha^*, A_{i1\alpha})} + \frac{1}{d_{\mathrm{U}}(A_\alpha^*, A_{i2\alpha})}} \tag{4}$$

While the linear KH interpolation uses only the two neighbouring fuzzy sets of the inspection, the stabilized KH rule interpolation uses all of the fuzzy sets of the given antecedent dimension of the rule base for the calculation of the result. Stabilizing the method requires raising the normalization factors to the power of the number of the input dimension k, as [26]

$$\inf\{B_\alpha^*\} = \frac{\sum_{i=1}^{2n} \left(\frac{1}{d_{\alpha L}(A^*, A_i)}\right)^k \inf\{B_{i\alpha}\}}{\sum_{i=1}^{2n} \left(\frac{1}{d_{\alpha L}(A^*, A_i)}\right)^k}$$

$$\tag{5}$$

$$\sup\{B_\alpha^*\} = \frac{\sum_{i=1}^{2n} \left(\frac{1}{d_{\alpha U}(A^*, A_i)}\right)^k \sup\{B_{i\alpha}\}}{\sum_{i=1}^{2n} \left(\frac{1}{d_{\alpha U}(A^*, A_i)}\right)^k}.$$

Here $d_{\alpha L/U}(A^*, A_i)$ denotes the upper/lower signed Minkovski distance between the observation and the α-cut of the ith antecedent, as given in the following equation

$$d_{\alpha L/U}(A_1, A_2) = \left(\sum_{i=1}^{k} \left(d_{i\alpha L/U}(A_1, A_2)\right)^k\right)^{1/k}. \tag{6}$$

2.2 Antecedents

In order to characterize the colonoscopy image segments regarding their polyp content, and thus helping the development of a computer aided diagnosing method, we built a fuzzy inference system for determining whether the segment contains polyp or not. Usually, those image segments (tiles) were well classifiable, which either did not have any polyp part, or a relatively large part was covered by a polyp. Also, the quality of the pictures influenced the classification success.

For antecedents we tried to use easily computable, mainly statistical parameters of the image segments. The reasons for this selection was the following. First, the program should be run on the fly, hence the easily computability. Second, the characteristics of the polyps visible for the human eyes. The polyps and the normal bowel wall differs by their shape, illumination, pattern, and sometimes in color. The polyps are usually brighter than the environment, have rather defined contour, expressed shadow around it, these are the characteristics that are almost always present in the case of polyps protruding into the bowel. The pattern of the veins or polyp surface is not always detectable in the images. The color of the polyp is usually almost the same as the healthy background, except in the case of some later phase malignant objects, but these are to be detected as well. Statistical parameters, such as mean or standard deviation of a matrix (i.e., an image segment) can be implemented very effectively, and the change of the standard deviation or the mean intensity in any of the colour channels, or the density of the pattern edges seemed to be promising candidates for being a basis of a classification scheme. The pattern edge density is calculated based on the Canny filtered version of the image segment: we simply calculated the rate of the white pixels (edges) within the segment. The structural entropy S_{str} and the spatial filling factor $\ln q$, which are two Rényi entropy based quantities that can characterize the shape of a distribution [29] are also easy to compute, has also the potential to characterize the roundish shapes [30]. The last group of parameters contains gradients of the image, as they are large in the case of rapid intensity changes, such as shadow behind the polyp to the bright surface of the polyp. Gradients are implemented as convolutional or block filters on the image, they are also easy to compute. We used the magnitude, the direction, the x component and the y component of the gradient images, and we will denote them by G_m, G_d, G_x and G_y.

The mean, the standard deviation, the edge density, the gradients and the structural entropies [29] of the tiles were also used for our previous methods [8, 31] as input parameters of the fuzzy inference system, together with the same parameters of the first wavelet transforms [32] of the tile (except for the gradients). The wavelet transform consists of a low-pass and a high-pass convolutional filter in both dimensions, thus there are 4 images arising, if 2D wavelet transform is carried out. The output after the two low-pass filter is very similar to the original image (only the size of its sides is reduced to their half): it shows the slowly varying part of the image. The output after the two high-pass filter shows the rapidly varying parts of the image.

Table 1 Original, 99 antecedents of the classification scheme. The first column gives the image type, i.e., the original image segment, its wavelet transforms, and its gradient filtered versions. The first row gives the statistical parameters calculated on the given image type. The shorthand notations R, G and B mean the colour channels red, green and blue

	R mean	R std	G mean	G std	B mean	B std	R edge	G edge	B edge	R S_{str}	R $\ln q$	G S_{str}	G $\ln q$	B S_{str}	$\ln q$
O^a	1	2	3	4	5	6	7	8	9	10	11	12	13	14	15
LL^b	16	17	18	19	20	21	22	23	24	25	26	27	28	29	30
HL^c	31	32	33	34	35	36	37	38	39	40	41	42	43	44	45
LH^d	46	47	48	49	50	51	52	53	54	55	56	57	58	59	60
HH^e	61	62	63	64	65	66	67	68	69	70	71	72	73	74	75
G_m	76	77	78	79	80	81									
G_d	82	83	84	85	86	87									
G_x	88	89	90	91	92	93									
G_y	94	95	96	97	98	99									

[a] original image, [b] Low-pass–Low-pass wavelet transform image, [c,d] High-pass–Low-pass and Low-pass–High pass wavelet transforms, [e] High-pass–High-pass wavelet transform image

The two mixed outputs of the wavelet transform produce such images that enhance either the horizontal, or the vertically rapidly varying components of the image.

The images have 3 colour channels, thus the total number of used antecedent dimensions is 99, arising the following way. 5 parameters, namely the mean, standard deviation, edge density, structural entropy, filling factor for each of the colour channels make 15, and these 15 parameters for both the original tile and the 4 first wavelet transforms make the number of the dimensions 75. The remaining 24 observables come from the mean and the standard deviation of 4 differently generated gradients. The parameters are summarized in Table 1.

3 Fuzzy Rules

Our method implements formula (5) of the stabilized KH rule interpolation. The membership functions are triangular with the support being the support of the measured data in the training set belonging to the given antecedent and consequent, while the peak (the $\alpha = 1$ cut point) was the mean of the measured values. The training set was generated as every second tile from the system, both in the no-polyp (0) and in the polyp (1) output cases. This very simple system gives very rough results, if the whole database is treated as one, however, if similar pictures are studied together as one group, the performance of their rulebases are very convincing: the ratio of the false positive classifications is usually very low, much below 10%, the ratio of true positive ones is above 70%. The conclusions were the same as in the case of the Mamdani inference if both were applicable. The results are shown in Fig. 1, together with the calculation consequences for a wavelet transform free antecedent set (with

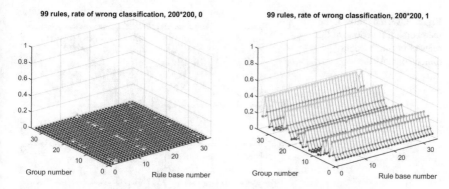

Fig. 1 Conclusion of the stabilized KH interpolation. L.h.s: no polyp misclassification rate, r.h.s: polyp misclassification rate. The antecedent number is 99, the tile size is 200 by 200

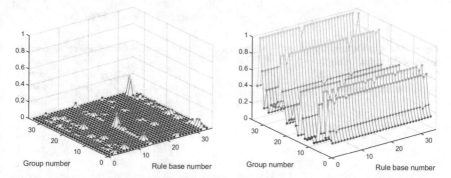

Fig. 2 Conclusion of the stabilized KH interpolation. L.h.s: no polyp misclassification rate, r.h.s: polyp misclassification rate. The antecedent number is 27 (no wavelet transforms), the tile size is 200 by 200

only 27 antecedents) in Fig. 2. It is clearly visible, that the wavelet analysis improves the result (practically, without wavelet transform, the method is not applicable).

In order to decrease the number of the antecedent dimensions, we studied, how large is the overlap between the two membership functions representing the rules of a given antecedent. As a first step, the antecedent measured data was linearly normalized so that the smallest of the measured data would be 0 while the largest 1. Sorted according to the antecedent number, the overlap of the supports of the membership functions in one antecedent dimension can be seen in Fig. 3. Most of the antecedents with small overlap belong to the wavelet-transformed image segments.

We also plotted the limits of the supports of the rules are plotted in Fig. 4.

It seems to be logical to omit those antecedents, where the overlap is large, e.g., above 90%. It also seems interesting, whether being partially ordered according to (1) plays role in the classification, thus in the next plot, in Fig. 5, the boolean values of being ordered are given, as well as of having large or small overlap.

Fig. 3 Overlap ratios of the membership functions of the rules of one antecedent dimensions. As the

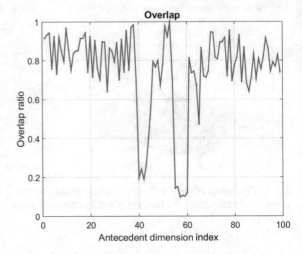

Fig. 4 The upper and lower limits of the membership functions of the rules of one antecedent dimensions

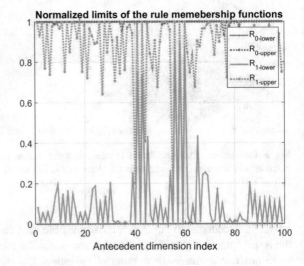

For different tile sizes similar result arise, if he tile size is large enough (usually tile sizes $n \times n$ were applied with n being larger than the tenth of the original picture's size).

4 Results

As a first, least drastic step, we omitted those input variables, that had larger than 90% overlap. From Fig. 6 it is visible, that even though the loss of the 25 antecedent decreases the precision of the conclusion, the decrement is very small. However, using

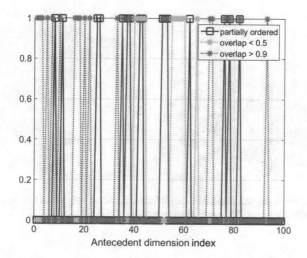

Fig. 5 Overlap ratios of the membership functions of the rules of one antecedent dimensions

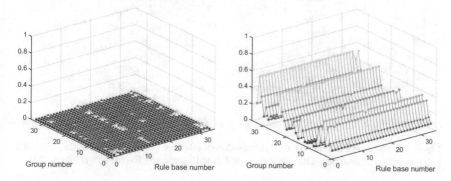

Fig. 6 Conclusion of the stabilized KH interpolation. L.h.s: no polyp misclassification rate, r.h.s: polyp misclassification rate. The antecedent number is 75 (no large overlap), the tile size is 200 by 200

only either the dimensions with small overlap, or the ones that are ordered, or both gives insufficient results, as Figs. 7, 8, and 9 shows. Although there are some image groups which give sufficiently precise conclusion with our method even with most of the original antecedents deleted, this is not a general rule, many images groups (especially if they contain blurred or distorted parts) are practically not classifiable in these cases.

The general success rate of this method is comparable to the ones presented in the Colonoscopy vision MICCAI challenge result summary [6], where the measure "recall rate" is the complementer to our misclassification rate for image segments with polyp (r.h.s pictures). In our case, the misclassification rate was between 0.4 and 0.03, depending on the picture type, which means an average recall rate better

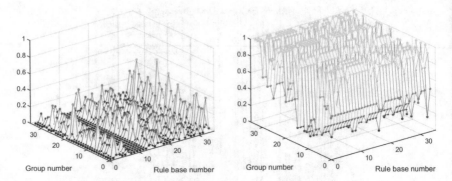

Fig. 7 Conclusion of the stabilized KH interpolation. L.h.s: no polyp misclassification rate, r.h.s: polyp misclassification rate. The antecedent number is 12 (small overlap), the tile size is 200 by 200

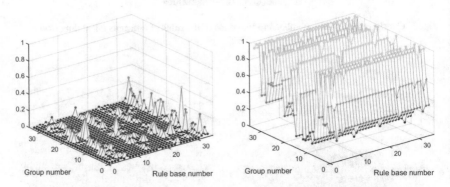

Fig. 8 Conclusion of the stabilized KH interpolation. L.h.s: no polyp misclassification rate, r.h.s: polyp misclassification rate. The antecedent number is 16 (ordered), the tile size is 200 by 200

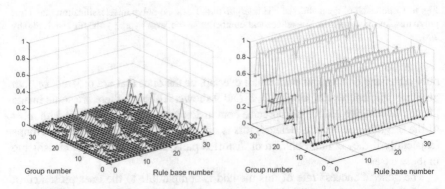

Fig. 9 Conclusion of the stabilized KH interpolation. L.h.s: no polyp misclassification rate, r.h.s: polyp misclassification rate. The antecedent number is 26 (small overlap or ordered), the tile size is 200 by 200

than 72%, for almost any of he rulebases. In [6], in the case of video frames with and without polyps the recall rate, i.e., the rate of the true positive compared to the number of all the frames with polyp was more than 71% with an end-to-end learning algorithm on convolutional neural network, while with their combined method they could get above 82%. The comparison is however a bit hard, as their method analyzed complete videos and video frames instead of image segments.

5 Conclusion

We applied stabilized KH interpolation for determining, whether a colorectal polyp or its segment is present on a colonoscopy image or tile. Wavelet analysis was proven to be an effective tool if the size of the observed vector seems to be too small: with wavelet analysis it is possible to regain part of the information that is lost if statistical parameters of the measured data is used as antecedent. Of the previously selected 99 potential antecedents we could sort out those that have rules with largely overlapping supports thus reducing the dimensionality to 75. We also demonstrated that it is not possible to omit too much inputs either on the basis of the overlap between the membership functions of a rule, or by their being ordered.

Acknowledgements The authors would like to thank the financial support of the project EFOP-3.6.1-16-2016-00017 and the ÚNKP-18-4 New National Excellence Programme of the Ministry of Human Capacities of Hungary. The research was also supported by National Research, Development and Innovation Office (NKFIH) K108405, K124055. The financial support of the Higher Education and Industry Cooperation Center at the Széchenyi University GINOP-2.3.4-15-2016-00003 is gratefully acknowledged.

References

1. Kudo, S., Hirota, S., Nakajima, T., et al.: Colorectal tumours and pit pattern. J. Clin. Pathol. **47**, 880–885 (1994)
2. Rácz, I., Horváth, A., Szalai, M., Spindler, Sz., Kiss, Gy., Regőczi, H., Horváth, Z.: Digital image processing software for predicting the histology of small colorectal polyps by using narrow-band imaging magnifying colonoscopy. Gastrointest. Endosc. **81**, 259 (2015)
3. Søreide, K., Nedrebø, B.S., Reite, A., et al.: Predicting cancer risk in colorectal adenoma: endoscopy morphology, morphometry and molecular markers. Expert Rev. Mol. Diagn **9**, 125–137 (2009)
4. Bernal, J., Sanchez, F.J., Vilariño, F.: Towards automatic polyp detection with a polyp appearance model. Pattern Recognit. **45**, 3166–3182 (2012)
5. Bernal, J., Sanchez, F.J., Fernández-Esparrach, G., Gil, D., Rodrígez, C., Vilariño, F.: WM-DOVA maps for accurate polyp highlighting in colonoscopy: validation versus saliency maps from physicians. Comput. Med. Imaging Graph. **43**, 99–111 (2015)
6. Bernal, J., et al.: Comparative validation of polyp detection methods in video colonoscopy: results from the MICCAI 2015 endoscopic vision challenge. IEEE Trans. Med. Imaging **36**, 1231–1249 (2017)

7. Silva, J.S., Histace, A., Romain, O., Dray, X., Granado, B.: Towards embedded detection of polyps in WCE images for early diagnosis of colorectal cancer. Int. J. Comput. Assist. Radiol. Surg. **9**, 283–293 (2014)

8. Nagy, S., Lilik, F., Kóczy, L.T.: Entropy based fuzzy classification and detection aid for colorectal polyps. In: IEEE Africon 2017, Cape Town, South Africa, 15–17 September 2017

9. Georgieva, V.M., Vassilev, S.G.: Kidney segmantation in ultrasound images via active contours. In: 11th International Conference on Communications. Electromagnetics and Medical Applications, Athens, Greece (2016)

10. Kass, M., Witkin, A., Terzopoulos, D.: Snakes: active contour models. Int. J. Comput. Vis. **1**, 321–331 (1988). https://doi.org/10.1007/BF00133570

11. Kóczy, L.T., Hirota, K.: Rule interpolation in approximate reasoning based fuzzy control. In: Proceedings of the 4th IFSA World Congress, Belgium, Brussels, pp. 89–92 (1991)

12. Zadeh, L.A.: Fuzzy sets. Inf. Control **8**, 338–353 (1965)

13. Zadeh, L.A.: Fuzzy algorithms. Inf. Control **12**, 94–102 (1968)

14. Zadeh, L.A.: Outline of a new approach to the analysis of complex systems and decision processes. IEEE Trans. Syst., Man Cybern., SMC-3, 28–44 (1973). https://doi.org/10.1109/TSMC.1973.5408575

15. Mamdani, E.H., Assilian, S.: An experiment in linguistic synthesis with a fuzzy logic controller. Int. J. Man-Mach. Stud. **7**, 1–13 (1975)

16. Hellendoorn, H., Thomas, C.: Defuzzification in fuzzy controllers. J. Intell. Fuzzy Syst. **1**, 109–123 (1993)

17. Sugeno, M.: An introductory survey of fuzzy control. Inf. Sci. **36**, 59–83 (1985)

18. Sugeno, M., Kang, G.T.: Structure identification of fuzzy model. Fuzzy Sets Syst. **28**, 15–33 (1988)

19. Takagi, T., Sugeno, M.: Fuzzy identification of systems and its applications to modeling and control. IEEE Trans. Syst., Man, Cybern., SMC-15, 116–132 (1985)

20. Kóczy, L.T., Hirota, K.: Rule interpolation by α-level sets in fuzzy approximate reasoning. J. BUSEFAL, Automne, URA-CNRS **46**, 115–123

21. Bouchon-Meunier, B., Dubois, D., Marsala, C., Prade, H., Ughetto, L.: A comparative view of interpolation methods between sparse fuzzy rules. In: Proceedings Joint 9th IFSA World Congress and 20th NAFIPS International Conference (Cat. No. 01TH8569). Vancouver. BC, Canada, IEEE, pp. 25–28 (2001)

22. Bouchon-Meunier, B., Dubois, D., Godo, L., Prade, H.: Fuzzy sets and possibility theory in approximate and plausible reasoning. In: Bezdek, J.C., Dubois, D., Prade, H. (eds.) Fuzzy Sets in Approximate Reasoning and Information Systems. Springer Science and Business Media, New York (1999)

23. Bouchon-Meunier, B., Esteva, F., Godo, L., Rifqi, M., Sandri, S.: A Principled Approach to Fuzzy Rule Base Interpolation Using Similarity Relations. EUSFLAT–LFA 2005, Barcelona, Spain. pp. 757–763 (hal-01072103) (2005)

24. Esteva, F., Rifqi, M., Bouchon-Meunier, B., Detyniecki, M.: Similarity-based fuzzy interpolation method. In: International Conference on Information Processing and Management of Uncertainty in Knowledge-Based Systems, IPMU 2004, Perugia, Italy, pp. 1443–1449 (hal-01072531) (2004)

25. Bouchon-Meunier, B., Marsala, C., Rifqi, M.: Interpolative reasoning based on graduality. In: Ninth IEEE International Conference on Fuzzy Systems, FUZZ-IEEE 2000 (Cat. No. 00CH37063), vol. 1, pp. 483–487. IEEE, San Antonio, TX, USA, 7–10 May 2000

26. Tikk, D., Joó, I., Kóczy, L.T., Várlaki, P., Moser, B., Gedeon, T.D.: Stability of interpolative fuzzy KH-controllers. Fuzzy Sets Syst. **125**, 105–119 (2002). https://doi.org/10.1016/S0165-0114(00)00104-4

27. Balázs, K., Kóczy, L.T.: Constructing dense, sparse and hierarchical fuzzy systems by applying evolutionary optimization techniques. Appl. Comput. Math. **11**, 81–101 (2012)

28. Dubois, D., Prade, H.: Gradual inference rules in approximate reasoning. Inf. Sci. **61**, 103–122 (1992)

29. Pipek, J., Varga, I.: Universal classification scheme for the spatial localization properties of one-particle states in finite d-dimensional systems. Phys. Rev. A **46**, 3148–3164. APS, Ridge NY-Washington DC (1992)
30. Nagy, S., Sziová, B., Pipek, J.: On structural entropy and spatial filling factor analysis of colonoscopy pictures. Entropy **21**, 256 (pp. 280–311 in printed version) (2019). https://doi.org/10.3390/e21030256
31. Nagy, S., Lilik, F., Kóczy, L.T.: On fuzzy classification with interpolation of the sparse rule bases. In: ESCIM 2017, Faro, Portugal, 6–8 October 2017
32. Daubechies, I.: Ten Lectures on Wavelets, CBMS-NSF Regional Conference Series in Applied Mathematics. SIAM, Philadelphia (1992)
33. Balázs, K., Kóczy, L.T.: Hierarchical-interpolative fuzzy system construction by genetic and bacterial memetic programming approaches. Int. J. Uncertainty, Fuzziness Knowl. Based Syst. **20**, 105–131 (2012)
34. Georgieva, V.M., Nagy, S., Kamenova, E., Horváth, A.: An approach for pit pattern recognition in colonoscopy images. Egypt. Comput. Sci. J. **39**, 72–82 (2015)
35. Nagy, Sz., Lilik, F., Kóczy, L.T.: Applicability of various wavelet families in fuzzy classification of access networks' telecommunication lines. In: FuzzIEEE 2017, Naples, Italy, 9–12 July 2017

Association Rule Mining for Unknown Video Games

Alexander Dockhorn, Chris Saxton, and Rudolf Kruse

Abstract Computational intelligence agents can reach expert levels in many known games, such as Chess, Go, and Morris. Those systems incorporate powerful machine learning algorithms, which are able to learn from, e.g., observations, play-traces, or by reinforcement learning. While many black box systems, such as deep neural networks, are able to achieve high performance in a wide range of applications, they generally lack interpretability. Additionally, previous systems often focused on a single or a small set of games, which makes it a cumbersome task to rebuild and retrain the agent for each possible application. This paper proposes a method, which extracts an interpretable set of game rules for previously unknown games. Frequent pattern mining is used to find common observation patterns in the game environment. Finally, game rules as well as winning-/losing-conditions are extracted via association rule analysis. Our evaluation shows that a wide range of game rules can be successfully extracted from previously unknown games. We further highlight how the application of fuzzy methods can advance our efforts in generating explainable artifical intelligence (AI) agents.

1 Introduction

Artificial intelligence (AI) in games is known to have beaten expert human players in a wide range of boardgames. A common example is the chess playing agent Deep Blue [4], which defeated the at the time reigning world champion Garry Kas-

A. Dockhorn · C. Saxton · R. Kruse (✉)
Institute for Intelligent Cooperating Systems, Otto-von-Guericke University,
Universitaetsplatz 2, 39106 Magdeburg, Germany
e-mail: rudolf.kruse@ovgu.de

A. Dockhorn
e-mail: alexander.dockhorn@ovgu.de

C. Saxton
e-mail: chris.saxton@ovgu.de

M.-J. Lesot and C. Marsala (eds.), *Fuzzy Approaches for Soft Computing and Approximate Reasoning: Theories and Applications*, Studies in Fuzziness and Soft Computing 394,
https://doi.org/10.1007/978-3-030-54341-9_22

parov. The system was able to rate millions of board positions per second and was tuned in cooperation with multiple chess grandmasters. Recently, GO was added to this list with AlphaGO's victory over the world's best human GO player Lee Sedol [28]. Despite the great success of AlphaGO its implications are unclear. The utilized black-box model, a neural network based board evaluation function trained by reinforcement learning [8], is hard to interpret and highly specialized on a single task.

Despite the strength in playing these games, humans find it nearly impossible to interpret the result of black-box systems such as deep neural networks [27]. Here, the knowledge is represented in high-dimensional spaces learned from tons of data. The decision-making process cannot be explained easily and be fooled by adversarial examples. See, for example, the deep convolutional neural network architecture AlexNet [11]. Researchers were able to classify one of the largest image libraries "ImageNet Large Scale Visual Recognition Challenge" (LSVRC-2010; 1.2 million images, 1000 object classes) [25] with an astonishing top-5 error rate of 17.0%. However, recent analysis on the best performing solutions showed that minor perturbations can drastically change the classification result [30].

In the context of game AI similar problems can lead to unexpected behavior of the developed agent. For example, the system might fail in unusual situations or perform unreliably. During the interaction with such systems, players or users can have difficulties in explaining the AI's behavior [5]. This can reduce the player's immersion and, therefore, decrease the player's enjoyment [29]. For this reason, the industry is in need of explainable AI systems, which act on the basis of comprehensible mechanisms. Crafting those systems by hand is an extremely costly and time-consuming task. Our work aims in developing such explainable AI systems by automatically adapting to the game environment.

While many proposed AI systems focus on generating an agent to play a single game, our work focuses on the task of creating an agent that is capable of learning to play multiple games. Thus, special attention goes to the generation of abstract learning systems, which are able to adapt to a wide range of game environments. Games provide a natural test bed for the task of general AI, because they exist in many variations and focus on providing a competing and balanced environment. Developing agents for games and especially for video games is of high interest due to the rapid growth of the (digital) entertainment industry. Learning AI agents would not just be able to play the game, but can adapt themselves to updates of the game throughout the development phase. Additionally, learning agents can be used for automatic game testing and balancing. Therefore, those systems would be able to drastically reduce time and costs of the development process.

The General Video Game AI (GVGAI) competition [21] provides games in a common description language, which eases the process of agent development and provides a unified benchmark for general purpose game AI. Previous results suggest that developed agents can achieve very good performance when they are provided with a forward model, which can be used to simulate the outcome of an agent's action. However, this forward model needs to be hand-crafted and may not be available during the development of a game. For this reason, the in 2017 introduced learning track added a new challenge, in which agents only have a few tries for training and understanding the game without receiving any information on the game's underlying rules.

In the learning track agents are provided with three levels during training, which introduce elements of the game being studied. The agent's performance is subsequently validated based on its performance in two unknown levels. Most recent results of the 2017's learning track competition show that developed agents performed barely better than an agent acting randomly. Proposed models include Monte Carlo Tree Search (MCTS) agents with various parameter estimation algorithms [10]. This might be due to a lack of an understanding of the game's general rules and conditions for winning or losing the game.

In this work our goal is to improve the generation of a suitable game model during the training phase. Our system especially aims for an interpretable approximation of the game's rules and its winning or losing conditions, also known as termination set. This is achieved using play-trace analysis based on frequent pattern mining and association rule analysis. The complexity of the approximated rule set is reduced using multiple rating criteria. Our work introduces new ways of developing and validating agents for previously unknown games and is especially interesting for the development of interpretable models of the agent's environment. We further discuss possible fuzzy extensions and their influence on the interpretability of generated models.

The remainder of this paper will be structured as follows: In Sect. 2 we provide a basic overview of the topic general game AI and the framework we used to simulate a wide range of games. Further in Sect. 3, frequent pattern and association rule mining methods are reviewed, since they are the key elements in our model induction process. After explaining the basic concepts, we introduce our play-trace generating mechanism in Sect. 4. On the basis of these play-traces, we first use association rule analysis to model each game. Special constraints of the termination set are discussed in Sect. 5. Furthermore, we propose various rule reduction mechanisms. These are necessary to filter a set of interesting rules, which describe the game reasonably well. In Sect. 6 we present how explainable agents can be learned to highlight the strengths of the induction process. Subsequently, we discuss fuzzy-based extensions for the proposed framework in Sect. 7. Final remarks and interesting implications for future work can be found in Sect. 8.

2 General Video Game AI

General Game AI is a popular area of research. In contrast to previous works in game AI an agent is evaluated on its capability in playing a diverse set of games reasonably well. While humans can pick up simple games and learn their rules in a matter of minutes, most programmed agents need a lot of data to play the game on a competitive level. The General Video Game AI (GVGAI) competition [21] promotes further research on general games AI. It provides a framework for the creation of simple games with a similar back-end. Thus, agents can be tested on a wide range of video games without additional coding.

The GVGAI-framework is based on the works of Tom Schaul, who adapted the game definition language to the context of video games. His Video Game Definition Language (VGDL) [26] allows to quickly design games by providing entities and interactions of the game in the game description, while providing the actual level in a separate level description. The game description consists of: an interaction set, a termination set, a sprite set, and a level mapping [21].

The interaction set provides the basic model of the game. It describes how single sprites behave, how the player input influences the games avatar, and how interactions between those sprites change the current state of the game. During a collision it is possible to change the players score, to move, destroy or add sprites to the current game state, or change inherent attributes of represented game-objects, such as health points, movement speed or resources [24]. The interaction set also provides information about the physics engine used by each sprite. Currently two modes of operation are available: either grid-based physics, or continuous physics [22]. Grid-based physics simulates a top-down view in which sprites move from one grid cell to another or pixel-wise. Continuous physics were recently added to analyze learning models under more complicated physic environments. Here, sprites can move continuously in between grid-cells and simulate basic laws of physics such as gravity.

In this work we focus on modeling the agent's control options, the interaction set, as well as the termination set. The termination set defines a set of rules in which the game is destined to end. Possible mechanics include:

1. the count of a sprite is less/greater than a specified threshold, e.g., the player sprite was removed
2. a variable exceeds a specified threshold
3. an interaction occurs (which most often deletes a sprite and triggers the first option).

The GVGAI-Framework provides two observation mechanisms. Either the pixel output of the game engine is used, or the observation set is provided as an array-like representation of the game environment. In our work we will focus on the latter, which removes the influence of the sprite set.

The level mapping defines relationships between characters used in the level definition and sprites to be used. The level definition contains the start configuration of the game and is used to easily change the level structure without changing the rules of the game.

The GVGAI-Framework currently provides a set of 90 grid-based and 10 continuous-physics games. Researchers can participate in various competition tracks. In the playing track agents can access full information on the game model and the current game state. The learning track limits the agents to a partial observable game-state and does not provide a game-model. Our work focuses on the learning-track, in which the agent cannot use a forward-model. Therefore, our methods aim in generating a forward model based on observations during multiple play-traces.

3 Frequent Pattern Mining (FPM) and Association Rule Analysis

In contrast to black-box models we aim for an interpretable system. The output should be in human-readable format. Especially interesting are well compressed systems such as decision trees [23] and association rules. The strength of these methods is their output in form of human-readable rules. Those are not just easy to interpret but also easy to adapt. Due to the complexity of the game environments, we choose to implement an association rules analysis. In contrast to decision trees, association rule analysis works efficiently with a high number of features [3, 18]. Learning a tree would have a large overhead during the feature selection phase and result in a tree with high width and depth [9]. Additionally to the simplicity of these rule based systems, both can be adapted to non-deterministic (game) environments by using fuzzy systems. Both, fuzzy decision trees as well as fuzzy rule sets, would be possible extensions to our proposed solution. The utility of fuzzy extensions for the proposed approach is discussed in Sect. 7.

Our work relies on frequent pattern mining and association rule analysis [1], which results in simple rules in the form of:

$$antecedence \rightarrow consequence$$

Specifically, we are interested in rules that map the current game-state and possibly the agents action to the next game-state. This allows the agent to anticipate enemy movements, tile interactions and avoid any of the losing conditions of the game.

We achieve this by extracting patterns from recorded play-traces. A play-trace is an ordered set of observations of every tick of the game. Two consecutive game-ticks (t_k, t_{k+1}) will be merged to a single transaction, in which the action and each change in an observable element is included as an item $i \in B$ in the transaction. For each set of items I we count the support. The support of an item set is the number of transactions $t \in T$ that include the specified item set.

$$\text{supp}_T(I) = \left| \{t \in T \mid I \subseteq t\} \right| \cdot |T|^{-1}$$

In Frequent Pattern Mining (also called Frequent Item Set Mining) we search for all item sets with a support higher than a specified threshold, the minimum support supp_{min}. Therefore, the set of frequent items F is given by:

$$F(\text{supp}_{min}) = \{I \subseteq B \mid \text{supp}_T(I) \geq \text{supp}_{min}\}$$

After all frequent item sets were extracted, we create rules of the form $X \rightarrow Y$ for each item set in the set of frequent item sets ($I \in F$), such that $X \cup Y = I$ and $X \cap Y = \emptyset$. The set of rules will be filtered based on support and confidence of a rule.

The support of an association rule indicates how often a rule can successfully be applied in the given transaction database.

$$\text{supp}(X \Rightarrow Y) = \text{supp}(X \cup Y) = |\{t \in T \mid X \cup Y \in t\}| \cdot |T|^{-1}$$

It can be interpreted as the relative frequency in which the rule could have been applied. The confidence of an association rule describes how often the rule can be correctly applied in the given transaction database.

$$\text{conf}(X \Rightarrow Y) = \frac{\text{supp}(X \cup Y)}{\text{supp}(X)}$$

It can also be interpreted as a conditional probability $P(E_Y|E_X)$, the probability of finding the consequence of the rule in a transaction in which the antecedence is known to be present.

4 Extracting Play-Traces Patterns and Game Rules

Our algorithm starts with zero knowledge and tries to approximate the games rules by analyzing the play-traces of previous runs. Extracting play-traces can be a cumbersome task. In general those play-traces should cover all possible interactions that can occur during the game. However, without any knowledge you cannot perform much better than acting randomly. This can be seen in the results of 2017's GVGAI-learning track in which only one agent performed slightly better than a random agent. For this reason we generate an initial set of play-traces using a random agent. During play-trace collection the random agent is not learning from any previous experience. Therefore, all transactions in the database will be based on the same knowledge of the game, more precisely, in this special case on the absence of knowledge. Further play-traces can be collected using the generated model for enhancing the planning capabilities of an agent.

The capabilities of our method will be largely restricted by the set of features we store during our play-trace generation. Table 1 lists the recorded features used in this study. This list was iteratively adjusted to the needs of our method.

We use Frequent Pattern Mining to find frequently occurring patterns between consecutive time-steps. Especially important is the game result, which can be either "*continuing*", "*won*", or "*lost*". In a play-trace that consists of n time-steps each time-step from $k = 1, \ldots, n - 1$ will be of the state "continuing" and only the last time-step is either marked with "won" or "lost". During longer games the support of only the ending time-steps will not reach a reasonable minimal support threshold. Therefore, no termination rules can be found in this case. Further, decreasing the minimal support to a level in which all termination rules can be found will result in far too many rules. However, we can make further use of special properties of the termination set for effective mining of such rules. These will be discussed in Sect. 5.

Table 1 Recorded attributes during the play-trace analysis

Feature	Range
Game tick	$\mathbb{N}$
Game score	$\mathbb{N}$
Game score change	$\mathbb{N}$
Game result	{Continuing, Won, Lost}
Player action	{Up, Down, Left, Right, Use, other}
Player grid-position	{$\mathbb{R}$, $\mathbb{R}$}
Position change X	$\mathbb{R}$
Position change Y	$\mathbb{R}$
No position change	*boolean*
Player collision	Object type
Changes in secondary attributes such as health, resources, etc.	
Object above	Object type
Object below	Object type
Object left	Object type
Object right	Object type
Object collision	{Sprite 1, Sprite 2}

All other rules will be filtered based on support and confidence. While the support is used to reduce the number of frequent item sets, the confidence limits the rule base to rules which have a high probability of being correct. This is especially important because mistakes in our model will accumulate over time and even further reduce the accuracy of our prediction.

5 Properties of the Termination Set

As a first example we consider the game depicted in Fig. 1. The agent (yellow character) can move to neighboring grid-cells in each of the cardinal directions. He wins the game when he is approaching the flag from the side and loses when coming

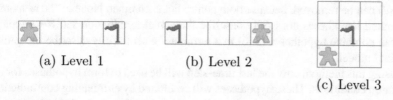

(a) Level 1   (b) Level 2

(c) Level 3

Fig. 1 Exemplary game where the player wins, when reaching the flag from the side and loses when he approaches from the top or the bottom

from the top or the bottom. In this game all winning play-traces in level 1 (Fig. 1a) reach this point by moving to the right during their last step, while all play-traces in level 2 (Fig. 1b) reach the target by moving to the left. By just looking at the final time-step-transition in all play-traces from the first and the second level five simple hypotheses are possible:

1. Moving right wins the game.
2. Moving left wins the game.
3. Reaching the flag wins the game.
4. Moving right and reaching the flag wins the game.
5. Moving left and reaching the flag wins the game.

Which of these is the most suitable hypothesis? Judging based on a single play-trace all seem equally valid. However, the hypothesis (1) and (2) can be excluded, in case we combine the observations of level 1 and level 2. Both levels will result in play-traces, which consist of multiple time-step-transitions in which a step to the right or to the left does not end the game. Deciding which of the other three hypotheses (3–5) is the most suitable cannot be done by just observing play-traces of a single level. Judging by complexity hypothesis (3) is the simplest and is correct for both levels. Both other hypotheses, (4) and (5), are valid for only one of the provided example levels. Therefore, we will return hypothesis (3) as a temporary result.

Nevertheless, in both levels there is only one way of fulfilling the general winning condition. In level 3, depicted in Fig. 1c, there are multiple options for winning and losing the game. The player will instantly lose in case he approaches the flag from the top or the bottom.

The following hypothesis can be generated from analyzing single play-traces of the third level:

1. Reaching the flag wins the game.
2. Reaching the flag loses the game
3. Moving right and reaching the flag wins the game.
4. Moving left and reaching the flag wins the game.
5. Moving up and reaching the flag loses the game.
6. Moving down and reaching the flag loses the game.

All hypotheses for "Moving in any direction to win or lose the game." were removed, since, by the logic presented above, some play-traces will contain moves in the specified direction that did not end the game. From the generated hypotheses, (1) and (2) can be removed, because both contradict each other. None of the remaining generated hypotheses does fully describe the actual termination set. Nevertheless, we can combine hypotheses (3–6) in a termination set to fully describe the game's original model.

Using this method, only the last time-step will be used to form hypotheses for termination conditions. These hypotheses will be filtered by eliminating contradictions using the transactions of all previous time-steps. For the need of simplicity more complex hypotheses are exchanged by more simple termination conditions. Finally, the termination set can consist of multiple simple termination criteria. Further work needs to be put into the automatic compression of the termination set. This will ensure a faster processing during run-time and an improved readability in more complex termination sets.

6 Applications of Association Rule Analysis for General Game AI

The strengths of the system will be further highlighted based on multiple games of the GVGAI-framework. In this paper we present a detailed analysis of the game "alien". For further results we refer to our current papers on the topic of forward model approximation [6, 7].

The game "alien" is a clone of the famous game space invaders. Here, the player controls a small spaceship and needs to defend itself from approaching aliens. Figure 2 shows a screenshot of the game. In the bottom row the player flies left or right. The spaceship can shoot a single bullet to destroy the alien spaceships before they arrive at the bottom of the screen. From time to time alien spaceships shoot as well. Their rockets and the player's bullets explode at the gray rocks in the centre of the game board. The player wins by destroying all aliens, but loses by either destroying all rocks, or in case a single alien reaches the bottom row.

Play-traces were collected by letting the random agent play 10 rounds of "alien". According to our proposed play-trace analysis we retrieved 8 rules. See Table 2 for the full list of rules.

Most of these rules are movement based, because movement occurs comparatively often. The lowered confidence of MoveRight appeared by coincidence, since the player always lost the game by moving right during the last time-step. During reset the game engine places the agent in the bottom left corner, therefore, a MoveLeft was detected. Interactions between sprites only occur rarely. For this reason, only one interaction rule can be found in the final rule set. The termination condition was

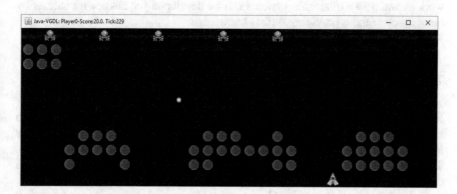

Fig. 2 Screenshot of the game alien of the GVGAI framework

Table 2 Association rules extracted from play-trace analysis

Rule			Support	Confidence
ActionLeft	→	MoveLeft	0.32	1.00
ActionRight	→	MoveRight	0.31	0.98
ActionUse	→	Stay	0.36	1.00
ObjectAbove, ActionUse	→	Stay	0.05	1.00
ObjectAbove, ActionLeft	→	MoveLeft	0.05	1.00
ObjectAbove, ActionRight	→	MoveRight	0.06	0.92
SpriteCollision	↔	HigherScore	0.06	1.00
PlayerCollision, ScoreDecrease	→	GameLost	1.00	1.00

successfully retrieved. However, since the agent did not win a game during 10 tries no rule for winning could have been retrieved. The final rule set describes the game reasonably well.

It is astonishing how well the systems grasps the true game model. Even without any further knowledge of the game, most necessary rules were detected. The resulting rule set is in human readable format and easy to understand. Agents can apply these rules in simulation-based search algorithms as approximation of the forward model. Chosen actions can later be justified by showing the user the related rules. Even the outcome of long action sequences can be explained by studying the agent's assumed outcome.

The result of this case-study also applies to many of the other games in the GVGAI-framework. A natural extension of this work, the Forward Model Approximation framework, can be found in [6, 7]. We also want to note that the development of game AI is just one of many interesting applications of this work. In general this work shows how explainable systems can be developed for unknown contexts by observation without providing further knowledge.

7   Analysis of Fuzzy Methods

In this section we will discuss opportunities for the application of fuzzy methods to improve the proposed approach. Fuzzy logic is a many-valued extension of binary logic, which can be exploited in cases that allow truth values between true and false. This could be a desirable characteristic for multiple components of the designed model.

First of all the agent described in the current framework is described to be perfect in its observation of the environment. While the computer game setting often allows to implement agents with perfect sensors, transferring this approach to the field of robotics would need adjustments to the sensor value processing. Especially the item set mining needs to be fault-tolerant in cases where sensor values can be missing or does only receive vague information. The mining of gradual patterns can be used to act in scenarios with missing sources of information [16]. Also using fault-tolerant frequent item set mining algorithms [20], such as SODIM [2], will allow the application of the proposed approach in an error-prone setting. Such a fuzzy item set matching cannot only be applied during the mining phase, but also during the association rule processing phase. This may allow matching incomplete patterns for acting in situations in which none of the crisp rules apply.

Next to a fuzzy item set mining, the association rule mining process needs to be adapted for situations in which multiple rules may apply. This is especially relevant in non-deterministic games, in which a single situation can yield differing results. The proposed approach will generate conflicting association rules, which need to be handled by the agent during the decision-making process. By making the rule-exploitation fuzzy we allow the agent to weigh the applicability of multiple rules and their outcomes. Aggregating the results of these rules will allow the agent to act in situations in which the outcome of a situation can only be described by multiple degrees of possibility.

While the extensions above describe how the agent will be able to act in vague situations, fuzzy methods can also be applied to reduce complex rule sets to a human understandable form. Describing the final agent using a small set of fuzzy rules may help to understand the reasoning process and its resulting behavior. For example, fuzzy summaries can be used to reduce the fuzzy rule set to a small set of linguistic terms [13]. Therefore, applying fuzzy concepts cannot just help to adapt the proposed method to vague scenarios, but can also help to increase the explainability of the overall system behavior.

A final aspect, which is often ignored, is the interaction between the developed agent and a human user. We already mentioned the industry's interest in the development of self learning systems, but after the agent is adapted to the developed system it will, ultimately, interact with the human user. Measuring the user's emotions will inherently lead to subjective information. Managing and interpreting vague data gained from user questionnaires can be made possible by applying fuzzy data mining [15]. Sentiment analysis and opinion mining [19] will be the major research areas for increasing the user's enjoyment.

8 Conclusions and Outlook

In this work we discussed a rule induction mechanism for modeling unknown video games. Our case-study shows that simple and accurate rules can be extracted from a small set of observations. The learned model results in an approximation of the forward model. Utilizing this model helps to explain the agent's behavior and enables it to provide explanations of its intentions.

Our work highlights the capabilities of explainable machine learning systems, which can be used in complex environments without further knowledge. In contrast to other popular machine learning algorithms our proposed method results in a rule set readable by humans. This offers users the chance to further analyze the strengths and weaknesses of the developed agent and fosters the possibility of a strong human-computer interaction.

With this in mind many new research opportunities arise. For example, the possibilities for a stronger human-computer interaction would motivate tutoring systems, in which a machine teaches a new user based on the true model of an application and the model that can be induced from the user's behavior. Especially interesting would be the generation of new learning examples, which help the user to get a grasp of the true model.

In contrast to teaching humans, the resulting model can also be used for a step wise increase in the agent's skill level. For this purpose, an agent would receive the induced rule set as an incomplete forward model. This would allow the agent to plan multiple steps ahead. Referring to our example game "alien", the agent would try to avoid collisions with other game entities to not lose the game. In combination with the extracted movement rules, the agent will be able to dodge bullets, thus, increasing its chances of winning the game. Collecting more play-traces with a partially informed agent will lead to an even better approximation of the true game model. Even without the true forward model, prediction of future game states could succeed with a reasonable enough approximation of the original model. This iterating approach will allow a step wise update of the agent's capabilities, ultimately allowing agents to learn how to be successful in a wide range of more complex applications, without any human interference.

Nevertheless, more complex models would also be in need of a more complex representation. Hierarchical rule representations and the mining of gradual patterns [16] would prove useful in preserving simplicity of the rule set in such scenarios. Another idea would be the development of local environment models. Especially multi-agent models might prove useful in modeling independent components of the environment. Such a distributed approach assures faster computation and higher memory efficiency. This could be achieved by applying the proposed approach to local observations of each entity of an environment. Selecting a scope of each local model and finding its relevant features will be a major focus due to the complexity of the environment.

Last but not least we see a special importance in the development of explainable models for non-deterministic applications. Those pose special constraints on the learning system, in which, for example, rules can trigger unreliably or have overlapping antecedences. Fuzzy models such as fuzzy decision trees [14, 17] and fuzzy rule sets [12, 16] may be the key to modeling such environments.

This selection of research opportunities does not aim to be complete, but it highlights the many directions in which explainable AI may advance during the next years. We are looking forward to all the upcoming applications, methods, and results that this field is about to offer.

References

1. Borgelt, C.: Frequent item set mining. Wiley Interdisc. Rev.: Data Min. Knowl. Discovery **2**(6), 437–456 (2012)
2. Borgelt, C., Kötter, T.: Mining fault-tolerant item sets using subset size occurrence distributions. In: Gama, J., Bradley, E., Hollmén, J. (eds.) Advances in Intelligent Data Analysis X, pp. 43–54. Springer, Berlin Heidelberg, Berlin, Heidelberg (2011)
3. Borgelt, C., Yang, X., Nogales-Cadenas, R., Carmona-Saez, P., Pascual-Montano, A.: Finding closed frequent item sets by intersecting transactions. In: Proceedings of the 14th International Conference on Extending Database Technology—EDBT/ICDT '11, ACM Press, New York, USA, p. 367 (2011)
4. Campbell, M., Hoane Jr. A.J., Hsu, F.H.: Deep blue. Artif. Intell. **134**(1–2), 57–83 (2002)
5. Core, M., Lane, H., Van Lent, M., Gomboc, D., Solomon, S., Rosenberg, M.: Building explainable artificial intelligence systems. In: Proceedings of the 21st National Conference on Artificial Intelligence and the 18th Innovative Applications of Artificial Intelligence Conference, AAAI-06/IAAI-06. vol. 2, pp. 1766–1773 (2006)
6. Dockhorn, A., Apeldoorn, D.: Forward model approximation for general video game learning. In: 2018 IEEE Conference on Computational Intelligence and Games (CIG) (2018), to be published
7. Dockhorn, A., Tippelt, T., Kruse, R.: Model decomposition for forward model approximation. In: 2018 IEEE Symposium Series on Computational Intelligence (SSCI) (2018), to be published
8. Giryes, R., Elad, M.: Reinforcement learning: a survey. In: European Signal Processing Conference, pp. 1475–1479 (2011)
9. Guyon, I., Elisseeff, A.: An introduction to variable and feature selection. J. Mach. Learn. Res. (JMLR) **3**(3), 1157–1182 (2003)
10. Ilhan, E., Etaner-Uyar, A.S.: Monte Carlo tree search with temporal-difference learning for general video game playing. In: 2017 IEEE Conference on Computational Intelligence and Games (CIG), IEEE, New York, pp. 317–324 (2017)
11. Krizhevsky, A., Sutskever, I., Hinton, G.E.: ImageNet classification with deep convolutional neural networks. In: Advances In Neural Information Processing Systems, pp. 1–9 (2012)
12. Kruse, R., Borgelt, C., Braune, C., Mostaghim, S., Steinbrecher, M.: Computational Intelligence. Texts in Computer Science, 2nd edn. Springer, London (2016)
13. Laurent, A., Marsala, C., Bouchon-Meunier, B.: Improvement of the interpretability of fuzzy rule based systems: quantifiers, similarities and aggregators. In: Lawry, J., Shanahan, J., Ralescu, L.A. (eds.) Modelling with Words: Learning, Fusion, and Reasoning within a Formal Linguistic Represntation Framework, Springer, Berlin, Heidelberg, pp. 102–123 (2003)
14. Marsala, C.: Application of fuzzy rule induction to data mining. In: Andreasen, T., Christiansen, H., Larsen, H.L. (eds.) Flexible Query Answering Systems, Springer, Berlin, Heidelberg, pp. 260–271 (1998)
15. Marsala, C., Bouchon-Meunier, B.: Fuzzy data mining and management of interpretable and subjective information. Fuzzy Sets Syst. **281**, 252–259 (2015)
16. Mellouli, N., Bouchon-Meunier, B.: Abductive reasoning and measures of similitude in the presence of fuzzy rules. Fuzzy Sets Syst. **137**(1 SPEC.), 177–188 (2003)
17. Mitra, S., Konwar, K., Pal, S.: Fuzzy decision tree, linguistic rules and fuzzy knowledge-based network: generation and evaluation. IEEE Trans. Syst., Man Cybern., Part C (Appl. Rev.) **32**(4), 328–339 (2002)
18. Pan, F., Cong, G., Tung, A.K.H., Yang, J., Zaki, M.J.: Carpenter. In: Proceedings of the Ninth ACM SIGKDD International Conference on Knowledge Discovery and Data Mining—KDD '03, ACM Press, New York, USA, p. 637 (2003)
19. Pang, B., Lee, L.: Opinion mining and sentiment analysis. Found. Trends® Inf. Retrieval **2**(1–2), 1–135 (2008)
20. Pei, J., Tung, A.K., Han, J.: Fault-tolerant frequent pattern mining: problems and challenges. DMKD **1**, 42 (2001)

21. Perez-Liebana, D., Samothrakis, S., Togelius, J., Schaul, T., Lucas, S.M., Couetoux, A., Lee, J., Lim, C.U., Thompson, T.: The 2014 general video game playing competition. IEEE Trans. Comput. Intell. AI Games **8**(3), 229–243 (2016)
22. Perez-Liebana, D., Stephenson, M., Gaina, R.D., Renz, J., Lucas, S.M.: Introducing real world physics and macro-actions to general video game ai. In: 2017 IEEE Conference on Computational Intelligence and Games (CIG), IEEE, pp. 248–255 (2017)
23. Quinlan, J.R.: Induction of decision trees. Mach. Learn. **1**(1), 81–106 (1986)
24. Rogers, S.: Level Up!: The Guide to Great Video Game Design. Wiley (2010)
25. Russakovsky, O., Deng, J., Su, H., Krause, J., Satheesh, S., Ma, S., Huang, Z., Karpathy, A., Khosla, A., Bernstein, M., Berg, A.C., Fei-Fei, L.: ImageNet large scale visual recognition challenge. Int. J. Comput. Vis. (IJCV) **115**(3), 211–252 (2015)
26. Schaul, T.: An extensible description language for video games. IEEE Trans. Comput. Intell. AI in Games **6**(4), 325–331 (2014)
27. Schmidhuber, J.: Deep learning in neural networks: an overview. Neural Netw. **61**, 85–117 (2015)
28. Silver, D., Huang, A., Maddison, C.J., Guez, A., Sifre, L., van den Driessche, G., Schrittwieser, J., Antonoglou, I., Panneershelvam, V., Lanctot, M., Dieleman, S., Grewe, D., Nham, J., Kalchbrenner, N., Sutskever, I., Lillicrap, T., Leach, M., Kavukcuoglu, K., Graepel, T., Hassabis, D.: Mastering the game of Go with deep neural networks and tree search. Nature **529**(7587), 484–489 (2016)
29. Sweetser, P., Wyeth, P.: GameFlow: a model for evaluating player enjoyment in games. Comput. Entertain. **3**(3), 3–3 (2005)
30. Szegedy, C., Zaremba, W., Sutskever, I., Bruna, J., Erhan, D., Goodfellow, I.J., Fergus, R.: Intriguing properties of neural networks. CoRR abs/1312.6199 (2013)

Semantic Web: Graphs, Imprecision and Knowledge Generation

Marek Z. Reformat

Abstract Growing interests in Artificial Intelligence generate expectations for constructing intelligent systems supporting humans in a variety of activities. Such systems would require the ability to analyze data and information, as well as to synthesize new elements of knowledge. To make it possible, there is a need to develop techniques and algorithms capable of extracting logic structures from data, processing and modifying these structures, and creating new ones. In this context, a graph-based representation of data—in particular a knowledge graph proposed by the Semantic Web initiative—is of a special interest. Graphs enable representing and defining semantics of data via utilization of multiple types of relations. Additionally, imperfections associated with processes of collecting and aggregating data and the imprecise nature of knowledge in variety of domains require data representation formats that can handle vagueness and uncertainty. In our opinion, knowledge graphs combined with elements of fuzzy set theory represent important building blocks of knowledge generation procedures. This chapter contains a brief overview of a framework enabling construction of systems capable of extracting logic structures from data and synthesizing knowledge. Utilization of fuzziness enables representation of imprecision and its inclusion in knowledge generation processes.

1   Introduction

The web represents an immense repository of information. A number of structured and unstructured data sources is constantly growing. Yet, the increased amount of data although recognized as a positive and beneficial fact creates challenges regarding its full utilization. This also leads to elevated expectations regarding development of comprehensive methods for processing data and information, and transforming them into formats suitable for knowledge generation. One can envision systems that

M. Z. Reformat (✉)
Electrical and Computer Engineering, University of Alberta, Edmonton, Canada
e-mail: Marek.Reformat@ualberta.ca

271

M.-J. Lesot and C. Marsala (eds.), *Fuzzy Approaches for Soft Computing and Approximate Reasoning: Theories and Applications*, Studies in Fuzziness and Soft Computing 394, https://doi.org/10.1007/978-3-030-54341-9_23

continuously monitor the web, collect information, extract elements and facts and enhance existing knowledge bases, build data models based on them, and enable knowledge creation processes via different synthesis mechanisms.

The presented above vision of processing data and converting it into knowledge becomes more realistic due to graph-based data representation formats. The intrinsic ability of graphs to express a variety of relations between entities creates conditions suitable for processing data in a semantically oriented way. Further, such a data format allows for constructing feature-based definitions of concepts.

We hypothesize that feature-based definitions are essential for systems aiming at knowledge generation. It is fundamental to know what features an entity has to have in order to belong to a particular category: which features are more important and which ones are optional. Further, a generation process can be perceived as a process of 'playing' with features. This can be interpreted as a 'permutation' or 're-representation' [23] of known concepts. We can create a new concept or a new instance of it via replacing features (concept level) or their particular values (instance level).

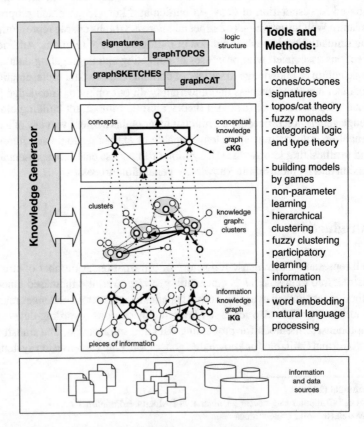

Fig. 1 Overview knowledge generation framework (*KGFrame*)

In this chapter, we propose a framework composed of methods and techniques suited for developing systems capable of semantic-based analysis of data and synthesis of knowledge, Fig. 1. It includes tools and algorithms for constructing definitions of concepts based on knowledge graphs, as well as formal-based mechanisms for extracting logic structures and their manipulation for the purpose of knowledge generation.

2 Knowledge Generation Framework: Introduction

The proposed framework for developing systems capable of knowledge generation relies on knowledge graphs, category theory, type theory, and fuzziness.

A graph-based data representation format called RDF (Resource Description Framework [1]) is a common denominator of Semantic Web [3], Linked Open Data [4] and Knowledge Graphs (KGs). Standardized by W3C, RDF provides a way of representing any piece of information as a triple—called an RDF triple—of the form: *subject-property-object*. A *subject* is an entity being defined, an *object* is a defining entity, and *property* is a relation between both entities (Sect. 3.1).

Category theory [2] is a discipline of mathematics dedicated to analysis of relationships between entities. The basic elements of category theory are *objects* that are entities of arbitrary complexity, and *morphisms* (*arrows*) that are connections— of any type—between objects. Mechanisms of category theory, such as limits and co-limits, as well as sketches, are tools suitable for manipulating and constructing new objects, and reasoning about them. An important categorical concept is called *monad*. It is a special functor (form of a transformation) from a category to itself. It is a mechanism of introducing new structures on the existing objects of category. Additionally, there is a special category called *topos* [18]. In a verbal form "topos can be thought of as a category theoretic 'generalization' (abstraction) of the structure of universe of sets and functions that removes certain logical and geometric restrictions (constraints) of this structure while maintaining its virtues" [17]. Its internal logic and language is viewed as a multi-sort type theory.

In type theory, a notion of *signature* is of special importance. A signature is specified by a set of types (*sorts*), and a collection of functions (*operations*) defined on the sorts [7]. We can think of signatures as basic building blocks that are used to compose other types. Its combination with categorical logic [15] is a primary example of formal processes for analyzing and synthesizing terms of different types.

Fuzzy sets have been introduced to capture imprecision of human expressions, and to translate them into a mathematical form [25]. They are used to define sets with elements that belong to them to a degree. Since their introduction, fuzzy sets have been investigated and used in many different domains of science and engineering. They are especially suitable for expressing imprecision associated with terms and statements. A substantial body of work is done on formalizing fuzzy sets and combining them with category theory and type theory [22]. Inclusion of fuzziness in definitions of sorts and operations leads to fuzzy signatures that are of high significance for knowledge creation processes.

Our goal is to combine these theories and their mechanisms to obtain a more comprehensive insight as well as abilities to generate new concepts and pieces of information associated with them. We consider a broad utilization of mechanisms of category theory [12, 24] as a promising strategy to accomplish this. As a result, we propose a data-driven and feature-oriented *Knowledge Generation Framework* (*KGFrame*). It comprises techniques for processing knowledge graphs and constructing feature-based definitions of concepts and dependencies between them. It contains formal mechanisms for extracting logic structures, and processes suitable for using these structures to synthesize new concepts. However, not all features of concepts are equally important, and not all of them contribute to concept definitions to the same degree. Thus, utilization of fuzziness in processes of constructing definitions and generating knowledge is natural as well as necessary.

In general, *KGFrame* is a collection of principles and ideas, mechanisms and algorithms from a variety of data and logic related disciplines. Its overview is presented in Fig. 1. It shows a number of layers of processed information from actual data to data models and logic structures, as well as a list of multiple processes for:

– building knowledge graphs and concept definitions from data;
– constructing different logic structures based on the constructed graphs;
– introducing and utilizing elements of fuzziness in graphs and logic structures;
– synthesizing new elements of knowledge based on graphs and logic structures.

3 Knowledge Graphs: Semantic Web Representation of Data

A graph-based representation of data provides a means to perceive data as a net of interconnected pieces of information. If such a format is combined with well defined relations existing between those pieces a semantically rich data structure is obtained.

3.1 Representation of Actual Data

Knowledge Graphs (KGs), as defined by W3C, are built from pieces of actual data represented as RDF triples of the form *subject-property-object* [1]. Triples can share their subjects and objects, while their properties represent relations existing between them. If we consider a single entity and a set of triples that have this entity as their subject then the properties of these triples together with their objects can be perceived as features of this entity. In other words, a single entity is described by multiple triples that are the entity's features. All interconnected entities constitute a net of inter-related nodes. Here, graphs that contain actual data will be called *data*

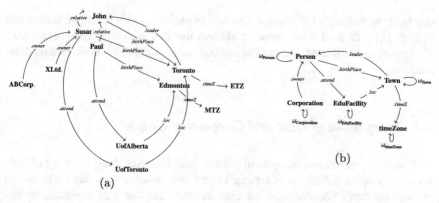

Fig. 2 Knowledge graphs: **a** a data knowledge graph—*dataKG*—representing relations between people, educational institutions, towns, corporations, and time zones; and **b** a conceptual knowledge graph—*conceptKG*—built based on this *dataKG*

Knowledge Graphs (*dataKG*s). A simple example of such a graph is shown in Fig. 2a: an entity **Susan** is described via a number of features, i.e., it is the subject of triples *<...-.birthPlace-Toronto>*, *<...-.attend-UofToronto>*, *<...-.relative-John>*, and the object of *<ABCorp.-.owner-...>* and *<XLtd.-.owner-...>*.

3.2 Representation of Concept Definitions

It is our belief that analysis of data and its synthesis is enabled when we operate on definitions of concepts. Such definitions can be built based on RDF triples that constitute a *dataKG*. A process of identifying features of a given concept, i.e., relations between this concept and other concepts together with these concepts, leads to generation of a feature-based definition of this concept. One way of accomplishing that is clustering of *dataKG*'s RDF triples. As a result, we obtain a graph of concept definitions, called hereafter a *conceptual Knowledge Graph* (*conceptKG*), Fig. 2b.

As it has been mentioned above, a process of constructing a *conceptKG* involves clustering *dataKG* triples. One of possible approaches is based on similarity between RDF triples estimated according to the process presented in [14]. A level of similarity between two entities, defined as sets of RDF triples, is estimated by comparing entities' triples, i.e., features on a one-by-one basis. This creates a similarity matrix between entities that is used in a *hierarchical clustering* process. Analysis of the obtained hierarchy provides information about: relations of inclusion between concepts; representative elements of each cluster at arbitrary levels of abstraction, and **degrees of belonging** individual pieces of information to particular clusters.

Another possible clustering process is based on representing RDF triples as propositions and applying *fuzzy clustering of categorical data* [16]. A process of translating triples into propositions has been applied to perform a participatory learning that leads to linking RDF triples—features of entities—with **levels of confidence** in them [6]. The final result includes clusters together with their **fuzzy centroids**. These centroids—treated as definitions of concepts—are fuzzy sets defined on other concepts.

3.3 Integration of Data and Conceptual Graphs

A clustering process and analysis of clusters lead to establishing a variety of connections between definitions of concepts and pieces of actual data that are instances of these concepts. We obtain a comprehensive structure that is an integrated formation of both *dataKG* and *conceptKG*. Such a structure gives us the ability to treat all data, information, and knowledge as a network of interconnected and interdependent concepts and their instances. In particular, we obtain:

- *vertical relations* between concepts in a form of super- and sub-concepts;
- *horizontal relations* between concepts and their instances (actual pieces of data);
- *conceptual relations* representing a variety of interactions between concept definitions; we can consider concepts as fuzzy sets with fuzzy relations between them, or as their crisp versions after a de-fuzzification process [25].

A graph obtained via integration of *dataKG* and *conceptKG*—called a *full Knowledge Graph (fullKG)*—is further processed in order to generate logic structures representing underlying data.

4 Logic Structure: Topos and Fuzziness

Different notions of category theory and type theory—such as *topos*, *signatures*, and *fuzzy term monads*—are used to process a *conceptKG*. Construction of a category of topos provides a different view of concept definitions, i.e., concepts are perceived as sets of features with logical relations identified between them. A *conceptKG* can be also recognized as a collection of signatures. It is a basis for constructing fuzzy term mondas. Here, we briefly illustrate a method of constructing topos and a logical structure for a crisp version of concepts (Sects. 4.1 and 4.2), while provide only an outline of an approach to deal with a fuzzy version of concepts (Sect. 4.3).

4.1 Logic of Topos

Mechanisms of category theory are used to process a *conceptKG* for the purpose of extracting a logic structure from RDF data. As a result, we obtain a representation of definitions of concepts as sets of concept features and relations between these sets. An internal logic of topos represents a logic embedded in RDF data [21].

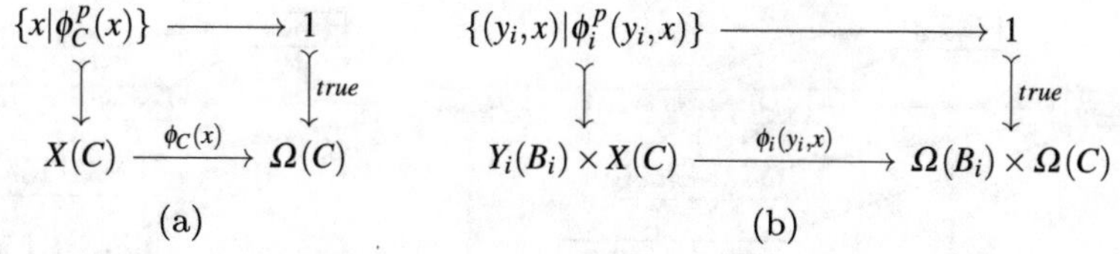

$$\{x|\phi_C^p(x)\} \longrightarrow 1 \qquad\qquad \{(y_i,x)|\phi_i^p(y_i,x)\} \longrightarrow 1$$

$$\downarrow \qquad\qquad \downarrow true \qquad\qquad \downarrow \qquad\qquad\qquad\qquad \downarrow true$$

$$X(C) \xrightarrow{\phi_C(x)} \Omega(C) \qquad Y_i(B_i) \times X(C) \xrightarrow{\phi_i(y_i,x)} \Omega(B_i) \times \Omega(C)$$

$$\text{(a)} \qquad\qquad\qquad\qquad\qquad \text{(b)}$$

Fig. 3 Subobject classifiers: **a** predicate ϕ_C defined on stage C; and **b** predicate defined on a cartesian product on stages C and B_i

A set of processes and algorithms is necessary for constructing topos. It includes procedures for: (1) converting a *conceptKG* into a category of concepts (in a sense of category theory); and (2) constructing, based on the category of concepts, topos specific elements—subobject classifiers—to formally express relations among sets of concept features, and to enable constructing a category of signatures.

Let us provide a few definitions. A category of concepts—named *conceptCAT*—is built based on a *conceptKG*. In this category, an *object* C represents a concept from a knowledge graph *conceptKG*. A *morphism*, on the other hand, represents an RDF property. Its domain is a subject of an RDF triple with this property, while its codomain is the triple's object. A *sieve* on C is defined as a set of morphisms that have C as their codomain and other objects B_i as their domains. In the sense of RDF triples, a sieve is a set of RDF properties that have C as their object, and B_i as their subjects. Similarly, a *cosieve* is composed of morphisms that have C as their domain and other B_j as codomains. So, a cosieve is a set of RDF properties that have C as their subject, and B_j as their objects. We also define ϕ_C^p as a *predicate* associated with an RDF property that has C as its object, $\{x|\phi_C^p(x)\}$ as a set of elements of $X(C)$ that satisfy ϕ_C^p, and $X(C)$ as a set of elements defined on C.

An subobject classifier $\Omega(C)$ built on C (see examples below) is an important building block of diagrams representing an internal logic of topos. Let us consider a canonical form of the diagram, Fig. 3a. A special morphism $\phi_C(x)$, called a *characteristic arrow*, takes elements of $X(C)$ identified by an inclusion $\{x|\phi_C^p(x)\} \rightarrowtail X(C)$ to components of $\Omega(C)$. In our case, components of $\Omega(C)$ are subsets of features of a concept from *conceptKG* represented by C. It means that the characteristic arrow identifies elements of $X(C)$ that satisfied the predicate $\phi_C^p(x)$ representing an RDF property. If we analyze a morphism $B_i \xrightarrow{\phi}_i C$, i.e., an RDF property associated with it, we can interpreted it as a predicate defined on a cartesian product of on two sets: elements of $Y_i(B_i)$ over B_i, and elements of $X(C)$ over C, Fig. 3b. We state that subobjects of $\Omega(C)$ and $\Omega(B_i)$ can be used to identify elements of $X(C)$ and $Y(B_i)$ that satisfy properties associated with their components.

A process of constructing subobject classifiers relies on the notion of sieves and cosieves. A subobject classifier $\Omega(C) = \{S|S \text{ is sieve on } C \in conceptCAT\}$ is a set of sieves defined on C, and $\Omega_{op}(C) = \{S|S \text{ is cosieve on } C \in conceptCAT\}$ is a set of cosieves defined on C. Based on the fragment of *conceptKG*, Fig. 2(b), we can construct two subobject classifiers on the concept **Town**, Fig. 4a, b: $\Omega(Town)$ is composed of power sets of $\{.loc, .birthPlace\}$, while $\Omega_{op}(Town)$ is composed of $\{.timeZ, .leader\}$.

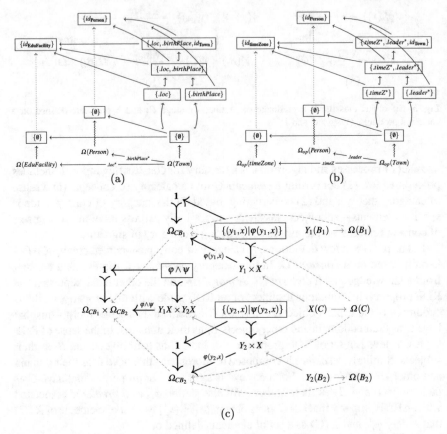

Fig. 4 Subobject classifiers on **Town**: **a** sieves; **b** cosieves; (* indicates an inverse morphism); and **c** predicates defined on **Town** where: $C = Town$; $B_1 = EduFacility$; $B_2 = Person$; $\varphi = .loc$; and $\psi = .birthPlace$ (**c**)

Let us focus on the node **Town**, Fig. 2b, and two predicates $.loc()$ and $.birth$-$Place()$. The diagram, shown in Fig. 4c, contains three loops (marked with bold lines) of the form presented in Fig. 3: one for $.loc() = \varphi$, one for $.birthPlace() = \psi$, and one for $.loc() \wedge .birthPlace() = \varphi \wedge \psi$. Each loop identifies elements x, y_1, and y_2 that satisfy predicates associated with the loops. So, $\Omega(Town)$, $\Omega(EduFacility)$ and $\Omega(Person)$ together with characteristic arrows constitute a structure that determines subsets of elements satisfying predicates created based on the morphisms that have **Town** as their codomain.

A similar process is applied for cosieves. Morphisms of the form $C \xrightarrow{\phi}_j B_j$ are turned into predicates defined on the cartesian products of elements over C, i.e., $X(C)$ and over B_j, i.e., $Y_j(B_j)$. As in the case of **Town** and sieves in Fig. 4a, we obtain a similar diagram when dealing with cosieves Fig. 4b. Now, the subobject classifiers $\Omega_{op}(Town)$, $\Omega_{op}(timeZone)$ and $\Omega_{op}(Person)$ with their *characteristic arrows* determine subsets of elements satisfying the predicates created based on the morphisms that have **Town** as their domain.

Sets of elements satisfying predicate formulas can be perceived as partially ordered sets. Consequently, if such a process is applied for a cartesian product of subobject classifiers $\Omega()$ and $\Omega_{op}()$, we obtain a definition of concept C based on morphisms between C and B's (in both directions). The result is a lattice-like structure of 'dependences' between subsets of elements from $X(C)$ and $Y(B)$'s satisfying different subsets of predicates defined by sieves of $\Omega()$ and cosieves of $\Omega_{op}()$.

4.2 Concept Signatures

Definitions of concepts, obtained using the process presented in the previous subsection, can be perceived as signatures. Each signature contains *sorts* that are other concepts defining a given concept, and *operations* that are relations defined on the sorts. Additionally, we are able to determine degree of importance of those relations (hierarchical clustering) [5, 20], and degree of importance of individual sorts (fuzzy clustering of categorical data). In this way our definitions of concepts have fuzzy relations to other concepts, and fuzzy sets as their sorts (Sect. 4.3).

From Sect. 4.1, we know that subobject classifiers are composed of multiple sieves/cosieves. Each sieve/cosieve contains one or more morphisms (RDF properties). In such a case, we state that a single sieve of $\Omega(C)$ is connected to one of sieves of $\Omega(B_i)$, and one cosieve of $\Omega_{op}(C)$ is connected to one cosieve of $\Omega_{op}(B_j)$. Thus, we have predicates of the form $\phi_i(y_i \in Y_i(B_i), x \in X(C))$ based on sieves' morphisms $B_i \xrightarrow{\phi}_i C$, and $\phi_j(x \in X(C), y_j \in Y_j(B_j))$ on cosieves' morphisms $C \xrightarrow{\phi}_j B_j$. All of them represent relations between elements of $X(C)$ defined on C and elements of $Y(B)$'s defined on B's.

Following that, we can interpret a structure composed of Ω and Ω_{op} as a reference for building signatures [19]. A process of transforming it into signatures, can be described by defining a signature $\Sigma_X = (S, F, R)$ where $S = \{X, Y_1, \ldots, Y_i, \ldots, Y_n\}$ is a set of sorts of the signature, $F = \{\phi_m(Y_m) \rightarrow X, \ldots, \phi_n(X) \rightarrow Y_n, \ldots\}$ is a set of functions defined on X and Y's, while $R = \{\phi_p(Y_p, X), \ldots, \phi_q(X, Y_q), \ldots\}$ is a set of relations.

For the illustration purposes, we construct a structure of signatures built based on the *concept KG* presented in Fig. 2b. Again, let us focus at the node **Town**. In the context of its 'neighbours' as seen in Fig. 2b, the following signature is created.

$$\Sigma_{Town} = (S_T, F_T, R_T)$$

Fig. 5 Category of signatures **Sign** constructed based on *concept KG* in Fig. 2

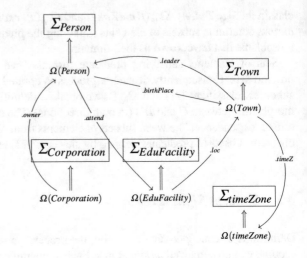

where

$$S_T = \{Town, Person, EduFacility, timeZone\}$$
$$F_T = \{.leader(Town) = Person,$$
$$.timeZ(Town) = timeZone,$$
$$.birthPlace(Person) = Town\}$$
$$.loc(EduFacility) = Town\}$$
$$R_T = \emptyset.$$

Based on this signature, we can define the following set of axioms:

$for\ y_1, y_4 : Person$ $.birthPlace(y_1) = .loc(y_2)$ $.leader(.birthPlace(y_1)) = y_4$
     $y_2 : EduFacility$ $.timeZ(.birthPlace(y_1)) = y_3$ $.leader(.loc(y_2)) = y_4$
     $y_3 : timeZone$ $.timeZ(.loc(y_2)) = y_3$ $.timeZ(x_2) = y_3$
     $x_1, x_2 : Town$ $.leader(x_1) = y_4$ $x_1 = x_2$

Following this procedure and making use of connections created at the time of building a topos category, we are able to create a category of signatures, Fig. 5.

4.3 Fuzzy Term Monads

The work of Eklund [8, 9, 11] constitutes a foundation, built on formal methods, that combines elements of fuzzy sets with elements of type theory. The work focuses on formal mechanisms suitable for introducing and working with fuzziness and signatures. In particular, it allows to deal with two aspects of signatures, that in our context represent concept definitions: fuzzy operations, and fuzzy sorts.

The first approach 'ties' fuzziness to *term monads*, and their compositions [8]. The term monad (Sect. 2) is a term functor $T_\Sigma : Set \to Set$ defined over a *Set*. It contains two transformations: $\eta_X : X \to T_\Sigma(X)$ that takes $x \in X$ to the term x; and $\mu_X : T_\Sigma(T_\Sigma(X)) \to T_\Sigma(X)$ that takes a term over terms over X and considers it a term over X. This means that $T_\Sigma(X)$ becomes a set of all terms over the set X of variables, and $X \subset Set$. Such defined term functor is suitable for handling **uncertainty of operations**.

The second approach 'introduces' fuzzy terms defined over $Set(L)$. It is based on setting fuzzy sets over a lattice L, represented as a category $Set(L)$ originally introduced by Goguen [13]. The further developments of that idea have led to the introduction of fuzzy term monads originally as single-sorted terms [11]. Later, it has been extended to multi-sorted ones [9] to handle multi-sort signatures. This approach deals with **uncertainty of sorts**.

Both of these approaches are essential for dealing with fuzziness in concept definitions for purposes of modifying and/or creating new definitions. Algorithms and methods for constructing term monads based on *fullKG* are essential for the proposed framework. Specific tasks can be identified to assemble a series of categorical steps to extract logical relations existing between signatures/definitions of concepts and their instances, and then to analyze them in order to form transformations that constitute monads.

5 Knowledge Generation Processes

A knowledge structure constructed using mechanisms of category theory is a structure equipped with multiple types of relations, including fuzzy ones, between specific pieces of information and definitions of concepts. All of this is integrated with elements of logic.

Further, the apparatus of category theory enables synthesis of new data and concepts associated with them. This can be accomplished using such constructors of category theory as pushouts, colimits, and sketches [2, 12, 18, 24]. Their usage ensures correctness of newly generated data elements. Application and integration of notions of Grothendieck topology [18] and categorical logic [15] together with the fuzzy-related concepts, such as fuzzy terms [9] and categorical fuzzy logic programming [10] create a sophisticated yet powerful framework enabling sensible generation of new concepts fully utilizing information represented in a form of knowledge graphs and results of their processing.

6 Conclusion

The chapter is an attempt to introduce and describe one of possible approaches to construct intelligent systems.

Topoi together with its algebraic structures and elements of fuzzy set theory lay out a formal way for expressing relations between data concepts, and provide tools for validating their correctness. Utilization of the proposed methodology will enable design context-aware intelligent components that are able to generate new concepts related to information and knowledge they possess, as well as their surroundings.

Furthermore, it is expected the presented framework will lead to significant contributions to methods suitable for building systems called Knowledge Generators. Such systems will work on the behalf of users to collect and analyze data, to monitor it for any changes, and to generate new knowledge based on it. This would represent a step towards customized systems helping users exploit and build knowledge from growing amount of information existing on the web.

References

1. https://www.w3.org/rdf/ (visited 2019)
2. Barr, M., Wells, C.: Category Theory for Computing Science. Les Publications CRM, Montreal (1999)
3. Berners-Lee, T., Hendler, J.: Scientific publishing on the semantic web. Nature **410**, 1023–1024 (2001)
4. Bizer, C., Heath, T., Berners-Lee, T.: Linked data: the story so far. Int. J. Semant. Web Inform. Syst. **5**(3), 1–22 (2009)
5. Chen, J.X., Reformat, M.Z.: Modelling experienced-based data in linked data environment. In: INCoS-2014 (2014)
6. Chen, J.X., Reformat, M.Z.: Learning processes based on data sources with certainty levels in linked open data. In: IEEE/WIC/ACM Web Intelligence (2016)
7. Crole, R.L.: Categories for Types. Cambridge University Press (1993)
8. Eklund, P.: Signatures for assessment. In: IPMU'10 (2010)
9. Eklund, P., Galan, M.A., Helgessona, R., Kortelainenc, J.: Fuzzy terms. Fuzzy Sets Syst. **256**, 211–235 (2014)
10. Eklund, P., Galan, M.A., Helgessona, R., Kortelainenc, J., Moreno, G., Vazquez, C.: Signatures for assessment. In: WILF 2013, LNAI 8256, pp. 109–121 (2013)
11. Eklund, P., Kortelainen, J., Stout, L.N.: Adding fuzziness to terms and powerobjects using a monadic approach. Fuzzy Sets Syst. **192**, 104–122 (2012)
12. Fiadeiro, J.L.: Categories for Software Engineering. Springer-Verlag, Berlin (2005)
13. Gougen, J.A.: L-fuzzy sets. J. Math. Anal. App. **18**, 145–174 (1967)
14. Hossein Zadeh, P.D., Reformat, M.Z.: Context-aware similarity assessment within semantic space formed in linked data. J. Ambient Intel. Human Comput. **4**, 515–532 (2013)
15. Jacobs, B.: Categorical Logic and Type Theory. Elsevier (1999)
16. Kim, D.W., Lee, K.H., Lee, D.: Fuzzy clustering of categorical data using fuzzy centroids. Pattern Recogn. Lett. **25**, 1263–1271 (2004)
17. Kostecki, R.P.: An introduction to topos theory. University of Warsaw, Technical report (2011)
18. Mac Lane, S., Moerdijk, I.: Sheaves in Geometry and Logic: A First Introduction to Topos Theory. Springer (1992)

19. Reformat, M.Z., D'Aniello, G., Gaeta, M.: Knowledge graphs, category theory and signatures. In: IEEE/WIC/ACM Web Intelligence (2018)
20. Reformat, M.Z., Hossein Zadeh, P.D.: Assimilation of information in RDF-based knowledge base. In: IPMU'12 (2012)
21. Reformat, M.Z., Put, T.: Constructing topos from RDF data. In: IEEE/WIC/ACM Web Intelligence (2015)
22. Rodabaugh, S.R., Klement, E.P., Hohle, U.: Sheaves in Geometry and Logic: A First Introduction to Topos Theory. Springer (1992)
23. Ward, T.B., Smith, S.M., Vaid, J.: Creative thought: an investigation of conceptual structures and processes. American Psychological Association (1997)
24. Williamson, K., Healy, M., Barker, R.: Industrial applications of software synthesis via category theory-case studies using specware. Autom. Softw. Eng. **8**, 7–30 (2001)
25. Zadeh, L.A.: Fuzzy sets. Inform. Control **8**, 338–353 (1965)

Z-Numbers: How They Describe Student Confidence and How They Can Explain (and Improve) Laplacian and Schroedinger Eigenmap Dimension Reduction in Data Analysis

Vladik Kreinovich, Olga Kosheleva, and Michael Zakharevich

Abstract Experts have different degrees of confidence in their statements. To describe these different degrees of confidence, Lotfi A. Zadeh proposed the notion of a Z-number: a fuzzy set (or other type of uncertainty) supplemented by a degree of confidence in the statement corresponding to fuzzy sets. In this chapter, we show that Z-numbers provide a natural formalization of the competence-vs-confidence dichotomy, which is especially important for educating low-income students. We also show that Z-numbers provide a natural theoretical explanation for several empirically heuristic techniques of dimension reduction in data analysis, such as Laplacian and Schroedinger eigenmaps, and, moreover, show how these methods can be further improved.

1 Introduction

Our knowledge is often imprecise.

In some cases, we only know the probabilities of different situations. For example, often, we do not know for sure whether it will rain tomorrow or not. To predict tomorrow's weather, we can look into similar past weather patterns. If it turns out that it rained in 60% of such cases, we can conclude that the rain will occur with probability 60%. In general, in such cases, to each possible situation, we assign a probability—a numerical degree describing to what extent this situation is possible.

V. Kreinovich (✉) · O. Kosheleva
University of Texas at El Paso, El Paso, TX 79968, USA
e-mail: vladik@utep.edu

O. Kosheleva
e-mail: olga@utep.edu

M. Zakharevich
SeeCure Systems, Inc., 1040 Continentals Way #12, Belmont, CA 94002, USA
e-mail: michael@seecure360.com

285

M.-J. Lesot and C. Marsala (eds.), *Fuzzy Approaches for Soft Computing and Approximate Reasoning: Theories and Applications*, Studies in Fuzziness and Soft Computing 394,
https://doi.org/10.1007/978-3-030-54341-9_24

Often, people describe their knowledge by using imprecise ("fuzzy") natural-language expressions like "we expect a heavy rain". To describe the meaning of words like "heavy", we can use fuzzy logic, in which, e.g., for each possible amount of rain we elicit, from the expert, a degree to which this particular amount corresponds to heavy rain.

In both cases, in addition to the uncertainty described by the corresponding degree, we may also not be absolutely sure about the related piece of knowledge. For example, in the probabilistic case, we may not be sure that the current weather pattern is indeed similar to the patterns based on which we estimates the probability of rain. In the fuzzy case, a meteorologist may not be absolutely sure that the rain will be heavy. To adequately describe this situation, Lotfi Zadeh proposed to supplement the usual degree of certainty with an additional degree describing to what extent the original piece of knowledge is reliable. The resulting pair of degrees is known as a *Z-number*.

In this chapter, we explain how to use Z-numbers in data processing, and we show, on a case study, that Z-numbers can help: their use explains the existing successful data processing heuristics, and it also provides ideas on how these heuristics can be further improved.

2 Need to Take into Account Accuracy and Reliability When Processing Data

2.1 Need for Data Processing

In many practical situations, we are interested in the values of the quantities $x_1, \ldots, x_n$ which are difficult (or even impossible) to measure directly. For example, in GPS-based localization, we want to find where different objects (and we) are, i.e., we want to find the coordinates of different objects. However, it is not possible to directly measure coordinates.

What we can measure in such situations is some auxiliary quantities $y_1, \ldots, y_m$ that depend on the desired quantities x_i in a known way, i.e., for which $y_j = f_j(x_1, \ldots, x_n)$ for known algorithms f_i. For example, to find the location of an object, we can measure distances between objects and/or angles between directions towards different objects.

Once we know the results $\widetilde{y}_j$ of measuring the quantities y_j, we need to reconstruct the desired quantities x_i from the corresponding system of equations:

$$\widetilde{y}_1 \approx f_1(x_1, \ldots, x_n), \quad \ldots, \quad \widetilde{y}_m \approx f_m(x_1, \ldots, x_n); \tag{1}$$

we write approximately equal, because measurements are never absolutely accurate, there is usually a difference $\Delta y_j \overset{\text{def}}{=} \widetilde{y}_j - y_j$ (known as *measurement error*) between the measurement result $\widetilde{y}_j$ and the actual value y_j of the measured quantity.

The process of reconstructing x_i from $\widetilde{y}_j$ is an important case of *data processing*.

In some applications—e.g., in many medical situations—it is difficult to find related easier-to-measure quantities y_j, but we *can* find related quantities that can be well estimated by an expert: e.g., by the patient's appearance or reaction to different tests. In such situations, to reconstruct the desired quantities x_i, instead of the measurement results, we can use the expert estimates $\widetilde{y}_j$.

2.2 How to Take Accuracy into Account: Probabilistic Case

In many practical situations, based on the previous experience of using the corresponding measuring instruments, we know the probabilities of different values of measurement errors. In precise terms, we know the corresponding probability density functions $\rho_j(\Delta y_j) = \rho_j(\widetilde{y}_j - f_j(x_1, \ldots, x_n))$.

Measurement errors corresponding to different measurements are usually independent. As a result, the overall probability of given observations is equal to the product of the corresponding probabilities:

$$\prod_{j=1}^{m} \rho_j(\widetilde{y}_j - f_j(x_1, \ldots, x_n)). \tag{1a}$$

For different values of x_i, this probability is different. It is therefore reasonable to select the most probable combination $(x_1, \ldots, x_n)$, i.e., the combination for which the product (1) attains the largest possible value. This quantity (1) is known as *likelihood*, and the above idea is known as the *Maximum Likelihood method*; see, e.g., [17].

The probabilities $\rho_j(\Delta y_j)$ can be reasonably small, and the number of measurements is often large. The product of a large number of small values is often too small, sometimes smaller than the smallest positive real number in a usual computer representation. To avoid this problem, practitioners use the fact that maximizing a function is equivalent to minimizing its negative logarithm

$$\sum_{j=1}^{m} \psi_j(\widetilde{y}_j - f_j(x_1, \ldots, x_n)), \tag{2}$$

where we denoted $\psi_j(z) \stackrel{\text{def}}{=} -\ln(\rho_j(z))$.

Often, the measurement error is the result of a joint effect of a large number of independent factors. In such situations, due to the Central Limit Theorem (see, e.g., [17]), the distributions are close to Gaussian—and, by re-calibrating the measuring instruments, we can usually safely assume that the mean of the measurement error is 0. In this case, $\rho_j(\Delta y_j) \sim \exp\left(-\dfrac{(\Delta y_j)^2}{\sigma_j^2}\right)$, where σ_j is the standard deviation of the j-th distribution. Thus, minimizing the expression (2) is equivalent to minimizing the sum

$$\sum_{j=1}^{m} \frac{(\widetilde{y}_j - f_j(x_1, \ldots, x_n))^2}{\sigma_j^2}. \tag{3}$$

This is known as the *Least Squares method*.

In particular, if we do not have any reason to believe that different measurements have different accuracy, it makes sense to assume that they all have the same accuracy $\sigma_1 = \sigma_2 = \ldots$. In this case, (3) becomes equivalent to minimizing the sum

$$\sum_{j=1}^{m} (\widetilde{y}_j - f_j(x_1, \ldots, x_n))^2. \tag{3a}$$

2.3 How to Take Accuracy into Account: Fuzzy Case

Often, instead of the probabilities of different values of the approximation error, we only have expert opinions about the possibility of different values. Describing these opinions in computer-understandable terms was one of the main motivations for fuzzy logic; see, e.g., [6, 10, 12, 14, 18]. It is therefore reasonable to describe these opinions in terms of the membership functions $\mu_j(\Delta y_j) = \mu_j(\widetilde{y}_j - f_j(x_1, \ldots, x_n))$.

In line with the general ideas of fuzzy logic, to describe the expert's degree of confidence that:

- the first approximation error is Δy_1, *and*
- the second approximation error is Δy_2,
- etc.,

we can apply the corresponding "and"-operation (t-norm) $f_\&(a, b)$, and get the value

$$f_\&(\mu_1(\widetilde{y}_1 - f_1(x_1, \ldots, x_n)), \ldots, \mu_m(\widetilde{y}_m - f_m(x_1, \ldots, x_n))). \tag{4}$$

It is thus reasonable to select the values x_i for which the degree (4) is the largest possible.

It is known (see, e.g., [13]) that for every $\varepsilon > 0$, each t-norm can be approximated by an Archimedean one, i.e., by a t-norm of the type $f_\&(a, b) = g^{-1}(g(a) \cdot g(b))$ for some increasing function $g(a)$. Thus, without losing generality, we can assume that our t-norm has this form. For such t-norms, the expression (4) takes the form

$$g^{-1}(g(\mu_1(\widetilde{y} - f_1(x_1, \ldots, x_n))) \cdot \ldots \cdot g(\mu_m(\widetilde{y} - f_m(x_1, \ldots, x_n))).$$

So, maximizing the expression (4) is equivalent to maximizing the product of type (1), where we denoted $\rho_j(z) \stackrel{\text{def}}{=} g(\mu_j(z))$, and is, hence, equivalent to minimizing the corresponding sum (2).

2.4   Need to Take Reliability into Account

In the above text, we implicitly assumed that every measuring instrument functions absolutely reliably and thus, every number $\widetilde{y}_j$ that we get comes from the actual measurement. In practice, measuring instruments are imperfect, sometimes they malfunction, and thus, once in a while, we get a value that has nothing to do with the measured quantity—i.e., an *outlier*.

Some outliers are easy to detect and filter out: e.g., if we measure body temperature and get 0 degrees, clearly the device is not working. In many other cases, however, it is not so easy to detect outliers. Similarly, some expert estimates can be way off.

In such cases, when processing data, we need to take into account that the values $\widetilde{y}_j$ are un-reliable: some of these values may be un-related to measurements.

3   Z-Numbers

3.1   What Do We Know About the Possible Outliers

Information about accuracy of measurements (or expert estimates) comes from our past experience:

- we know how frequent were different deviations between the measured and actual values, and
- we can thus estimate the probabilities of different deviations Δy_j.

  Similarly, based on our past experience:

- we can determine how frequently the values produced by the measuring instrument (or by an expert) turned out to be outliers, and
- thus, estimate the probability p_j that a given value $\widetilde{y}_j$ is an outlier.

  In both cases, we arrive at the following description.

3.2   Probabilistic Case

In the probabilistic case, for each j:

- we know the probability distribution function $\rho_j(\Delta y_j)$, and
- we know the corresponding probability p_j.

3.3 Fuzzy Case

In the fuzzy case, for each j:

- we know the corresponding membership function $\mu_j(\Delta y_j)$—or, equivalently, the corresponding function $\rho_j(\Delta y_j)$—and
- we also know the corresponding probability p_j.

3.4 General Case

L. Zadeh called such a pair (ρ_j, p_j) or (μ_j, p_j)—that describes both the accuracy and the reliability—a Z-number; see, e.g., [1, 19].

3.5 Problem

How can we process data in the presence of possible outliers?

If we knew which values $\tilde{y}_j$ are outliers, we could simply ignore these values and process all others. In practice, however, we do not know which measurement results are outliers, we only know the probabilities of each of them being an outlier. In principle, we could consider all possible outlier subsets—but since there are exponentially many such possible subsets, this would require an un-feasible exponential time. So what can we do?

3.6 Idea

We do not know which values $\tilde{y}_j$ are outliers, but knowing the probability p_j means that we know that if we repeat the measurements N times, than in approximately $p_j \cdot N$ cases we will have accurate estimates—and in the remaining $N - p_j \cdot N$ cases, we will have outliers.

To utilize this information, let us consider an imaginary situation in which each value $\tilde{y}_j$ is repeated N times.

Good news is that if all values $\tilde{y}_j$ were absolutely reliable, and simply repeat each value $\tilde{y}_j$ the same number of times N, the result of data processing will not change. Indeed, e.g., in the minimization formulation (2), repeating each value N times simply increases the minimized expression by a factor of N—and, of course, both the original expression (2) and the same expression multiplied by N attains their minimum on the exact same tuple x_i.

So, it makes sense to consider repetitions. But once we have many (N) repetitions, we kind of know which values are outliers—namely, we know that only $N \cdot p_1$ of copies of $\widetilde{y}_1$ are accurate estimates, etc. So, in processing data, we take into account:

- only $N \cdot p_1$ copies of the value $\widetilde{y}_1$,
- only $N \cdot p_2$ copies of the values $\widetilde{y}_2$,
- etc.

When we apply the expression (2) to these values, we end up with selecting the tuple $(x_1, \ldots, x_n)$ that minimizes the sum

$$\sum_{j=1}^{m} (N \cdot p_j) \cdot \psi_j(\widetilde{y}_j - f_j(x_1, \ldots, x_n)).$$

Strictly speaking, this expression depends on the unknown number of repetitions N, but good news is that if we divide the above expression by N, we get a new expression that no longer depends on N—but which attains its minimum at exactly the same tuple $(x_1, \ldots, x_n)$. Thus, we arrive at the following recommendation.

3.7   Resulting Algorithm

When we know the reliability p_j of each value $\widetilde{y}_j$, then we should select the tuple $(x_1, \ldots, x_n)$ that minimizes the expression

$$\sum_{j=1}^{m} p_j \cdot \psi_j(\widetilde{y}_j - f_j(x_1, \ldots, x_n)). \tag{5}$$

In particular, in the case of normal distribution, applying the same idea to formula (3) leads to the need to minimize the expression

$$\sum_{j=1}^{m} p_j \cdot \frac{(\widetilde{y}_j - f_j(x_1, \ldots, x_n))^2}{\sigma_j^2} = \sum_{j=1}^{m} \frac{(\widetilde{y}_j - f_j(x_1, \ldots, x_n))^2}{(\sigma_j')^2}, \tag{6}$$

where we denoted $\sigma_j' \stackrel{\text{def}}{=} \dfrac{\sigma_j}{\sqrt{p_j}}$.

3.8 How Good Is This Algorithm?

To check whether this algorithm is good, we will show, on the case study of dimension reduction, that the ideas behind this algorithm provide a natural explanation for an empirically successful heuristic approach.

4 Examples and Case Study

4.1 Z-Numbers and Teaching

Up to know, we considered the case when Z-numbers describe measurements or expert estimates, but there is another important area where Z-numbers are useful: teaching.

Namely, usually, the success of teaching is gauged by how accurate are the students' answers. However, it is important to also take into account how confident the students are in their answers:

- if a student gives the right answer, but he or she is not confident, this means there is still room for improvement,
- on the other hand, if a student gives the wrong answer, but he or she is not sure, the situation is not so bad: it means that in a similar future real-life case, the student will probably doublecheck or consult someone else and thus, avoid making a wrong decision.

In [11], we showed how to take both accuracy and reliability into account when gauging the result of teaching.

The need to take both accuracy and confidence into account is especially important for female students, low-income students, and students from under-represented minority groups, since these students typically show decreased confidence—even when their accurate answers show that they have reached a high level of competence; see, e.g., [8].

4.2 Case Study: Dimension Reduction

In many practical situations, we analyze a large number of objects of a certain type. For example, in medical research, we study all the patients that suffer from a given disease.

In many such situations, we do not know which quantities will turn out to be relevant. Thus, not to miss any relevant quantity, we measure as many quantities as possible. As a result, for each object, we have a large number of measurement results and/or expert estimates. In other words, each object is represented by a point in a very high-dimensional space.

Processing such high-dimensional data is often very time-consuming. It is therefore desirable to reduce the amount of data. Good news—coming from our experience—is that in most practical situations, most of the collected data is irrelevant, that there are usually a few important combinations of the original parameters that are relevant for our specific problem. In other words, with respect to the corresponding problem, we can as well use a low-dimensional representation of the data.

To use this possibility, we need to be able to reduce the data dimension.

4.3 Reformulating the Problem in Terms of Z-Numbers

We want to assign, to each point s_i in the multi-D space, a point q_i in the lower-dimensional space. The main criterion that we want to satisfy is that if s_i and s_j are close, then the corresponding points q_i and q_j should also be close.

If we had a clear (crisp) idea of which pairs (s_i, s_j) are close and which pairs are not close, we would simply require that the values q_i and q_j corresponding to these pairs are close, i.e., that

$$q_i \approx q_j$$

for all such pairs. By applying the Least Squares approach to this situation, we would then arrive at the problem of minimizing the sum $\sum \|q_i - q_j\|^2$, where the sum is taken over all such pairs. Of course, to avoid the trivial and useless solution $q_1 = q_2 = \ldots$, we need to "normalize" these solutions: e.g. by requiring that $\sum_i \|q_i\|^2 = 1$.

In practice, we usually do not have an absolutely clear idea of which points are close to each other and which are not. A reasonable idea is to describe closeness in probabilistic terms. Since there can be many different reasons why objects are somewhat different, it makes sense to apply the same Central Limit theorem argument that we used before and conclude that closeness corresponds to a normal distribution.

Since we do not have a priori knowledge of which components of the original vectors s_i are more relevant and which are less relevant, it is therefore reasonable to assume that the corresponding Gaussian distribution is invariant with respect to all permutations of these components (and changing their signs), and thus, that the normal distribution has the form const $\cdot \exp\left(-\frac{\|s_i - s_j\|^2}{2\sigma^2}\right)$ for some $\sigma > 0$. Thus, we arrive at the following Z-number-type problem:

$$q_i \approx q_j \text{ with probability } p_{ij} = \text{const} \cdot \exp\left(-\frac{\|s_i - s_j\|^2}{2\sigma^2}\right). \tag{7}$$

4.4   Our Algorithm Leads to a Known Successful Heuristic

If we apply the above algorithm to this problem, we arrive at the need to minimize
the expression

$$\sum_{i,j} p_{ij} \cdot \|q_i - q_j\|^2, \tag{8}$$

where the values p_{ij} are defined by the formula (7). (Of course, some normalization
like $\sum_i \|q_i\|^2 = 1$ is needed.) This is equivalent to minimizing the sum

$$\sum_{i,j} w_{ij} \cdot \|q_i - q_j\|^2, \tag{8a}$$

where we denoted

$$w_{ij} = \exp\left(-\frac{\|s_i - s_j\|^2}{2\sigma^2}\right). \tag{7a}$$

This is indeed one of the most successful heuristic methods for dimension
reduction—it is known as the *Laplacian eigenmap*, since its solution can be described
in terms of eigenvectors of the corresponding Laplacian operator $\nabla^2 \varphi = \sum_{i=1}^{d} \frac{\partial^2 \varphi}{\partial^2 x_i}$;
see, e.g., [2–5, 9, 15, 16].

So, *Z-numbers provide a theoretical explanation for the empirical success of
Laplace eigenmaps—a heuristic approach to dimension reduction.*

4.5   Taking into Account that Some Objects May Be Not
Relevant

In the above analysis, we assumed that for each object, we are 100% sure that this
object belongs to the desired class. In practice, we are often not fully confident about
this. For example, when we study a certain disease, we are not always sure that a
patient suffers from this very disease—and not from some similar one.

In general, the further away the object from the "typical" (average) situation—
which, by shifting, we can always assume to be 0—the less probable it is that this
object actually belongs to the desired class. In making this conclusion, we should
not take into account irrelevant components of the points s_i. Thus, this conclusion
should be based only on the values q_i—which contain only relevant combinations.

Similar to the above argument, we can safely assume that the corresponding
distribution is Gaussian, with probability P_i proportional to $\exp\left(-\frac{\|q_i\|^2}{2\sigma_i^2}\right)$. Here,
different values σ_i correspond to different degrees of confidence that this object
belongs to the class:

- when $\sigma_i^{-2} = 0$, this means that the probability does not depend on q_i at all: in other words, we are so confident, that no matter how big the deviation from the typical object, our degree of confidence does not change;
- on the other hand, if σ_i^{-2} is large, then even a small deviation from the typical value will make us conclude that this object does not belong to the desired class.

In this case, to get a more adequate description of the situation, to the product (1), we need to add the factors corresponding to different objects. After taking negative logarithm, these terms are equivalent to adding terms proportional to $V_i \cdot \|q_i\|^2$ to the sum (2), where we denoted $V_i \overset{\text{def}}{=} \sigma_i^{-2}$.

In particular, for the dimension reduction problem, this means that instead of minimizing the expression (8), we minimize a more complex expression

$$\sum_{i,j} w_{ij} \cdot \|q_i - q_j\|^2 + \alpha \cdot \sum_i V_i \cdot \|q_i\|^2. \tag{9}$$

This expression has indeed been proposed and successfully applied—on a heuristic basis—in [7]. This approach is known as *Schroedinger eigenmap*, since it corresponds to using eigenvectors of the operator $\nabla^2 \varphi + \text{const} \cdot V \cdot \varphi$ from Schroedinger's equations in quantum physics.

Thus, *Z-numbers provide a theoretical explanation for the empirical success of this a heuristic approach as well.*

4.6 Can We Go Beyond Justification of Existing Approaches?

A theoretical justification of known heuristic approaches is nice, but can we learn something new from this approach? Yes, we can.

While, as we have shown, the Schroedinger approach is well-justified for the case when we are not sure whether objects belongs to the class, this approach is also used in a completely different situation: when:

- we have an additional discrete value V_i characterizing each object, and
- we want to require $q_i \approx q_j$ only for objects that have close values of V_i and V_j.

For this situation, the Schroedinger approach is not perfect: indeed, even in the simplest case when V_i takes two possible values—which we can describe as 0 and 1—the result of minimizing the expression (9) depends on which of the two possible value we associate with 0 and which with 1.

In view of our analysis, it is more adequate to add the similarity between the value V_i and V_j to the description of closeness, i.e., to use an expression

$$w_{ij} = \exp\left(-\frac{\|s_i - s_j\|^2}{2\sigma^2} - \frac{(V_i - V_j)^2}{2\sigma_0^2}\right), \tag{10}$$

for some $\sigma_0 > 0$. The resulting probabilities does not change if we swap 0 and 1 value of V_i—thus, the resulting minimized expression (8) will not change after this swap, and hence, the produced optimizing arrangement q_i will not change—which is exactly what we wanted.

5   Conclusion

To more adequately describe expert knowledge, Lotfi Zadeh proposed to supplement the usual degree of certainty with an additional degree describing to what extent the corresponding piece of knowledge is reliable. The resulting pair of degrees is known as a *Z-number*.

In this chapter, we explain how we can use Z-numbers in data processing, and we show, on the example of dimension reduction problem in data analysis, that Z-numbers can indeed be very helpful; namely, Z-numbers:

- enable us to explain the existing successful heuristics, and
- provide us with ideas on how these heuristics can be improved.

Acknowledgements This work was supported in part by the National Science Foundation grant HRD-1242122 (Cyber-ShARE Center of Excellence). The authors are greatly thankful to the anonymous referees for valuable suggestions.

References

1. Aliev, R.A., Huseynov, O.H., Aliyev, R.R., Alizadeh, A.A.: The Arithmetic of Z-Numbers: Theory and Applications. World Scientific, Singapore (2015)
2. Belkin, M., Niyogi, P.: Laplacian eigenmaps for dimensionality reduction and data representation. Neural Comput. **15**, 1373–1396 (2003)
3. Belkin, M., Niyogi, P.: Convergence of Laplacian eigenmaps. In: Koller, D., Schuurmans, D., Bengio, Y., Bottou, L. (eds.) Advances in Neural Information Processing Systems 21 (NIPS'2008). Springer Verlag, Berlin, Heidelberg, New York (2008)
4. Belkin, M., Niyogi, P.: Towards a theoretical foundation for Laplacian-based manifold methods. J. Comput. Syst. Sci. **74**(8), 1289–1308 (2008)
5. Belkin, M., Niyogi, P., Sindhwani, V.: Manifold regularization: a geometric framework for learning from labeled and unlabeled examples. J. Mach. Learn. Res. **7**, 2399–2434 (2006)
6. Belohlavek, R., Dauben, J.W., Klir, G.J.: Fuzzy Logic and Mathematics: A Historical Perspective. Oxford University Press, New York (2017)
7. Czaja, W., Ehler, M.: Schroedinger eigenmaps for the analysis of bio-medical data. IEEE Trans. Pattern Anal. Mach. Intell. **35**(5), 1274–1280 (2013)
8. Frisby, C.L.: Meeting the Psychoeducational Needs of Minority Students: Evidence-Based Guidelines for School Psychologists and Other School Personnel. Wiley, Hoboken, New Jersey (2013)
9. Izenman, A.J.: Spectral embedding methods for manifold learning, In: Ma, Y., Fu, Y. (eds.) Maniforld Learning: Theory and Applications, pp. 1–36. CRC Press, Boca Raton, Florida (2012)

10. Klir, G., Yuan, B.: Fuzzy Sets and Fuzzy Logic. Prentice Hall, Upper Saddle River, New Jersey (1995)
11. Kosheleva, O., Lorkowski, J., Felix, V., Kreinovich, V.: How to take into account student's degree of confidence when grading exams. In: Proceedings of the 5th International Conference "Mathematics Education: Theory and Practice" MATHEDU'2015, Kazan, Russia, November 27–28, 2015, pp. 29–30 (2015)
12. Mendel, J.M.: Uncertain Rule-Based Fuzzy Systems: Introduction and New Directions. Springer, Cham, Switzerland (2017)
13. Nguyen, H.T., Kreinovich, V., Wojciechowski, P.: Strict Archimedean t-norms and t-conorms are universal approximators. Int. J. Approx. Reason. **18**(3–4), 239–249 (1998)
14. Nguyen, H.T., Walker, E.A.: A First Course in Fuzzy Logic. Chapman and Hall/CRC, Boca Raton, Florida (2006)
15. Saul, L.K., Weinberger, K.Q., Sha, F., Ham, J., Lee, D.D.: Spectral methods for dimensionality reduction. In: Chapelle, O., Scholkopf, B., Zien, A. (eds.) Semi-Supervised Learning. MIT Press (2013)
16. Shaw, B.: Graph Embedding and Nonlinear Dimensionality Reduction, Ph.D. Dissertation, Columbia University (2011)
17. Sheskin, D.J.: Handbook of Parametric and Nonparametric Statistical Procedures. Chapman and Hall/CRC, Boca Raton, Florida (2011)
18. Zadeh, L.A.: Fuzzy sets. Inf. Control **8**, 338–353 (1965)
19. Zadeh, L.A.: A note on Z-numbers. Inf. Sci. **181**, 2923–2932 (2011)

Printed in the United States
by Baker & Taylor Publisher Services